FLORA

OF

TROPICAL EAST AFRICA

INDEX OF COLLECTING LOCALITIES

by

DIANA POLHILL

ROYAL BOTANIC GARDENS
KEW

Printed in Great Britain by
Whitstable Litho Printers Ltd.

ISBN 0 947643 09 5

INTRODUCTION & ACKNOWLEDGEMENTS

Since the issue of the first edition in 1970 the number of entries has increased by 50% to nearly 12,000. These include the majority of localities likely to be encountered on the labels of herbarium specimens from East Africa (Uganda, Kenya and Tanzania).

Localities of earlier collectors have been traced from itineraries derived from their travelogues and from the citation of their specimens in subsequent botanical works. A synopsis, with references, will be found in J.B. Gillett, The history of botanical exploration of the area of the Flora of Tropical East Africa, in Junta de Investigações do Ultramar, Compt. Rend. IVe Réunion A.E.T.F.A.T. (1962). So long as journeys were restricted to foot marches, the few localities not traced from contemporary maps can be localised fairly precisely from the more complete itineraries. In such cases the co-ordinates are inserted with an approximation sign or an indication is given of the nearest recorded place.

For the major collectors since the introduction of motor vehicles localities have been extracted from the field note books where available. Otherwise records are taken from the citations in published parts of the Flora or from enquiries from authors.

Spellings adopted reflect current usage as far as possible. Obsolete and less preferred spellings have been included in square brackets. Acceptable alternatives are shown with an equals sign. The collectors should be listed against whichever variant spelling they used, but over the years there has been some unevenness of treatment and they may be found under the entry with the accepted spelling.

The geographical divisions of the Flora are the administrative provinces and districts in effect at the time the project was started. They are shown on Uganda Survey Land and Mines Department, 1 : 1,000,000 map, No. A.1005 (1948), an untitled Government administrative 1: 1,000,000 map of Kenya (1948), and the Tanganyika Survey Division, 1: 2,000,000 map. It is unfortunate that most of the political divisions have changed subsequently, but it is not practical to make changes to the Flora districts at this stage of the production. Nonetheless the names of the places on which the old districts were based are still familiar and provide a convenient reference point.

Approximations for areas are indicated by "c.". Where the locality has not been traced, but the next one is known, the co-ordinates of the latter are given and qualified by "near".

ACKNOWLEDGEMENTS. The gazetteer was based originally on a card-index started with the Flora project. The development of the card-index and the preparation of itineraries became largely the responsibility of Mr J.B. Gillett until his appointment as Botanist-in-Charge at the East African Herbarium in 1964. I am very grateful for his continued support and enthusiasm. I am also very grateful for the help from colleagues at Kew and at the East African Herbarium, Nairobi. I extend special thanks to the Director and staff of the Royal Geographical Society of London for access to their extensive map collection. Furthermore I am privileged for access to the following map collections: British Museum (Natural History), London; the Commonwealth Forestry Herbarium, Oxford; the Department of Overseas Surveys, Southampton; Foreign and Commonwealth Office, London; Land Resources Development Centre, Tolworth; Royal Commonwealth Society, London; School of Oriental and African Studies, London University and the Public Record Office, Kew. I am very grateful for the help and kindness extended to me by the staff of all these institutes. Lastly, but not least, I wish to thank the many individuals who have written in response to specific enquiries or have made available their field note books or other sources of information.

My special thanks go to Mrs Eileen Attwood for her personal interest in type-setting the book with such speed and accuracy and to my husband Roger for all his help and encouragement.

1

ABBREVIATIONS

admin.	=	administrative
Agric.Dept.	=	Agricultural Department
agric. stn.	=	agricultural station
bdg.	=	bridge
b.h.	=	bore-hole
b.p.	=	boundary post
c.	=	approximately
chnl.	=	channel
co.	=	county
coll.	=	college
distr.	=	district
escarp.	=	escarpment
est.	=	estate
exper. fm.	=	experimental farm
exper. stn.	=	experimental station
For. Dept.	=	Forestry Department
f.r.	=	forest reserve
fm.	=	farm
for.	=	forest
for. stn.	=	forest station
ft.	=	fort
fy.	=	ferry
govt. stn.	=	government station
g.r.	=	game reserve
H.Q.	=	Headquarters
I.	=	island
I.R.L.C.S.	=	International Red Locust Control Service
Is.	=	islands
L.	=	lake
ln.	=	location
l.s.	=	landing stage
mkt.	=	market
mssn.	=	mission
mt.	=	mountain
Nat. Pk.	=	National Park
pen.	=	peninsula
pl.	=	plantation
plat.	=	plateau
popl.	=	populated place (i.e. village, town)
p.p.	=	police post
pt.	=	point
prov.	=	province
P.W.D.	=	Public Works Department
R.	=	river
r.c.	=	rest camp
res.	=	reserve
res. stn.	=	research station
r.h.	=	rest house
r.p.	=	ranger's post
rsta.	=	railway station
sch.	=	school
spr.	=	spring
str.	=	stream
str./s.	=	seasonal stream
sw.	=	swamp
t.a.	=	tribal area
t.c.	=	trading centre
tech. sch.	=	technical school
val.	=	valley
vet. stn.	=	veterinary station
w.h.('s)	=	water-hole(s)
w.h./s.	=	seasonal waterhole
w.p.	=	water pan

GEOGRAPHICAL DIVISIONS OF THE FLORA

Province	District
Uganda	
U1 Northern	West Nile, Acholi, Lango, Karamoja
U2 Western	Bunyoro, Toro, Ankole, Kigezi
U3 Eastern	Teso, Busoga, Mbale
U4 Buganda	Masaka, Mengo, Mubende
Kenya	
K1 Northern Frontier	Northern Frontier
K2 Turkana	Turkana, West Suk
K3 Rift Valley	Trans-Nzoia, Uasin Gishu, Nandi, Elgeyo, Baringo, Laikipia, Ravine, Nakuru, Naivasha
K4 Central	North Nyeri, South Nyeri, Fort Hall, Kiambu, Nairobi, Meru, Embu, Machakos, Kitui
K5 Nyanza	North Kavirondo, Central Kavirondo, South Kavirondo, Kisumu-Londiani, Kericho
K6 Masai	Masai
K7 Coast	Teita, Kwale, Mombasa, Kilifi, Tana River, Lamu

Tanganyika (mainland Tanzania)	
T1 Lake	Bukoba, Biharamulo, Ngara, Mwanza, Kwimba, Shinyanga, Maswa, Musoma, North Mara
T2 Northern	Masai, Mbulu, Arusha, Moshi
T3 Tanga	Pare, Lushoto, Handeni, Tanga, Pangani
T4 Western	Buha, Kahama, Nzega, Kigoma, Tabora, Mpanda, Ufipa
T5 Central	Singida, Kondoa, Dodoma, Mpwapwa
T6 Eastern	Kilosa, Morogoro, Bagamoyo, Uzaramo, Rufiji, Ulanga
T7 Southern Highlands	Chunya, Mbeya, Iringa, Rungwe, Njombe
T8 Southern	Songea, Kilwa, Tunduru, Masasi, Lindi, Newala, Mikindani

Zanzibar	
Z	Zanzibar Island
P	Pemba Island

ABBREVIATIONS FOR ADJACENT COUNTRIES

Sudan	= Sud.	Mozambique	= Moz.
Ethiopia	= Eth.	Rwanda	= Rw.
Zambia	= Zam.	Burundi	= Bur.
Malawi	= Mal.	Somali Republic	= Som.

Name		Region	Coordinates	Collector
Agago, co.	U1	Acholi	c.2 50 N 33 20 E	
Aghondi, popl.	T5	Dodoma	5 45 S 34 43 E	Polhill & Paulo
[Agili] see Ogili				
Agilo, hill	U1	Acholi	3 11 N 33 17 E	
[Agoe] see Agoi				
Agoi, str.	U1	Lango	2 22 N 32 54 E	
	U1	West Nile	c.2 53 N 31 02 E	
	U1	West Nile	c.2 53 N 31 02 E	
[Agora] see Agoro				
Agoro, popl., t.a.	U1	Acholi	3 45 N 33 02 E	Eggeling, Greenway & Hummel
	U1	Acholi	3 45 N 33 02 E	Eggeling, Greenway & Hummel, A.S. Thomas, Purseglove
Agoro-Agu, f.r.	U1	Acholi	c.3 52 N 33 02 E	Eggeling, Chandler
Agu, sw.	U3	Teso	c.1 28 N 33 42 E	Bagshawe
[Agua, R.] see Aswa	U1	Acholi	c.2 57 N 32 03 E–2 35 N 31 27 E	
Agule, popl., r.h.	U3	Mbale	1 17 N 33 42.5 E	Eggeling
Agur, popl.	U1	Lango	2 26 N 32 56 E	
Agwata, popl.	U1	Lango	1 58.5 N 32 59 E	Mukasa
Agwera ? = Awere	U1	Acholi		
Ahero, popl., p.p.	K5	Central Kavirondo	0 10 S 34 55 E	McMahon, Davidson
[Ai-ivu] see Ayivu	U1	West Nile	c.3 05 N 30 55 E	
Ailsa = Kitekelo, old farm	T7	Mbeya	8 54 S 33 59 E	
Ainabkoi, rsta.	K3	Ravine	0 10 N 35 31 E	Bickford, Tweedie
Ainabkoi, ln.	K5	Kericho	0 28 S 35 11 E	
Airport Swamp	K3	Trans-Nzoia	0 58 N 34 57 E	Tweedie
Aitcho, popl., escarp., str.	T2	Mbulu	3 41.5 S 35 32.5 E	Bally
Aitibu = Itibo, popl.	K5	Kericho	0 33 S 35 00 E	Dale
Aitjo = Aitcho, popl., escarp., str.	T2	Mbulu	3 41.5 S 35 32.5 E	Williams
Aitong, Tsetse Survey and Control	K6	Masai	1 11 S 35 15 E	Glover et al.
[Aiyu] see Ayo	U1/Sudan	West Nile	3 46 N 31 39 E–3 51 N 31 49 E	
Ajao, popl.	K1	Northern Frontier	2 57.5 N 39 42 E	
Ajao, hill	K1	Northern Frontier	2 56.5 N 39 42.5 E	Dale
Ajao, w.h.	K1	Northern Frontier	2 57 N 39 42.5 E	Dale, Bally & Smith

Name	Code	Region	Coordinates	Authority
Ajaumarka, mt.	K1	Northern Frontier	2 16 N 37 54 E	Jonsell
[Ajugopi] see Adzugopi	U1	West Nile	3 23 N 31 58 E	Hancock
[Ajumani] see Adjumani	U1	West Nile	3 22 N 31 47 E	A.S. Thomas
Aka, hill	U1	West Nile	c.2 25 N 31 01 E	Chancellor
Akarat = Akoret	K2	Turkana	c.2 20 N 35 10 E	Pratt
Akiriamet, str.	K3	Baringo	c.1 20 N 35 43 E	
Akisim, mt.	U1/3	Karamoja/Teso	2 07 N 34 17 E	Eggeling, A.S. Thomas
[Akobi] see Aboki	U1	Lango	2 21 N 32 41 E	A. S. Thomas
Akokoro, popl. t.c.	U1	Lango	1 42 N 32 23.5 E	
Akoret = Akarat	K2	Turkana *not* Karamoja	c.2 20 N 35 10 E	Philip
Alango, w.h.	K1	Northern Frontier	0 19 N 38 27 E	
Alap, hill	K2	Turkana	prob. c. 1 55 N 35 20 E	J. Wilson
Alaska Highway, road	K3	Nakuru/Naivasha	c.0 40 S 36 07 E	Pegler
Albert, L. = Mobutu Sese Seko	U1, 2,4/Zaire	several	c.1 00—2 25 N 30 20—31 30 E	Numerous
[Albert Edward, L.] see Edward	U2/Zaire	Toro/Ankole/Kigezi	c.0 25 S 29 30 E	
Albert Nile, R.	U1	West Nile/Acholi	c.3 15 N 31 20 E—3 36 N 32 02 E	Mildbraed, Scott Elliot
Alebtong, popl.	U1	Lango	2 16 N 33 14 E	
[Alecelec] see Alekilek	U3	Teso	2 06 N 34 11 E	Harker
Alego, t.a.	K5	Central Kavirondo	c.0 06 N 34 21 E	Davidson
Alekilek, hill	U3	Teso	2 06 N 34 11 E	
Aler, dam	U1	Lango	Not traced	Eggeling
[Alhi] see Athi	K4/6	Nairobi/Masai	c.1 25 S 36 52 E	Pospischil
[Alia] see Allia	K1	Northern Frontier	c.3 45 N 36 16 E	
Allia, bay	K1	Northern Frontier	c.3 45 N 36 16 E	
[Allui] see Alui	U1	West Nile	2 24 N 31 22 E	Eggeling
Aloi, popl.	U1	Lango	2 17 N 33 10 E	
Alori (Ndoto Mts.)	K1	Northern Frontier	c.1 45 N 37 08 E	Newbould
Althohovu	T3	Lushoto/Tanga	Not traced	Peter
[Alt Langenburg] see Lumbila	T7	Njombe	9 34.5 S 34 07.5 E	
Alui, f.r.	U1	West Nile	2 24 N 31 22 E	Eggeling

Name	Region code	Region	Coordinates	Collector
Amahenge, popl.	U2	Kigezi	0 54 S 29 44 E	Purseglove
Amaich, ginnery	U1	Lango	2 07 N 32 57 E	Eggeling
Amakita = Amakita's, old popl.	T8	Songea	c.10 50 S 35 03 E	Busse
Amala, R.	K5/6	Kericho/Masai	c.0 33 S 35 43 E–1 02 S 35 14.5 E	Evans, Pfennig
Amala Bridge	K5/6	Kericho/Masai	0 54 S 35 26 E	
Amaler, popl.	U1	Karamoja	1 47 N 34 41 E	Eggeling
Amani, popl., mssn., r.h.	T3	Lushoto	5 06 S 38 38 E	Numerous
Amani, f.r.	T3	Lushoto	5 08 S 38 39 E	
Amani West, f.r.	T3	Lushoto	5 06 S 38 37 E	Semsei
Ambangulu, est.	T3	Lushoto	5 05 S 38 26 E	Wallace, Peter
[Ambanguru] see Ambangulu	T3	Lushoto	5 05 S 38 26 E	
Amboni = Honi (in part)	K4	N. & S. Nyeri	0 21 S 36 44 E–0 24 S 36 59.5 E	Hooper & Townsend
Amboni, caves	T3	Tanga	5 04 S 39 03 E	Bogner, Eggeling
Amboni, popl., est.	T3	Tanga	5 03 S 39 03 E	Numerous
Amboseli, g.r.	K6	Masai	c.2 30 S 37 00 E	Numerous
Amboseli, L.	K6	Masai	2 37 S 37 08 E	Greenway, Verdcourt
Amgamwa, popl.	U1	Karamoja	2 51 N 34 13 E	Eggeling, Tothill
Amiel, popl.	U1	Acholi	2 58 N 33 27 E	Dawkins
[Amolatar] see Molitar	U1	Lango	1 38 N 32 50 E	Mukasa
Amor Hill ? = Omoro Hill	U1	Acholi	?2 42.5 N 32 44.5 E	Brasnett, Eggeling, A.S. Thomas
Amua, popl., r.h.	U1	West Nile	3 37 N 31 49 E	J. Wilson, A.S. Thomas
Amuda, popl.	U1	Karamoja	2 12 N 34 36 E	Numerous
Amudat, popl., r.c.	U1	Karamoja	1 57 N 34 57 E	Kimani
Amukura, popl., mssn.	K5	North Kavirondo	0 34 N 34 16 E	A.S. Thomas
Amuria, co.	U3	Teso	c.2 05 N 33 45 E	A.S. Thomas
Amuria, r.h.	U3	Teso	2 02 N 33 38 E	A.S. Thomas
Amuria, t.a.	U3	Teso	c.2 07 N 33 40 E	
Andamaksiu (Nou Forest)	T2	Mbulu	c.4 05 S 35 30 E	Carmichael
[Andarer] see Ndareda	T2	Mbulu	4 13 S 35 33 E	Carmichael
Angai, R.	T8	Kilwa	10 08 S 37 33 E–9 58 S 38 08 E	Nicholson

Ang'ata Kiti, area	T2	Masai	c.2 46 S 35 25 E	Oteke
Angata Opiri	T2	Masai	c.2 55 S 36 00 E	Richards
Angata Salei, area	T2	Masai	c.2 40 S 35 40 E	Newbould
Angata Sere (near Kijungu)	T2	Masai	near 5 22 S 37 11 E	Vesey-FitzGerald
[Angoniland] see Ungoni	T8	Songea	c.10 45 S 36 00 E	Busse
Ankole, distr.	U2	Ankole	c.0 30 S 30 30 E	
Ankole Kingdom = Ankole District	U2	Ankole	c.0 30 S 30 30 E	Purseglove, Dawe
Anmeri, tea est.	T7	Iringa	8 36.5 S 35 21.5 E	Carmichael
Anri, ln.	K1	Northern Frontier	c.2 38 N 40 53 E	Gillett
[Anyek] see Anyeke	U1	Lango	2 23 N 32 31 E	
Anyeke, popl., r.h.	U1	Lango	2 23 N 32 31 E	
Apala, popl.	U1	Lango	2 24 N 33 02 E	
Apedet, hill	U1	Karamoja	1 39 N 34 36 E	Dyson-Hudson
Apelapong, Upe co.	K2	Turkana	2 25 N 35 05 E	Philip
Api, popl.	U1	Lango	2 19 N 32 47 E	
Apoli, popl.	U3	Mbale	0 36 N 34 05 E	Katende
Apule, str.	U1	Karamoja	c.2 44 N 34 45 E–2 42 N 34 20 E	A.S. Thomas
Araba ? see Aropa	U1	West Nile	?3 02 N 31 03 E	Purseglove
Arabaka, dam	U3	Teso	1 51 N 33 40 E	Kabuye
Arabuko, popl.	K7	Kilifi	3 15 S 39 58 E	R.M. Graham
Arabuko Sokoke, for.	K7	Kilifi	c.3 20 S 39 52 E	Numerous
Arapai, r.h.	U3	Teso	1 47.5 N 33 37 E	
Arawa, area	K3	Trans-Nzoia	1 09.5 N 34 42.5 E	Tweedie
Archer's Post, popl.	K1	Northern Frontier	0 39 N 37 41 E	D.C. Edwards, Gardner, Shantz
Ardai Plains	T2	Masai	c.3 21–3 26 S 36 20–36 25 E	Fuggles Couchman, Greenway, Bally
Ardai Ranch	T2	Masai	c.3 21 S 36 20 E	Kisaka
Aremo, popl.	U1	Karamoja	2 35 S 33 47 E	Tweedie
Arganu, R.	K5	South Kavirondo	prob. c.0 45 S 34 28 E	Jarrett
[Aridai] see Ardai	T2	Masai	c.3 21 S 36 20 E	Kisaka
Aringa, co.	U1	West Nile	3 35 N 31 15 E	Langdale-Brown, Eggeling

Name	Code	Region	Coordinates	Source
Aringa, str.	U1	Acholi	c.3 20 N 32 57 E	Eggeling, Greenway & Hummel
Arivu, popl., mssn., r.h.	U1	West Nile	2 52 N 30 59 E	
Ark, game lodge	K4	South Nyeri	0 21 S 36 48 E	Hooper & Townsend
Armfield, Mt.	T5	Mpwapwa	c.6 17 S 36 29.5 E	Hornby
Aropa, popl.	U1	West Nile	3 02 N 31 03 E	
Aror = Arror, R.	K3	Elgeyo	1 16 N 35 29 E–0 58 N 35 38 E	Mabberley & McCall, Tweedie
Arror, R.	K3	Elgeyo	1 16 N 35 29 E–0 58 N 35 38 E	
Arua, hill	U1	West Nile	3 01 N 30 55 E	
Arua, popl., mssn., r.h.	U1	West Nile	3 01.5 N 30 55 E	Numerous
Aruba, dam	K7	Teita	3 21 S 38 49 E	Greenway
Aruba, hill	K7	Teita	3 15 S 38 53 E	
Aruba, lodge	K7	Teita	3 21 S 38 49 E	Jeffrey et al., Greenway & Kanuri
Arusha, distr.	T2	Arusha	c.3 20 S 36 45 E	
Arusha, popl.	T2	Arusha	3 22 S 36 41.5 E	Numerous
Arusha Chini, popl.	T2	Moshi	3 35.5 S 37 21 E	Numerous
Arusha Juu, old area	T2	Arusha	3 20 S 36 42 E	Fischer
Arusha National Park	T2	Arusha	c.3 15 S 36 45 E	Numerous
(comprised of Meru Game Reserve and Ngurdoto Nat. Park)				
Asagari = Esaigeri, area, w.h.	K6	Masai	1 22 S 36 32 E	C.G. Rogers
Ashuwei, popl.	K7	Lamu	1 56 S 41 19 E	Rawlins
[Ashwe] see Ashuwei	K7	Lamu	1 56 S 41 19 E	Dale
[Asi] see Athi	K4/5	several	c.1 25 S 36 53.5 E–2 59 S 38 31 E	Fischer*
*(most likely Machakos Dist.)				
[Assua] see Amua	U1	West Nile	3 37 N 31 49 E	Eggeling
[Assua] see Aswa	U1/Sudan	Acholi/Lango	2 22 N 33 32 E–3 44 N 31 55 E	
Aswa, co.	U1	Acholi	3 00 N 32 30 E	
Aswa, R.	U1	Acholi	c.2 57 N 32 03 E–2 35 N 31 27 E	
Aswa, R.	U1/Sudan	Acholi/Lango	2 22 N 33 32 E–3 44 N 31 55 E	Eggeling
[Aswe] see Ashuwei	K7	Lamu	1 56 S 41 19 E	Rawlins
Atheno, hills	K2	Turkana	c.3 40 N 35 40 E	

9

Athi, R.	K4/6	several	c.1 25 S 36 53.5 E−2 59 S 38 31 E	Numerous
Athi Plains	K4/6	Nairobi/Masai	c.1 25 S 36 52 E	Numerous
Athi River, popl., rsta.	K4	Nairobi/Machakos	1 26.5 S 36 59 E	Numerous
Atiak, popl., mssn., t.c.	U1	Acholi	3 16 N 32 07 E	Chorley, Eggeling
[Atiek] see Atyak	U1	West Nile	2 35 N 30 54 E	Eggeling
Atira, popl., mssn., t.c.	U3	Teso	1 37 N 33 31 E	
[Attiak] see Atyak	U1	West Nile	2 35 N 30 54 E	Eggeling
[Atumatak] see Atumutaok	U1	Karamoja	2 14 N 34 39 E	Kerfoot
Atumutaok, hill	U1	Karamoja	2 14 N 34 39 E	
Atura, popl.	U1	Lango	2 07 N 32 20 E	A.S. Thomas
Atyak, popl.	U1	West Nile	2 35 N 30 54 E	
Austin, L.	T2	Mbulu	4 02.5 S 35 16.5 E	
Avisana	K7	Kwale	3 53 S 39 29 E	Kassner
Awach, popl., r.h.	U1	Acholi	2 59 N 32 24 E	Eggeling
Awach, popl.	K5	Central Kavirondo	0 05.5 S 34 29 E	Agnew et al.
Awal Abdulla, ln.	K1	Northern Frontier	3 56 N 41 21 E	Bally & Carter
Awara Plain, area	K1	Northern Frontier	3 45 N 41 07 E	Bally & Smith
Awasi, mssn.	K5	Central Kavirondo	0 10 S 35 05 E	Bjornstad
Awere, hill	U1	Acholi	2 42 N 32 46 E	
Awere, r.h.	U1	Acholi	2 43 N 32 50 E	
Ayangyangi, sw.	K1	Northern Frontier	1 55 N 36 05 E	Mathew, Mwangangi & Gwynne
Ayer, popl.	U1	Lango	2 16.5 N 32 41 E	Eggeling, A.S. Thomas
Ayipe, mt., f.r.	U1	West Nile	3 40 N 31 46 E	
Ayivu, co.	U1	West Nile	c.3 05 N 30 55 E	
Ayivu, popl.	U1	West Nile	3 05 N 30 54 E	Brasnett, Oakley
Ayo, str.	U1/Sudan	West Nile	3 46 N 31 39 E−3 51 N 31 49 E	
Ayweri Cwero, popl.	U1	Acholi	2 56 N 32 32 E	
Babati, popl.	T2	Mbulu	4 13 S 35 45 E	Numerous
[Babay] see Rabai	K7 *not* Z	Kilifi	3 56 S 39 34 E	Sacleux

Name		Area	Coordinates	Collector
[Babungi] see Bubungi				
[Bachuma] see Buchuma				
Badyang, popl.	U3	Mbale	1 05 N 34 20 E	Nganga
Baga, pl.	K7	Teita	3 38.5 S 38 54 E	Eggeling
Baga, f.r.	U1	Lango	2 02 N 32 54 E	Drummond & Hemsley, Braun
Baga II, f.r.	T3	Lushoto	4 47 S 38 25 E	Markham
Bagaga = Ugaga, area	T3	Lushoto	4 46.5 S 38 26.5 E	Peter
Bagala, rsta.	T3	Lushoto	4 48 S 38 28.5 E	Peter
[Bagalla] see Bagala				
Bagamoyo, popl.	T4	Buha	c.4 35 S 30 05 E	Peter
Bagamoyo, popl.	T6	Uzaramo	6 48 S 38 34.5 E	Zimmermann
Bagamoyo, distr.	T6	Uzaramo	6 48 S 38 34.5 E	Chabwela, Yohannes, Sayalel
Bagamoyo, popl., r.h.	T3	Tanga	5 11 S 38 51 E	Numerous
Bagara, area	T5	Dodoma	6 31 S 34 04.5 E	
Bagara = Baghara, str.	T6	Bagamoyo	c.6 15 S 38 30 E	
[Bagare] see Bagara				
Bagoyo, ridge	T6	Bagamoyo	6 25.5 S 38 54 E	Carmichael
Bagoyo, str.	T2	Mbulu	c.4 25 S 35 27 E	Richards
[Baha] see Kibaha				
[Bahadale] see Mbalambala				
Baharine Plains	T2	Mbulu	4 26 S 35 24 E–4 28 S 35 32.5 E	Greenway
Bahati, escarp.	T2	Mbulu	3 29.5 S 35 46.5 E	Peter
Bahati, for.	T2	Mbulu	c.3 30 S 35 47.5 E	Richards
Bahati, popl.	T2	Mbulu	6 46 S 38 55 E	Piers
Bahati (near Morogoro)	T6	Uzaramo	0 02.5 S 39 03.5 E	Bogdan, Bruce
Bahi, popl., rsta.	K1	Northern Frontier	prob. c.7 55 S 31 52 E	C.G. Rogers
Baio, mt.	T4	Ufipa	c.0 10 S 36 11 E	E.M. Bruce, Eggeling
Bajo, popl.	K3	Nakuru	c.0 06 S 36 12 E	Peter, Polhill & Paulo
Bakari-Kipindima, popl.	K3	Nakuru	1 17 S 36 51 E	Evans
Bakari-Rondo, popl.	K4	Nairobi	near 6 49 S 37 40 E	Dummer
	T6	Morogoro	5 59 S 35 19 E	Busse
	T5	Dodoma	1 46 N 37 32.5 E	Braun
	K1	Northern Frontier	0 23 N 32 43.5 E	
	U4	Mengo	10 12 S 38 55 E	
	T8	Lindi	10 08 S 39 15 E	
	T8	Lindi		

[Bakeria] see Bukuria				
Baker's Camp (Fatiko)				
Bakers View, popl., hill				
[Bakira] see Bwakira Chini				
Bakitabuk = Barkitabu, popl.				
Bako, sw.				
[Bakora] see Bukora				
Bakwa, area				
[Balambala] see Mbalambala				
Balangai, popl.				
Balangai East, f.r.				
Balangai West, f.r.				
[Balangati] see Mbalageti				
[Balangeti] see Mbalageti				
Balangida, L.				
Balawoli, popl.				
[Balbal] see Duluti				
[Balbal See] see Duluti Lake				
Balbala, w.h.				
[Balballa] see Balbala				
Bale, popl., mssn., t.c.				
Baleni, popl.				
Balenje, area, road				
Balesa Kulal, wells				
Balmoral (near Busingiro)				
Baloble, hill, area, w.h.				
[Balobleh] see Baloble				
[Baloboleh] see Baloble				
[Balogonja] see Bolgonja				
[Baloledi] see Bololedi				

Name	Code	Region	Coordinates	Collector
[Bakeria] see Bukuria	K5	South Kavirondo		
Baker's Camp (Fatiko)	U1	Acholi	3 01 N 32 19 E	Napier
Bakers View, popl., hill	U2	Bunyoro	1 14 N 30 47 E	Maxwell Forbes
[Bakira] see Bwakira Chini	T6	Morogoro	7 24 S 37 45 E	Lind
Bakitabuk = Barkitabu, popl.	K6	Masai	1 29 S 35 34.5 E	Schlieben
Bako, sw.	T6	Bagamoyo	6 29 S 38 53.5 E	Vesey-FitzGerald
[Bakora] see Bukora	U4	Masaka	0 49 S 31 29 E–0 50 S 31 41 E	Wingfield
Bakwa, area	T3	Handeni	5 57 S 37 18 E	Bagshawe
[Balambala] see Mbalambala	K1	Northern Frontier	c.0 05 S 39 07 E	J. Adamson, Bally
Balangai, popl.	T3	Lushoto	4 55.5 S 38 37 E	Peter, Faulkner
Balangai East, f.r.	T3	Lushoto	4 57 S 38 31 E	Hughes
Balangai West, f.r.	T3	Lushoto	4 56 S 38 25 E	
[Balangati] see Mbalageti	T1	Musoma/Maswa	c.2 53 S 34 54 E–2 12 S 33 49 E	Greenway
[Balangeti] see Mbalageti	T1	Musoma/Maswa	c.2 53 S 34 54 E–2 12 S 33 49 E	Greenway
Balangida, L.	T2	Mbulu	4 20 S 35 20 E	Hornby
Balawoli, popl.	U3	Busoga	1 02 N 33 06 E	
[Balbal] see Duluti	T2	Arusha	3 23 S 36 47 E	Uhlig
[Balbal See] see Duluti Lake	T2	Arusha	3 23 S 36 47 E	Uhlig
Balbala, w.h.	K1	Northern Frontier	1 51.5 N 40 44 E	
[Balballa] see Balbala	K1	Northern Frontier	1 51.5 N 40 44 E	Langdale-Brown, Pfennig
Bale, popl., mssn., t.c.	U4	Mengo	1 06 N 32 53 E	Greenway
Baleni, popl.	T6	Rufiji	7 51 S 39 48.5 E	Rodgers
Balenje, area, road	T8	Kilwa	c.8 41 S 38 32 E	Bally
Balesa Kulal, wells	K1	Northern Frontier	2 32 N 37 05 E	
Balmoral (near Busingiro)	U2	Bunyoro	near 1 44 N 31 28 E	Eggeling
Baloble, hill, area, w.h.	K1	Northern Frontier	c.3 16 N 40 01 E	Dale
[Balobleh] see Baloble	K1	Northern Frontier	c.3 16 N 40 01 E	
[Baloboleh] see Baloble	K1	Northern Frontier	c.3 16 N 40 01 E	Dale
[Balogonja] see Bolgonja	T1	Musoma	c.1 47 S 35 12 E–1 34 S 34 58 E	Greenway
[Baloledi] see Bololedi	K6/T1/T2	Masai/Musoma/Masai	c.1 52 S 35 40 E–2 13 S 35 08 E	Greenway

12

Balolo, gorge (Yaida Valley)	T2	Mbulu	c.4 00 S 35 05 E	Richards
Baloti, popl.	T2	Moshi	3 15 S 37 07 E	
[Balupa] see Buluba	U3	Busoga	0 30 N 33 23 E	G.H. Wood
[Balwada] see Bulwada	U4	Mengo	0 11 N 31 47 E	
Bamanda	T4	Tabora	Not traced	Boehm
Bamasuta (Kajara County)	U2	Ankole	c.1 00 S 30 10 E	Snowden
Bamba, popl.	K7	Kilifi	3 32 S 39 31 E	Bally, Dale, Moomaw
Bamba, t.a.	T3	Lushoto/Tanga	c.5 00 S 38 50 E	Peter, Bogner
[Bamba Bay] see Mbamba Bay	T8	Songea	11 17 S 34 46 E	Zimmer
Bamba Ridge, f.r.	T3	Lushoto	4 57 S 38 47 E	
Bamboo Forest, for., road	K4	Kiambu	c.0 55 S 36 36 E	Verdcourt
Bamburi, beach, popl.	K7	Mombasa	4 00 S 39 43.5 E	Drummond & Hemsley, Bally, Tweedie
Bamburi Cement Works, Baobab Ranch	K7	Mombasa	4 00 S 39 42 E	Bally & Carter
Bamgodjo	T1	Bukoba	Not traced	Holtz
Bamunanika, popl.	U4	Mengo	0 41 N 32 36 E	
Bana, f.r.	T6	Bagamoyo	c.6 36 S 38 57 E	Mgaza, Ruffo, Procter
Banagi, hill	T1	Musoma	2 18 S 34 50 E	Numerous
Banagi, popl., r.c.	T1	Musoma	2 16 S 34 51 E	Numerous
[Banani] see Benane	K1	Northern Frontier	0 40 N 40 30 E	J. Adamson
[Banas] see Banisa	K1	Northern Frontier	3 56.5 N 40 21 E	
Banda, popl., est.	U4	Mengo	0 17 N 32 49.5 E	Dummer
Banda, popl.	U4	Mengo	0 21 N 32 38 E	Pool, A.S. Thomas
Banda, f.r.	T6	Uzaramo	6 58 S 38 50 E	Procter, Shabani
Banderisoheria	T3	Pangani	c.5 23 S 38 52 E	Zimmermann, Tanner
Bandimi = Baudimi, mt.	T7	Rungwe	9 18 S 33 23 E	
[Banessa] see Banisa	K1	Northern Frontier	3 56.5 N 40 21 E	Gillett
Bangala, popl.	T7	Chunya	prob.c.8 22 S 32 55 E	
[Bangala] see Mbangala	T8	Masasi	c.10 40 S 38 20 E–11 09 S 38 55 E	Milne-Redhead & Taylor
Bangayega, popl.	T5	Dodoma	5 43 S 34 33 E	
Bangayega, rsta.	T5	Dodoma	5 43 S 34 38 E	Peter

13

Name	Div.	District	Coordinates	Collectors
[Bangoni] see Bargoni	K7	Lamu		J. Adamson
[Baninga] see Buninga	U4	Masaka	c.0 17 S 32 18 E	Maitland
Banisa, w.h., pan	K1	Northern Frontier	3 56.5 N 40 21 E	Delamere, Gillett, Riva
[Banissa] see Banisa	K1	Northern Frontier	3 56.5 N 40 21 E	
Banya, air strip	K1	Northern Frontier	4 26.5 N 36 15 E	
Baomo, popl.	K7	Tana River	1 56 S 40 08 E	
Bar, popl.	U1	Lango	2 15 N 33 02 E	
Barabili, popl.	U1	Acholi	2 42 N 32 19 E	
Barafu Kopjes, hills	T2	Masai	c.2 30 S 35 25 E	Greenway
Baragi, hills	T2	Masai	c.2 46 S 36 48 E	
Baragoi, popl., air strip, w.h.	K1	Northern Frontier	1 47 N 36 47.5 E	Bally, Carter & Stannard, J. Adamson
[Baragoli] see Baragoi	K1	Northern Frontier	1 47 N 36 47.5 E	
[Baraguess] see Uaraguess	K1	Northern Frontier	0 57 N 37 23.5 E	Dale
[Baranga] see Buranga	U2	Toro	0 50 N 30 10 E	Bagshawe
Barankata, popl., f.r.	T2	Moshi	3 11 S 37 15 E	Steele
Barata = Kuru Barata, popl., w.h.	K1	Northern Frontier	0 43 N 38 21 E	J. Adamson
[Barceloi] see Barsaloi	K1	Northern Frontier	1 20 N 36 52 E	
Barclay's Farm (12.8 km. N. of Njoro)	K3	Nakuru	c.0 13 S 35 57 E	Bogdan
Bargera, R.	K3	Baringo	c.0 11 N 35 39 E–0 32 N 36 05 E	
Bargoni, area	K7	Lamu	1 56 S 40 40 E	Kuchar
Bargoni, popl.	K7	Lamu	2 02 S 40 47 E	Bagshawe, Dawe
Bari, area	U1	Sudan not U1	c.4 00 N 31 30 E	
Barigoni, plains	K7	Lamu	c.1 58 N 40 46 E	
[Barikiva] see Barikiwa	T8	Kilwa	9 30 S 37 54 E	
Barikiwa, popl.	T8	Kilwa	9 30 S 37 54 E	Busse
Barikiwa, valley, str.	T8	Kilwa	9 30 S 37 52 E–9 28 S 38 06 E	Crosse-Upcott
Baringo, distr.	K3	Baringo	c.0 45 N 35 55 E	
Baringo, L.	K3	Baringo	c.0 38 N 36 05 E	Numerous
[Bariti] see Bwiti	T3	Lushoto	4 47 S 38 49 E	Peter
Barkitabu, popl.	K6	Masai	1 29 S 35 34.5 E	

Barsalinga, p.p.	K1	Northern Frontier	0 47 N 37 06 E	
Barsaloi, popl.	K1	Northern Frontier	1 20 N 36 52 E	
[Barsoloi] see Barsaloi	K1	Northern Frontier	1 20 N 36 52 E	
Bartagwet, hill	K1	Northern Frontier	1 55 N 36 56 E	Carter & Stannard
Bartolimo, popl.	K3	Baringo	0 41 N 35 49 E	Wimbush
Barton's Estate, fm.	K3	Naivasha	c.0 48 S 36 18 E	Pegler
[Barumba] see Burumba	U2	Ankole	0 58 S 30 49 E	Bagshawe
Bashay, popl., mssn.	T2	Mbulu	4 00 S 35 20.5 E	Richards
[Basi] see Bassi	T3	Tanga	5 09 S 38 56 E	Mgaza
Basodarer, area, dam, str.	T2	Mbulu	c.4 05 S 35 13 E	Carmichael
Basodesh, L., popl.	T2	Mbulu	4 18 S 35 07.5 E	B.D. Burtt
Basotu, L.	T2	Mbulu	4 22.5 S 35 05 E	B.D. Burtt, van Rensburg
Basotu, popl.	T2	Mbulu	4 22 S 35 04.5 E	B.D. Burtt, Bally
Bassanyek, spr.	T2	Mbulu	c.4 30 S 35 24 E	Carmichael
Bassi, f.r.	T3	Tanga	5 09 S 38 56 E	Mgaza, Semsei
[Bassoda-Merka] see Tlawi	T2	Mbulu	3 55 S 35 29 E	
[Basuto] see Basotu	T2	Mbulu	4 22 S 35 04.5 E	Carmichael
Batai = Batei, popl.	K2	West Suk	1 25.5 N 35 19 E	Meyerhoff
Baudimi = Bandimi, mt.	T7	Rungwe	9 18 S 33 23 E	Goetze
[Baumo] see Baomo	K7	Tana River	1 56 S 40 08 E	Sampson
[Bavanye] see Buanji	T7	Njombe/Mbeya	c.8 55 S 34 05 E	Stolz
[Bawanje] see Buanji	T7	Njombe/Mbeya	c.8 55 S 34 05 E	
[Bawanya, near Entebbe] ? see Buwaya	U4	Mengo	?0 08 N 32 24 E	Maitland
[Baya] see Mbeya	T7	Mbeya	c.8 50 S 33 20 E	
[Bayango] see Buyango	T1	Bukoba	1 09 S 31 38 E	Haarer
[Baya Nyundo] see Nyundo	K7	Kilifi	3 49 S 39 35 E	Thorold
[Baymo] see Baomo	K7	Tana River	1 56 S 40 08 E	
Beda, popl.	T1	Mwanza	2 36 S 33 00 E	Tanner
Beda, area	T1	Shinyanga	3 34 S 33 17.5 E	B.D. Burtt
[Begolo] ? see Bukulu	T5	Kondoa	?4 35 S 35 45 E	Peter

15

Begu Kwa Mlindi, popl.	T3	Handeni	5 28 S 37 44 E	Busse
Behobeho, str., area	T6	Morogoro/Rufiji	7 40 S 37 49 E – 7 37 S 37 52 E	Nicholson
[Beho-Beho] see Kwa Mhinda	T6	Rufiji	7 37 S 37 58 E	Goetze
Behungi, popl., r.c., sw.	U2	Kigezi	1 17 S 29 48 E	Numerous
Beila, w.h.	K1	Northern Frontier	3 59 N 41 42 E	Riva
Bela = Beila	K1	Northern Frontier	3 59 N 41 42 E	Riva
Belam, str.	T2	Arusha	c.3 09 S 36 51 E	Richards
Belam ma larg, farm	T2	Arusha	c.3 10 S 36 51 E	Richards
Belazoni, canal, popl.	K7	Lamu	c.0 31 S 40 21 E	Sampson
Belengi, ln.	T4	Mpanda	c.6 00 S 29 50 E	Harley
[Belessa Kulal] see Balesa Kulal	K1	Northern Frontier	2 32 N 37 05 E	Bally
Belgut, ln.	K5	Kericho	0 20 S 35 10 E	Kerfoot
Belkros, t.a.	K1	Northern Frontier	c.1 25 N 39 52 E	Pratt
Belo = Bilo, for., str.	T2	Arusha	c.3 15 S 36 48 E	Richards
[Beloble] see Baloble	K1	Northern Frontier	c.3 16 N 40 01 E	J. Adamson
Benane, w.h.'s, pans	K1	Northern Frontier	c.0 40 N 40 30 E	J. Adamson
[Benani] see Benane	K1	Northern Frontier	0 40 N 40 30 E	
Bendera, popl.	T8	Songea	10 52 S 34 58 E	Busse
Benet, area	U3	Mbale	c.1 22 N 34 33 E	Eggeling, A.S. Thomas, J. Wilson
Berega, mssn.	T6	Kilosa	6 11.5 S 37 08 E	B.D. Burtt, Hornby
Bereku, popl., area	T5	Kondoa	4 27 S 35 44 E	Numerous
[Beréu] see Bereku	T5	Kondoa	4 27 S 35 44 E	Peter
Berkeley, bay	U3	Mbale	c.0 14 N 33 58 E	Scott Elliot
[Beru] see Bereku	T5	Kondoa	4 27 S 35 44 E	Peter
[Beruli] see Buruli	U4	Mengo	c.1 25 N 32 20 E	Eggeling
[Besereku] see Biseruka	U2	Bunyoro	1 33 N 31 10 E	
Beta, popl., mssn.	U4	Masaka	0 18 S 32 08 E	
Bet el Ras, ruin	Z	Zanzibar	6 07 S 39 12.5 E	Oxtoby
[Beya] see Mbeya	T7	Mbeya	c.8 50 S 33 20 E	
Bibisande, popl.	T5	Dodoma	5 52 S 34 15 E	Stuhlmann

Bicha, popl., L.	T5	Kondoa	4 56 S 35 49 E	Hazel
Bigando, popl.	U2	Bunyoro	1 09 N 31 36 E	Bagshawe
Bigera, str.	U2	Ankole	c.0 04 N–0 06 S 30 26 E	A.S. Thomas
[Biggo] see Bigo	U4	Masaka	0 09 N 31 15 E	Greenway
Big Momela = Momela Kubwa, L.	T2	Arusha	3 13 S 36 54.5 E	Hedberg, Osmaston
Bigo, hut	U2	Toro	0 23.5 N 29 55 E	
Bigo, ruin	U4	Masaka	0 09 N 31 15 E	
Bigoro, popl.	T1	Bukoba	1 38 S 31 03 E	Ford
Bigwa, popl.	T6	Uzaramo	7 12.5 S 39 10 E	Procter
Bihanga, rsta.	U2	Toro	0 12 N 30 38 E	
Biharamulo, distr.	T1	Biharamulo	c.2 35 S 31 25 E	Carmichael
Biharamulo, f.r.	T1	Biharamulo	3 05 S 31 30 E	Gane, Procter, Verdcourt
Biharamulo, popl., r.h., mssn.	T1	Biharamulo	2 38 S 31 18 E	Polhill & Paulo
Bihawana, mssn.	T5	Dodoma	6 16 S 35 38 E	Godman
[Bihungi] see Behungi	U2	Kigezi	1 17 S 29 48 E	Eggeling
[Biiso] see Biso	U2	Bunyoro	1 46 N 31 25 E	Peter
[Bikare] see Kwa Bikare	T4	Buha	near 4 27 S 29 58 E	Lye
Biko, hill	U4	Mengo	0 46 N 31 56 E	G. Taylor
Bikoni, hill	U2	Toro	0 21.5 N 30 02.5 E	
Bikonzi, popl.	U2	Bunyoro	1 36 N 31 39 E	Gillett
Bil Bil, w.h., area, str./s	K7	Tana River	c.1 02 S 39 48 E	Greenway
Bilo = Belo, R.	T2	Arusha	c.3 15 S 36 48 E	B.D. Burtt
[Bincha] see Bicha	T5	Kondoa	4 56 S 35 49 E	Procter
Binguni, R.	T6	Uzaramo	c.6 59 S 39 17 E	
Binyin = Binyiny	U3	Mbale	1 25 N 34 32 E	Eggeling
[Binyinyi] see Binyiny	U3	Mbale	1 25 N 34 32 E	A.S. Thomas
[Birinze] see Birinzi	U4	Masaka	0 15 S 31 53 E	A.S. Thomas
Birinzi, popl., L.	U4	Masaka	0 15 S 31 53 E	Peter
[Birira] see Mbirira	T4	Buha	4 21 S 30 10 E	
Biseruka, popl.	U2	Bunyoro	1 33 N 31 10 E	Eggeling

Name	Region	District	Coordinates	Collector
Bololedi, R.	K6/T1/T2	Masai/Musoma/Maswa	c.1 52 S 35 40 E–2 13 S 35 08 E	Greenway
[Bolongonja] see Bolgonja	T1	Musoma	c.1 47 S 35 12 E–1 34 S 34 58 E	Haarer
[Boloti] see Baloti	T2	Moshi	3 15 S 37 07 E	Drummond & Hemsley
Boma, pen.	T3	Tanga	4 53 S 39 11 E	Trapnell
[Bomala] see Bumala	K5	North Kavirondo	0 18 N 34 12 E	Stolz
Bomalakitana (near Kyimbila)	T7	Rungwe	near 9 17 S 33 39 E	Richards
Boma la Megi	T2	Arusha	3 15 S 36 52 E	
Bomalandani, popl.	T3	Tanga	4 52 S 39 12 E	Uhlig
Boma la Ngombe, popl.	T2	Moshi	3 20 S 37 10 E	McLoughlin, Mwazumbi
Boma la Mzinga, for.	T7	Iringa	8 07 S 36 05 E	Drummond & Hemsley
[Bomandani] see Bomalandani	T3	Tanga	4 52 S 39 12 E	Gillespie
Bomani, I.	K7	Lamu	c.1 46 S 41 30 E	Richards
Bomariva, est., area	T7	Mbeya	c.9 11 S 32 52 E	Numerous
Bombo, popl.	U4	Mengo	0 35 N 32 32 E	Greenway
Bombo, hill	T3	Lushoto	4 25 S 38 22.5 E	Peter
Bombo, hill	T3	Lushoto	5 07.5 S 38 33 E	
Bombo, popl.	T3	Lushoto	4 55 S 38 10 E	Greenway
Bombo, popl.	T3	Lushoto	4 51 S 38 41 E	Holst
Bombo, R.	T3	Lushoto/Tanga	c.4 46 S 38 42 E–4 31 S 38 47 E	Cribb & Grey-Wilson
[Bomboni] see Bumbuli	T3	Lushoto	4 52 S 38 28 E	Milne-Redhead & Taylor, Faulkner, Holst
Bombwera, popl., t.c., mssn.	T3	Tanga	5 12 S 38 43 E	Kassner
Bome, R.	K7	Kwale	4 06 S 39 29 E	Numerous
Bomole, for.	T3	Lushoto	c.5 06 S 38 37 E	Numerous
Bomole, hill	T3	Lushoto	5 06 S 38 37 E	
[Bomwera] see Bombwera	T3	Tanga	5 12 S 38 43 E	Holst
[Bonda (W. Usambaras)] ? see Bondei	T3	Lushoto	4 44.5 S 38 29.5 E	
Bondei, popl.	T3	Lushoto	4 44.5 S 38 29.5 E	
Bondei, t.a.	T3	Tanga/Pangani	c.5 15 S 38 50 E	
Bondo, popl.	K5	Central Kavirondo	0 06 S 34 16.5 E	Agnew et al.
Bondoni, hill	K4	Machakos	1 39 S 37 14.5 E	Prescott Decie, Trump

20

Name		Region	Coordinates	Collectors
[Buangai] see Bwanjai	T1	Bukoba	1 13 S 31 42 E	Mildbraed
Buanji, t.a.	T7	Njombe/Mbeya	c.8 55 S 34 05 E	
[Bubande] see Bubandi	U2	Toro	0 39 N 29 59 E	Liebenberg, Eggeling
Bubandi, popl.	U2	Toro	0 39 N 29 59 E	Maitland
Bubembe, I.	U4	Masaka	0 27 S 32 20 E	Numerous
Bubu, R., sw.	T2/5	Mbulu/Končoa/Dodoma	4 15 S 35 29 E – 6 03 S 35 19 E	Busse
Bubu, R.	T5	Dodoma	6 18 S 35 54 E – 7 03 S 35 49 E	Oxtoby, Vaughan, R.O. Williams
Bububu, popl.	Z	Zanzibar	6 06 S 39 13 E	
Bububu, popl.	Z	Zanzibar	5 56 S 39 14 E	Tothill
Bubulo, popl.	U3	Mbale	0 57 N 34 17 E	Harris, Tothill
[Bubulu] see Bubulo	U3	Mbale	0 57 N 34 17 E	Maitland
Bubungi, popl., r.c.	U3	Mbale	1 05 N 34 20 E	Eggeling
Bubwe, str., for.	U2	Bunyoro	1 46 N 31 28 E – 1 51 N 31 23 E	Kennedy
Bucharwe, hill (S. of Mbarara)	U2	Ankole	c.0 48 S 30 37 E	B.D. Burtt
Buchosa, t.a.	T1	Mwanza	c.2 30 S 32 15 E	Bogdan, Polhill & Paulo, Verdcourt
Buchuma, rsta.	K7	Teita	3 38.5 S 38 54 E	Greenway
Buchuma Gate, ln.	K7	Teita	3 40 S 38 37.5 E	Numerous
Budadiri, popl.	U3	Mbale	1 10 N 34 20 E	Dale, Drummond & Hemsley, Napper
Buda Forest, f.r.	K7	Kwale	c.4 27 S 39 24 E	
Budaka, popl.	U3	Mbale	1 01 N 33 57 E	
Budaka Bugwere, co.	U3	Mbale	c.1 05 N 33 45 E	Snowden, Maitland
[Budama, old distr.] see Bukedi distr. S. of Namatala R.	U3	Mbale	c.0 40 N 34 00 E	
Budda, popl.	U4	Masaka	0 18 S 31 42 E	H.B. Johnston
[Buddo] see Buddu	U4	Masaka	c.0 30 S 31 45 E	
Buddu, co.	U4	Masaka	c.0 30 S 31 45 E	Dawe, Fyffe, Eggeling
[Buddu Wald] see Minziro Forest	U4/T1	Masaka/Bukoba	c.1 00 S 31 50 E	Mildbraed
Budo, mssn., for.	U4	Mengo	0 16 N 32 29 E	Eggeling, Chandler & Hancock
Budongo, f.r.	U2	Bunyoro	c.1 47 N 31 35 E	Numerous
Budongo, popl.	U2	Bunyoro	1 39 N 31 35 E	

Bududa, popl.	U3	Mbale	1 01 N 34 20 E	Snowden, Maitland
Budutu, ln.	T1	Mwanza	2 48.5 S 33 07 E	Tanner
Buekeragi Ridge	U2	Toro	0 11 N 29 53 E	Ross
Bueni, popl.	T7	Iringa	8 24 S 35 06 E	Goetze
Buffalo Point, ln.	T2	Arusha	3 18 S 36 56 E	Greenway
Buffalo Ridge, ln.	K7	Kwale	4 14 S 39 26 E	Bally & Smith
Buffalo Springs, spr.	K1	Northern Frontier	0 36.5 N 37 39 E	Polhill
[Bufindi] see Bufundi				
Bufumbira, co.	U2	Kigezi	1 17 S 29 54 E	Linder
Bufumbira, co.	U2	Kigezi	c.1 12 S 29 40 E	Purseglove, Eggeling
Bufumbo, popl.	U3	Mbale	1 06 N 34 15 E	Numerous
Bufumira, I.	U4	Masaka	c.0 21 S 32 24 E	Purseglove, A.S. Thomas, Maitland
Bufundi, popl. r.h.	U2	Kigezi	1 17 S 29 54 E	Numerous
Buga, popl. area	T5	Mpwapwa	c.6 58 S 36 36 E	Schmidt
Bugabe, popl.	U4	Mengo	0 32.5 N 33 00 E	Dummer
Bugabo (near L. Nabugabo)	U4	Masaka	near 0 22 S 31 54 E	Tallantire, Lye
Bugabula, co.	U3	Busoga	c.1 05 N 33 10 E	G.H. Wood
Bugadi, popl.	U3	Busoga	0 18 N 33 31 E	
Bugaga = Ugaga, area	T4	Buha	c.4 35 S 30 05 E	Peter
Bugahya, co.	U2	Bunyoro	c.1 30 N 31 20 E	
Bugaia, I.	U4	Mengo	c.0 03 N 33 16 E	
Bugala, I.	U4	Masaka	c.0 20 S 32 10 E	Maitland, Symes, A.S. Thomas
Bugala, I.	U4	Masaka	c.0 40 S 32 20 E	
Bugamba, mt., for.	U2	Ankole	0 45 S 30 27.5 E	Eggeling
Bugamba, popl., r.h.	U2	Ankole	0 44.5 S 30 31 E	Numerous
Bugamba, mssn.	U2	Ankole	0 44 S 30 29 E	
Bugambe, popl., mssn.	U2	Bunyoro	1 23 N 31 15 E	
Bugambe, popl.	U2	Bunyoro	1 26 N 31 13 E	
Bugambe, area	U2	Bunyoro	c.1 26 N 31 12 E	
Buganda, prov., kingdom	U4	Masaka/Mengo/Mubende	c.0 20 N 32 15 E	Maitland

Name		District	Coordinates	Collectors
Bugandika, popl.	T1	Bukoba	1 15 S 31 38 E	Gillman, Haarer, Ford
Bugando, area	T1	Mwanza	c.2 30 S 32 03 E	Stuhlmann, B.D. Burtt
[Bugangadzi] see Bugangazi				
Bugangari, popl.	U4	Mubende	c.1 00 N 31 20 E	Dawe, Snowden
Bugangazi, popl.	U2	Kigezi	0 43 S 29 51 E	Purseglove, Thornton
Bugangazi, co.	U4	Mubende	c.1 00 N 31 20 E	
[Buganzo] see Bugonzo				
Bugara, mt.	U3	Busoga	0 52 N 33 24 E	G.H. Wood
Bugarama, t.c.	T1	Bukoba	1 14 S 31 02 E	Watkins
Bugaya, popl.	T1	Ngara	2 52.5 S 30 32 E	Tanner
[Bugaya] see Bugaga	U3	Busoga	1 06 N 33 15 E	
[Bugembe] see Bubembe	T4	Buha	c.4 35 S 30 05 E	Peter
Bugene, popl.	U4	Masaka	0 27 S 32 20 E	A.S. Thomas
[Bugenue] ? see Bugene	T1	Bukoba	1 35 S 31 08 E	Haarer, Ford, Procter
Buget, popl.	T1	Bukoba	?1 35 S 31 08 E	Stuhlmann
[Bugera] see Bugara	T2	Mbulu	3 39,5 S 35 38.5 E	
Bugerere, co.	T1	Bukoba	1 14 S 31 02 E	
[Bugesege] see Bugusege	U4	Mengo	c.1 00 N 32 50 E	Numerous
[Buginyanga] see Buginyanya	U3	Mbale	1 09 N 34 15 E	
Buginyanya, popl. r.c.	U3	Mbale	1 17 N 34 21 E	
Bugiri, popl., r.h.	U3	Mbale	1 16.5 N 34 22 E	Hancock, A.S. Thomas, Snowden
	U3	Busoga	0 34 N 33 45 E	Drummond & Hemsley, G.H. Wood, Eggeling
Bugiri-Busoga, f.r.	U3	Busoga	c.0 34 N 33 45 E	Chandler, Hedberg
[Bugishu] see Bugisu	U3	Mbale	c.1 00 N 34 20 E	Peal, Lye
Bugisu, distr.	U3	Mbale	c.1 00 N 34 20 E	
(considered as part of Mbale District for F.T.E.A.)				
Bugolora, popl.	T1	Bukoba	1 13,5 S 31 34 E	Ukiriguru Res. Centre
Bugoma, f.r.	U2	Bunyoro	c.1 20 N 31 05 E	Numerous
Bugoma, mssn.	U4	Masaka	0 15 S 32 04 E	Maitland
Bugombe, popl., mssn.	U4	Mengo	0 05 S 32 44 E	Lye
Bugomolo, popl.	U2	Bunyoro	1 29,5 N 31 26 E	Eggeling

23

Place	Region	Code	Coordinates	Collectors
Bugona, popl.	Masaka	U4	0 31 S 31 16 E	Purseglove
Bugondo, fy., t.c.	Teso	U3	1 37 N 33 17 E	Chandler, Maitland
Bugongi, popl.	Ankole	U2	0 38 S 30 15 E	Eggeling, Snowden
Bugongo = Bugonzo, popl.	Busoga	U3	0 52 N 33 24 E	
Bugonzi, popl.	Masaka	U4	0 15 S 31 52 E	Lye & Katende
Bugonzo, popl.	Busoga	U3	0 52 N 33 24 E	G.H. Wood
[Bugorora] see Bugolora	Bukoba	T1	1 13.5 S 31 34 E	Pitt-Schenkel
Bugoye, r.h., f.r.	Toro	U2	0 18.5 N 30 06 E	Eggeling
Bugoye, popl.	Mengo	U4	0 12.5 N 32 46 E	Dummer
[Buguba Rock] ? see Buguda Hill	Teita	K7	?3 42 S 38 38 E	Bally
Buguda, hill	Teita	K7	3 42 S 38 38 E	
Bugufi, distr.	Ngara	T1	2 30 S 30 37 E	Chambers, Rounce, Tanner
[Bugumbe] see Bubembe	Masaka	U4	0 27 S 32 20 E	A.S. Thomas
Bugungu, area	Bunyoro	U2	c.2 10 N 31 40 E	Dummer, Eggeling
Bugusege, popl., t.c.	Mbale	U3	1 09 N 34 15 E	
Bugwe, West, f.r.	Mbale	U3	c.0 32 N 34 00 E	Dawkins, G.H. Wood
Bugwere, old distr.	Teso	U3	c.1 15 N 34 00 E	Snowden
Bugweri, co.	Busoga	U3	c.0 40 N 33 35 E	H.B. Johnston, Osmaston, Snowden
Buha, distr.	Buha	T4	c.4 00 S 30 30 E	
Buhaguzi, co.	Bunyoro	U2	c.1 20 N 31 05 E	
Buhama, mssn.	Mwanza	T1	2 21 S 32 29 E	Carmichael
Buhamba, near Kigoma	Kigoma	T4	near 4 53 S 29 38.5 E	Ross
Buhamila, popl.	Bukoba	T1	1 35.5 S 31 16 E	
[Buhamira] see Buhamila	Bukoba	T1	1 35.5 S 31 16 E	Haarer
Buhanika, popl.	Bunyoro	U2	1 25 N 31 25 E	
Buhara, popl.	Kigezi	U2	1 22 S 30 02 E	Purseglove, Norman
Buharati = Mburahati (near Dar es Salaam)	Uzaramo	T6	near 6 48 S 39 15 E	Peter
Buhemba, area	Musoma	T1	c.1 45 S 34 09 E	Grant
Buhemba, mine, p.p.	Musoma	T1	1 48 S 34 06 E	
Buhemba, popl., sch.	Musoma	T1	1 46 S 34 05 E	

Name		Region	Coordinates	Authority
Buhimba, popl.	U2	Bunyoro	1 20 N 31 19 E	Carmichael
Buhindi, f.r.	T1	Mwanza	c.2 25 S 32 15 E	Eggeling
Buhinga, popl.	T4	Buha	4 36.5 S 30 01 E	G.H. Wood
[Buholi] see Bukoli	U3	Busoga	c.0 20 N 33 50 E	Carmichael
Buholoholo, t.a.	T4	Mpanda	c.6 00 S 30 00 E	
Buhoro, popl.	T4	Buha	4 25 S 30 10 E	
Buhoro, popl.	T4	Buha	4 32 S 30 47.5 E	Disney, Procter
Buhoro Flats, plains	T7	Mbeya	c.8 30 S 34 30 E	Tanner
[Buhumbi] see Bukumbi	T1	Mwanza	c.2 40 S 33 00 E	Tanner
Buhunda, ln.	T1	Mwanza	c.2 55 S 32 51 E	Eggeling
Buhundu, popl.	U2	Toro	0 46 N 30 05.5 E	Rounce
Buhungukira, t.a.	T1	Kwimba	c.3 12 S 32 56 E	Synnott, Snowden
Buhweju, co.	U2	Ankole	c.0 18 S 30 18 E	Purseglove, Snowden
[Buhwezu] see Buhweju	U2	Ankole	c.0 18 S 30 18 E	Numerous
Buiko, popl.	T3	Lushoto	4 40 S 38 03 E	
Buiko, rsta.	T3	Lushoto	4 39.5 S 38 03 E	Osmaston
Buikwe, popl., rsta.	U4	Mengo	0 20 N 33 02 E	Holst, Cribb & Grey-Wilson
[Buiti] see Bwiti	T3	Lushoto	4 47 S 38 49 E	Hildebrandt
[Buityuma] see Buchuma	K7	Teita	3 38.5 S 38 54 E	Eggeling
Bujawe, f.r.	U2	Bunyoro	1 32 N 31 12 E	Dawe
Bujeju, for.	U4	Masaka	0 19 S 32 02 E	Purseglove
Bujenje, co.	U2	Bunyoro	c.1 50 N 31 30 E	
[Bujenzi] see Buyenzi	T4	Buha	4 27 S 29 58 E	Tanner
Bujingwa, area	T1	Mwanza	c.2 45 S 32 56 E	Duke of Abruzzi
Bujongolo, camp	U2	Toro	0 20 N 29 55 E	Dawkins, Lye
Bujuko, popl.	U4	Mengo	0 20 N 32 22 E	Eggeling
Bujuku, L.	U2	Toro	0 22.5 N 29 53 E	Numerous
Bujuku, str.	U2	Toro	0 22.5 N 29 53.5 E–0 21.5 N 29 58 E	
[Bujuni] see Buyuni	T3	Pangani	5 57 S 38 47.5 E	Numerous
Bukakata, mssn.	U4	Masaka	0 18 S 32 00 E	

Name	Region	District	Coordinates	Authority
Bukakata, pier	U4	Masaka	0 18 S 32 02 E	
Bukalasa, popl., farm inst.	U4	Mengo	0 43 N 32 30 E	Maitland, Snowden, Staples
Bukama, popl.	T4	Nzega	4 09.5 S 33 49 E	
[Bukarungu] see Bukurungu	U2	Toro	0 01 S 30 17.5 E	Bagshawe
Bukasa, I.	U4	Masaka	c.0 26 S 32 30 E	Numerous
Bukasa, popl.	U4	Mengo	0 19 N 32 40.5 E	Lind
Bukasa, f.r.	U4	Mengo	c.0 18 N 32 40.5 E	Eggeling
[Bukasa] see Magyo	U4	Mengo	0 10 N 33 17.5 E	Eggeling
[Bukassa] see Bukasa	U4	Masaka	c.0 26 S 32 30 E	
Bukedea, co.	U3	Teso	c.1 25 N 34 10 E	Snowden
Bukedea, popl., r.h.	U3	Teso	1 21 N 34 03 E	A.S. Thomas
Bukedi, distr.	U3	Mbale	c.0 40 N 34 00 E	Peal & Sakwa, Maitland
		(considered as part of Mbale District for F.T.E.A.)		
Bukenga	T1	Bukoba	Not traced	Ford
Bukeri, popl., mssn.	U4	Masaka	0 27 S 31 46 E	
[Bukeria] see Bukuria	K5	South Kavirondo	c.1 12 S 34 35 E	Napier
Bukiberu, ln.	U4	Mengo	0 02 N 32 26 E	Lye
Bukiga, popl.	U3	Mbale	1 03 N 34 21 E	Snowden
Bukimbiri, popl.	U2	Kigezi	1 11.5 S 29 42 E	Purseglove
Bukinda, popl., mssn.	U2	Kigezi	1 12 S 30 08 E	Leakey
Bukindo, popl.	T1	Mwanza	2 30 S 32 00 E	Stuhlmann
Bukindo, popl.	T1	Mwanza	2 02 S 33 06 E	
[Bukingukira] see Buhungukira	T1	Kwimba	c.3 12 S 32 56 E	Rounce
Bukiriro, popl.	T1	Ngara	2 47 S 30 33 E	Tanner
[Bukkol] see Bokkol	K1	Northern Frontier	1 52 N 37 01.5 E	
Bukoba, distr.	T1	Bukoba	c.1 35 S 31 30 E	Numerous
Bukoba, popl.	T1	Bukoba	1 20 S 31 49 E	Numerous
Bukoli, co.	U3	Busoga	c.0 20 N 33 50 E	Snowden, G.H. Wood
Bukoloto, popl.	U4	Mengo	0 41 N 32 55 E	
Bukome, area	T1	Biharamulo	c.2 40 S 31 40 E	Stuhlmann

Name	Code	District	Coordinates	Collector
Bukomero, popl., r.h.	U4	Mengo	0 41.5 N 32 02 E	Eggeling
Bukonde, popl.	U3	Mbale	1 05 N 34 13 E	Chandler, Snowden
[Bukora] see Bokora	U1	Karamoja	c.2 20 N 34 20 E	
Bukora, R.	U4	Masaka	0 49 S 31 29 E–0 50 S 31 41 E	
Buku, popl.	U4	Mengo	0 03.5 N 32 26 E	Eggeling
[Bukuloto] see Bukoloto	U4	Mengo	0 41 N 32 55 E	Pfennig
Bukulu, popl.	T5	Kondoa	4 35 S 35 45 E	
Bukumbi, est.	U2	Toro	0 40 N 30 11.5 E	Hazel
Bukumbi, area	T1	Mwanza	c.2 40 S 33 00 E	Tanner, Procter
Bukumbi, mssn.	T1	Mwanza	2 43 S 32 55 E	Stuhlmann
Bukumi, popl., r.h.	U2	Bunyoro	1 47 N 31 23 E	Dawkins, Eggeling, Purseglove
[Bukungukira] see Buhungukira	T1	Kwimba	c.3 12 S 32 56 E	
Bukura, popl., Agr. Dept.	K5	North Kavirondo	0 13.5 N 34 37 E	Barney, M.D. Graham
Bukuria, area	K5	South Kavirondo	c.1 12 S 34 35 E	Napier
Bukurungu, popl.	U2	Toro	0 01 S 30 17.5 E	
Bukurungu, str.	U2	Toro	c.0 24 N 29 55 E	
Bukuya, popl., r.h., r.c., t.c.	U4	Mengo	0 41 N 31 50 E	H.D. Johnston
Bukuzi (Bugala I.)	U4	Masaka	c.0 20 S 32 10 E	A.S. Thomas
Bukwa, popl., r.c.	U3	Mbale	1 18 N 34 45 E	Norman, Snowden, Eggeling
Bukwali, popl.	U2	Toro	0 39 N 30 17 E	Eggeling
Bukwaya, area	T1	Musoma	1 34 S 33 46 E	Tanner
Bukwimba, popl.	T1	Kwimba	2 48 S 33 26 E	
Bula ? see Buna	K1	Northern Frontier	?2 47 N 39 31 E	Dale
[Bulaga] see Bulago	U3	Mbale	1 15 N 34 21 E	Snowden
Bulago, popl., r.c.	U3	Mbale	1 15 N 34 21 E	Forbes, A.S. Thomas, Snowden
Bulamagi, popl., hill	U4	Mengo	0 25 N 33 07 E	Dummer
[Bulamajo] see Bur Maiyo	K1	Northern Frontier	2 58.5 N 40 16 E	Dale
[Bulambia] see Ulambya	T7	Rungwe	c.9 20 S 33 10 E	Stolz
Bulambuli, popl.	U3	Mbale	1 10 N 34 23 E	Numerous
[Bulambya] see Ulambya	T7	Rungwe	c.9 20 S 33 10 E	Stolz

27

28

Name		Region	Coordinates	Authority
Buluba, leper colony	U3	Busoga	0 30 N 33 23 E	G.H. Wood
Bulucheke, popl., mssn., r.h.	U3	Mbale	1 02.5 N 34 22 E	Forbes, Dale
Buluganya, popl.	U3	Mbale	1 13 N 34 22 E	A.S. Thomas, Peal
Bululu, R.	T3	Lushoto	c.5 03 S 38 23 E–5 04.5 S 38 33 E	Peter
[Bulumagi] see Bulamagi	U4	Mengo	0 25 N 33 07 E	Numerous
[Bulumezi] see Bulemezi	U4	Mengo	c.1 00 N 32 20 E	Numerous
Bulwa, tea est.	T3	Lushoto	5 02.5 S 38 38 E	Eggeling
Bulwada, popl., r.h.	U4	Mengo	0 11 N 31 47 E	Carmichael
Bulyahilo, ln.	T1	Mwanza	c.2 25 S 32 15 E	Eggeling
Bulyango, popl.	U2	Bunyoro	1 37.5 N 31 33 E	
Bumala, popl.	K5	North Kavirondo	0 18 N 34 12 E	Peal, A.S. Thomas
Bumasifwa, r.h., f.r.	U3	Mbale	1 11 N 34 22 E	Stuhlmann
[Bumbide] see Bumbiri	T1	Bukoba	1 40 S 31 52 E	Stuhlmann
[Bumbinde] see Bumbiri	T1	Bukoba	1 40 S 31 52 E	
Bumbiri, I.	T1	Bukoba	1 40 S 31 52 E	Drummond & Hemsley, Peter
Bumbuli, popl., mssn., r.h., t.c.	T3	Lushoto	4 52 S 38 28 E	?Peter
Bumbuli, popl. ?	T3	Lushoto	prob. c.5 07 S 38 40 E	Stuhlmann
Bumpeke, popl.	T1	Mwanza	3 05 S 32 28 E	Dawkins
Bumpenje, hill	U4	Mengo	0 11 N 32 25.5 E	Dale, Hemming, Bally, Greenway
Buna, popl., t.c.	K1	Northern Frontier	2 47 N 39 31 E	Lye
Bunado, popl.	U4	Masaka	0 19 S 31 59 E	B.D. Burtt, Eggeling
Bunagana, popl., r.c.	U2	Kigezi	1 18 S 29 36 E	Gillman
Bunazi, popl.	T1	Bukoba	1 13 S 31 24 E	Gillett
Bundali = Undali, t.a., hills	T7	Mbeya/Rungwe	c.9–10 30 S 33 15–30 E	R.M. Davies, Stolz
Bundali, mt.	T7	Rungwe	9 30 S 33 27 E	Ross, A.S. Thomas
Bundibugyo, popl., r.h.	U2	Toro	0 43 N 30 04 E	Osmaston
Bundimbere, for.	U2	Toro	0 41 N 30 01 E	Greenway
[Bundugi] see Dundugi	T7	Iringa	7 34 S 34 49 E	Numerous
Bunduki, popl., r.h.	T6	Morogoro	7 01.5 S 37 38 E	Tanner
Bunegeji, ln.	T1	Mwanza	c.2 45 S 32 56 E	

29

Name	Region	District	Coordinates	Authority
[Bura L.] see Bicha				
Burahya, co.	U2	Toro	c.0 40 N 30 20 E	
Buramayo, hill (near Mandera)	K1	Northern Frontier	near 3 56 N 41 52 E	Dale
Buranga, hot springs	U2	Toro	0 50 N 30 10 E	Maitland, A.S. Thomas
Buranga, pass	U2	Toro	c.0 50 N 30 12 E	
[Burdali] see Mbalambala	K1	Northern Frontier	0 02.5 S 39 03.5 E	Sampson
[Burduras] see Bur Duras	K1/Ethiopia	Northern Frontier	3 52 N 39 52 E	Gillett
Bur Duras, mt.	K1/Ethiopia	Northern Frontier	3 52 N 39 52 E	
[Burgone] see Bargoni	K7	Lamu		
Burguret Track	K4	North Nyeri	0 08 S 37 07 E–0 09 S 37 19 E	J. Adamson
Buri, hill	K1	Northern Frontier	3 14 N 40 00 E	Hedberg
Burigi, L.	T1	Bukoba	2 07 S 31 16 E	Bally & Smith
Burko, mt., f.r.	T2	Masai	3 19 S 36 13 E	Procter
Bur Maiyo, hill	K1	Northern Frontier	2 58.5 N 40 16 E	Carmichael
Burnt Forest p.p.	K3	Uasin Gishu	0 13 N 35 26 E	Dale, Webster, Tweedie
Burole, hill	K1	Northern Frontier	3 33 N 38 38 E	Gillett, Bally & Smith
[Burroli] see Burole	K1	Northern Frontier	3 33 N 38 38 E	Bally
Burrungat, plain	K6	Masai	1 27 S 35 09 E	Taiti
Buru, peak	K3	Naivasha	0 38 S 36 16 E	
Buruli, co.	U2	Bunyoro	c.1 35 N 31 50 E	Eggeling, H.B. Johnston, A.S. Thomas
Buruli, co.	U4	Mengo	c.1 25 N 32 20 E	Tanner
Buruma, area, popl.	T1	Musoma	c.1 39 S 33 50 E	
Burumba, popl	U2	Ankole	0 58 S 30 49 E	Richards
Burungi, L.	T2	Mbulu	3 53 S 35 53 E	
[Burwa] see Bulwa	T3	Lushoto	5 02.5 S 38 38 E	J. Adamson
Bur Wein, hill	K1	Northern Frontier	c.2 30 N 40 55 E	Maitland
Busaba, popl.	U3	Mbale	0 53 N 33 51 E	
Busabaga, popl.	U4	Mengo	0 18 N 32 54 E	Dept. Agric.
Busabara, prob. = Busabaga	U4	Mengo	prob.0 18 N 32 54 E	
Busagami, hill	T1	Mwanza	1 58 S 32 56 E	

31

Name		Region	Coordinates	Collector(s)
[Busamusi] see Busamuzi	U4	Mengo	0 15 N 33 15 E	
Busamuzi, area, popl.	U4	Mengo	0 15 N 33 15 E	Maitland
Busana, popl., r.h.	U4	Mengo	0 46 N 32 59 E	Eggeling, Liebenberg, A.S. Thomas
[Busanga] ? see Busanza	U2	Kigezi	?1 12 S 29 37 E	Norman
Busankara ? = Usankara	T4	Tabora	?5 04 S 31 52 E	Swynnerton
Busano, popl., r.h.	U3	Mbale	1 01 N 34 16 E	Snowden, Lye
[Busanu] see Busana	U4	Mengo	0 46 N 32 59 E	A.S. Thomas
Busanza, popl., mssn., r.h.	U2	Kigezi	1 12 S 29 37 E	Liebenberg, Eggeling
[Busaro] see Busaru	U2	Toro	0 40 N 30 02 E	Eggeling
Busaru, popl., r.h.	U2	Toro	0 40 N 30 02 E	
Busega, area	T4	Mpanda	c.6 30 S 31 45 E	
Buseke (near Keza)	T1	Ngara	near 2 47 S 30 42 E	Tanner
Busembatia, popl., rsta.	U3	Busoga	0 46 N 33 37.5 E	Osmaston, G.H.S. Wood
Busenya = Bushenya	T1	Bukoba	Not traced	Procter
Buseresere, popl.	T1	Biharamulo	3 03 S 31 53 E	Procter
Busesa, popl.	U3	Busoga	0 38 N 33 36 E	W.H. Lewis, Harker
Bushasha, popl., sw.	T1	Bukoba	1 05 S 31 48 E	Gillman
Bushenya, for.	T1	Bukoba	Not traced	Ford
Bushenyi, popl., r.h.	U2	Ankole	0 32 S 30 11 E	Cree, Purseglove, Eggeling
Bushiri, popl.	T3	Pangani	5 22 S 38 57.5 E	Numerous
Bushiri, sisal est.	T3	Pangani	5 20 S 38 57 E	Faulkner
Bushubi = Busubi, area	T1	Ngara	c.2 50 S 30 40 E	Tanner
Bushwhackers, r.c.	K4	Machakos	2 19 S 38 07 E	Napper
Busi, I.	U4	Mengo	0 02 N 32 20 E	Eggeling
Busia, distr.	K5	N. & Central Kavirondo	c.0 20 N 34 15 E	Numerous
Busia, popl.	U3	Mbale	0 28 N 34 05 E	
Busia, mkt.	K5	North Kavirondo	0 28 N 34 06 E	Brunt
Busiki, co.	U3	Busoga	c.0 50 N 33 45 E	
Busikimbi (Maisome I.)	T1	Mwanza	c.2 18 S 32 02 E	Carmichael
Busingiro, popl.	U2	Bunyoro	1 44 N 31 28 E	Numerous

Name		Region	Coordinates	Authority
Busingiro, hill	U2	Bunyoro	1 44 N 31 28 E	Numerous
[Busiogami] see Busagami	T1	Mwanza	1 58 S 32 56 E	Conrads
[Busira] see Musira	T1	Bukoba	1 21 S 31 51 E	Gillman
Busiro, co.	U4	Mengo	c.0 10 N 32 25 E	Dawe, Maitland
Busisi, fy.	T1	Mwanza	2 44 S 32 52 E	Verdcourt, Stuhlmann
Busiu, popl.	U3	Mbale	0 55.5 N 34 09 E	Maitland
[Buskenzi] see Bushenyi	U2	Ankole	0 32 S 30 11 E	Snowden
Busoga, distr.	U3	Busoga	c.0 45 N 33 30 E	Numerous
Busolwa, area, hill	T1	Mwanza	c.2 59 S 32 36 E	Carmichael
Busolwe, popl.	U3	Mbale	0 50 N 33 55 E	
Busondo, popl.	T4	Kigoma	5 20 S 30 25 E	Verdcourt, Moreau
[Busongola] see Bujongolo	U2	Toro	0 20 N 29 55 E	
Busongora, co.	U2	Toro	c.0 05 N 30 00 E	
Busowa, hill	U4	Mengo	0 46.5 N 32 13 E	Eggeling
Bussi, I.	U4	Mengo	0 02 N 32 20 E	Eggeling
[Bussiru] see Musira	T1	Bukoba	1 21 S 31 51 E	Mildbraed
[Bussissi] see Busisi	T1	Mwanza	2 44 S 32 52 E	Stuhlmann
Bustani = Garden (Swahili)	T3	Lushoto		Verdcourt
Busubi, distr.	T1	Ngara	c.2 50 S 30 40 E	Tanner
Busuju, co.	U4	Mengo	c.0 20 N 32 05 E	
Busunga, crater	U2	Toro	0 03.5 S 29 55 E	Lock
Busunju, popl.	U4	Mengo	0 34 N 32 12 E	G.H. Wood, Eggeling
Buswale, popl.	U3	Busoga	0 22 N 33 54 E	
[Butadiri] see Budadiri	U3	Mbale	1 10 N 34 20 E	Hedberg
[Butagu] see Butahu	U2/Zaire	Toro	0 21 N 29 53 E—0 31 N 29 38 E	Mildbraed
Butahu, str.	U2/Zaire	Toro	0 21 N 29 53 E—0 31 N 29 38 E	Salt, Stuhlmann
Butale, mssn., t.c.	U2	Ankole	0 19.5 S 30 24 E	A.S. Thomas
Butale, mt.	U4	Masaka	0 24 S 31 38 E	
[Butale] see Butali	U2	Kigezi	1 22 S 30 02 E	Synge
Butaleja, popl., r.h., t.c.	U3	Mbale	0 55 N 33 57 E	Maitland

Name		District	Coordinates	Authority
Butali, popl., r.h.	U2	Kigezi	1 22 S 30 02 E	
Butambala, co.	U4	Mengo	c.0 10 N 32 05 E	Karani
Butamira, f.r.	U3	Busoga	0 37 N 33 14 E	Numerous
Butandiga, popl., r.h.	U3	Mbale	1 12.5 N 34 22 E	
Buteba, popl.	U3	Mbale	0 32 N 34 08 E	
Butemba, popl., r.h.	U4	Mengo	1 09 N 31 37 E	Langdale-Brown, Harker
Butembe Bunya, co.	U3	Busoga	c.0 15 N 33 30 E	Webb
Butende, ln.	K5	South Kavirondo	c.1 10 S 34 27 E	?Napier
Butende, mssn.	K5	South Kavirondo	1 14 S 34 36 E	
Butengesa, area	U4	Mengo	0 59 N 32 03 E	J.P. Kennedy
Butenzi, sch.	U4	Masaka	0 23 S 31 49 E	Lye
Butere, popl., rsta.	K5	North Kavirondo	0 13 N 34 29.5 E	Graham
Butiaba, popl., popl, port, r.h., Flats	U2	Bunyoro	1 49 N 31 19 E	Numerous
[Butiba] see Buteba	U3	Mbale	0 32 N 34 08 E	G.H. Wood
Butimba, popl.	T1	Mwanza	2 34 S 32 53 E	Tanner
Butiru, popl.	U3	Mbale	0 49 N 34 17 E	Snowden
Butiti, popl., mssn., r.h.	U2	Toro	0 38 N 30 32 E	A.S. Thomas, Mukibi, Eggeling
Butiti, popl.	U2	Toro	0 39 N 30 32 E	
Butogo, popl.	U2	Toro	0 44 N 29 59 E	Eggeling
Butu = Buto	U4	Mengo	0 20 N 32 22.5 E	Styles
Butu, popl., mkt.	T3	Pare	3 41 S 37 42 E	Haarer
Butumbi (near Kantanda)	U2	Kigezi	near 0 55 S 29 59 E	Stuhlmann
Butungama (Wasa R.)	U2	Toro	prob. 0 56 N 30 22 E	Buechner
Butuntumula, r.h.	U4	Mengo	0 52 N 32 29 E	Langdale-Brown
Butwa, I.	T1	Mwanza	c.2 27 S 31 59 E	Carmichael
Buvuma, co.	U4	Mengo	c.0 05 N 33 15 E	
Buvuma, I.	U4	Mengo	c.0 15 N 33 20 E	Numerous
[Buvuma] see Boruma	T3/6	Handeni/Morogoro	c.6 00 S 37 25 E–5 48 S 37 27 E	Busse
[Buwaga] see Buyaga	U4	Masaka	0 30 S 31 48 E	Maitland
Buwalasi, popl., mssn., r.c.	U3	Mbale	1 11 N 34 14 E	Snowden

Name	Code	District	Coordinates	Collector
Buwalasi, tech. sch.	U3	Mbale	1 09 N 34 13 E	
Buwangire, popl.	U4	Masaka	0 13 S 31 58 E	Lye
Buwaya, popl.	U4	Mengo	0 08 N 32 24 E	
Buwekula, co.	U4	Mubende	c.0 30 N 31 15 E	
Buwenda, popl., hill	U3	Busoga	0 28 N 33 11 E	Osmaston, G.H.S. Wood
Buwunga, popl.	U4	Masaka	0 23 S 31 48 E	A.S. Thomas
Buyaga, mssn.	U4	Masaka	0 30 S 31 48 E	
Buyaga, co.	U4	Mubende	c.0 55 N 31 00 E	Dawe, Kauma
[Buyama I.] see Bunyama	U4	Masaka	c.0 23 S 32 18 E	Maitland
Buyamba, popl.	U4	Masaka	0 39 S 31 23.5 E	Purseglove
[Buyamtole] ? see Bunyantole	U3	Busoga	0 34.5 S 33 36 E	Harris
Buyango, popl.	T1	Bukoba	1 09 S 31 38 E	Gillman, Haarer, Procter
Buyangu (Kakamega Forest)	K5	North Kavirondo	c.0 15 N 34 52 E	Hooper & Townsend
Buyayu, popl.	U2	Toro	0 45 N 30 07 E	Maitland, Liebenberg
Buye, f.r.	U4	Masaka	0 31 S 31 49 E	Eggeling
Buyende, popl., r.h., t.c.	U3	Busoga	1 10 N 33 09.5 E	G.H.S. Wood
Buyenzi, t.a.	T4	Kigoma	c.5 20 S 30 38 E	Procter
Buyenzi, popl., mssn.	T4	Buha	4 27 S 29 58 E	Procter, Peter
Buyonga, popl.	T4	Buha	4 38 S 29 52 E	
Buyovu, I.	U4	Masaka	c.0 22 S 32 27 E	A.S. Thomas
Buyuni, ln., hill	T3	Pangani	5 57 S 38 47.5 E	Peter
Buyuni, popl.	T3	Pangani	5 22.5 S 38 59 E	
Buzini, area	T6	Bagamoyo	c.6 12 S 38 00 E	Hannington
Bwabya, hill	U2	Toro	0 43.5 N 30 20.5 E	
Bwaga, plain	T5	Mpwapwa	c.6 11 S 36 35 E	Hornby
Bwakira Chini, popl.	T6	Morogoro	7 24 S 37 45 E	
Bwakira Juu, popl.	T6	Morogoro	7 18 S 37 42 E	
[Bwakya] see Bwabya	U2	Toro	0 43.5 N 30 20.5 E	Thornton
Bwamba, co.	U2	Toro	c.0 50 N 30 06 E	Numerous
Bwamba, f.r.	U2	Toro	c.0 50 N 30 05 E	Greenway & Eggeling, Paulo

36

Name	Code	Region	Coordinates	Collector
Bwizi, popl.	U2	Toro	0 24 N 30 44 E	Brockington
Bwongera, popl.	U2	Ankole	0 49 S 30 06 E	Eggeling
[Byabakara] see Kyabakara	U2	Ankole	c.0 11 S 30 13 E	Synnott
Byante, f.r.	U4	Masaka	c.0 40 S 31 47 E	Philip
Caesar's Seat, ln.	K4	North Nyeri	c.0 09 S 37 18 E	C.G. Rogers
Calini, tea est. (Monga)	T3	Lushoto	c.5 05 S 38 35 E	Ali Omare
[Camba] see Kamba	K4	Machakos	c.1 15 S 38 15 E	Pole Evans & Erens
Campi ya Bibi = Kampi ya Bibi, ln.	K6	Masai	c.1 33 S 36 32 E	Harding
Campi ya Chui, ln.	K7	Tana River	0 03 S 38 33 E	R.G.S./Nat. Mus. Kenya Exped.
Campi ya Fisi, ln.	T2	Arusha	c.3 15 S 36 52 E	Greenway
Campi ya Mawi, ln.	T1	Musoma	prob. c.2 10 S 35 00 E	Greenway & Turner
[Campi ya Mpofu] see Campi ya Pofu	T1	Musoma	prob. c.2 10 S 35 00 E	Greenway
[Campi ya Ndege] see Kampi ya Ndege	K4	Machakos	c.1 12 S 37 33 E	Bally
Campi ya Nyama Yangu = Neumann's Camp	K1	Northern Frontier	0 34 N 37 34.5 E	
Campi ya Pofu, ln.	T1	Musoma	prob. c.2 10 S 35 00 E	Greenway
Campi ya Samaki, ln.	K3	Baringo	0 37 N 36 01.5 E	Tweedie
Campi ya Simba, ln.	T3	Pare	3 36 S 37 42 E	
Campi ya Swala, ln.	K7	Teita	3 25 S 38 41 E	Greenway
Cangort Farm	K3	Trans-Nzoia	0 56 N 34 53 E	Tweedie
Cantalla	?K1	?Northern Frontier	?near Marsabit	Delamere
Capri point (Mwanza)	T1	Mwanza	c.2 32 S 32 53 E	Eggeling, Procter
Castle Forest Station	K4	South Nyeri	0 22.5 S 37 18 E	Numerous
Cathedral Rock = Kapchok	K2	Turkana	1 50 N 35 07 E	
Causeway Rocks, ln.	K3	Trans-Nzoia	1 07 N 34 50.5 E	Tweedie
Cave Kopje, hill	K3	Trans-Nzoia	1 08.5 N 34 49 E	Tweedie
Caves of Elgon, fm.	K3	Trans-Nzoia	c.0 52.5 N 34 50 E	Bally, Tweedie
Cave Waterfall = Queen's Cave Waterfall	K4	South Nyeri	0 29 S 36 42 E	Coe, Polhill
Cawente, popl.	U1	Lango	1 47 N 32 39.5 E	
[Central Bugishu] see Central Bugisu	U3	Mbale	c.1 00 N 34 15 E	

Name	Code	District/Area	Coordinates	Collector
Central Bugisu, co.	U3	Mbale	c.1 00 N 34 15 E	
Central Island, I.	K2	Turkana	c.3 30 N 36 03 E	Padwa, Worthington
Central Kavirondo, distr.	K5	Central Kavirondo	c.0 00 34 30 E	
[Central Nyanza] see Central Kavirondo	K5	Central Kavirondo	c.0 00 34 30 E	
Chaani, plain	Z	Zanzibar	c.5 57 S 39 18 E	Greenway, Vaughan
[Chabia] see Chawia	K7	Teita	3 29 S 38 21.5 E	
Chada, L.	T4	Mpanda	6 58 S 31 16 E	Richards
Chafundika, str.	T7	Mbeya	c.8 20 S 32 32 E	Bullock
Chagga, t.a.	T2	Moshi	c.3 04 S 37 22 E	Carmichael
[Chagoria] see Chogoria	K4	Embu	c.0 15 S 37 35 E	Moreau
Chagu, L., popl.	T4	Kigoma	4 59 S 31 12 E	
[Chagwe] see Kyagwe	U4	Mengo	c.0 20 N 32 50 E	Hooper & Townsend
Chahafi, L.	U2	Kigezi	c.1 21 S 29 46.5 E	E. Brown, Maitland
Chahi, popl.	U2	Kigezi	1 17 S 29 42 E	Purseglove, Snowden, Gilbert Rogers
Chahi, for.	U2	Ankole	c.0 36 S 30 45 E	Purseglove
Chai, est., spr.	K4	Machakos	c.2 21 S 38 01 E	J. Brown
[Chajafa] see Chasafa	T5	Kondoa	4 52.5 S 35 30.5 E	
Chakachani, popl.	T3	Tanga	4 56 S 39 06 E	B.D. Burtt
Chake Chake, creek	P	Pemba	c.5 15 S 39 45 E	Greenway
Chake Chake, popl.	P	Pemba	5 15 S 39 46 E	Vaughan
Chake Mitatuni (Chake Chake area)	P	Pemba	? near 5 15 S 39 46 E	Vaughan
[Chakenge] see Chakengi	T6	Bagamoyo	6 39 S 38 51 E	Vaughan
Chakengi, popl.	T6	Bagamoyo	6 39 S 38 51 E	Holtz
[Chakwa] see Chwaka	Z	Zanzibar	6 10 S 39 26 E	
Chakwale, popl., mssn., t.c.	T6	Kilosa	6 04 S 36 58 E	Hornby
Chakwi, R.	U1	Karamoja	?c.1 52 N 34 45 E	Symes
Chala, L., crater	K7/T2	Teita/Moshi	3 19 S 37 42 E	Numerous
Chala, popl., mssn.	T4	Ufipa	7 35.5 S 31 16.5 E	Richards, Bullock
Chala North, mt.	T4	Ufipa	7 33 S 31 17 E	Richards
Chalangwa, mt., ridge	T7	Chunya/Mbeya	8 41 S 33 36 E	Richards, Eggeling

Name	Code	Region	Coordinates	Collector
Chala South, mt.	T4	Ufipa	7 36 S 31 18 E	
[Chalbe Desert] see Chalbi Desert				
Chalbi Desert	K1	Northern Frontier	3 00 N 37 20 E	Bally
Chale = Chali, popl.	K1	Northern Frontier	3 00 N 37 20 E	Bally
Chali, popl.	T5	Dodoma	6 15 S 35 15 E	
Chaliwindi = Chariwindi, Mt.	T5	Dodoma	6 15 S 35 15 E	
Chamabanda (Geita area)	T7	Iringa	7 49 S 35 00 E	Greenway
Chamarange	T1	Mwanza	Not traced	B.D. Burtt
Chamazoze (Bukwaya)	T1	Bukoba	prob. c.1 04 S 31 48 E	Gillman
Chamba, popl.	T1	Musoma	c.1 34 S 33 46 E	Tanner
Chambezi, agric. stn.	T8	Tunduru	11 34 S 36 58 E	
Chambura, str.	T6	Bagamoyo	6 34 S 38 55 E	S.A. Robertson
Changore, w.h.	U2	Ankole	c.0 07 S 30 04 E	Dawe
[Chamhilo] see Chamliho	T6	Kilosa	7 10 S 37 05 E	Procter
Chamliho, hill	T1	Musoma	1 57 S 34 09 E	Tanner
Chamoto, hill	T1	Musoma	1 57 S 34 09 E	
Chamozoze (Bukwaya)	T7	Mbeya	8 45.5 S 33 48.5 E	Wingfield
[Chamtei] see Maji ya Chumvi	T1	Musoma	c.1 34 S 33 46 E	Tanner
Chandamara, hill	K7	Kwale/Kilifi	3 48 S 39 23 E	Hildebrandt
[Chandani] see Chundani	T8	Songea	c.10 38 S 35 42 E	Milne-Redhead & Taylor
Changa, I.	P	Pemba	5 16 S 39 49 E	Greenway
Changamwe, popl.	Z	Zanzibar	6 07 S 39 10 E	Numerous
Changana, tea est.	K7	Mombasa	4 01 S 39 38 E	Mearns, Napier, Bally
[Changanwe] see Changamwe	K5	Kericho	0 27 S 35 18 E	Jex-Blake, Gray
[Chang'hwale] see Chakwale	K7	Mombasa	4 01 S 39 38 E	Kassner
Changombe, popl.	T6	Kilosa	6 04 S 36 58 E	
Changu, popl.	T6	Uzaramo	6 50 S 39 16 E	Peter
[Changwe] see Kyagwe	U4	Mengo	c.0 01 S 32 05 E	Scott Elliot
Chania, falls	U4	Mengo	c.0 20 N 32 50 E	Ussher
Chania, R.	K4	South Nyeri	0 27.5 S 36 43.5 E	
	K3/4	Naivasha/Fort Hall/Kiambu	0 39 S 36 41 E–1 02 S 37 04 E	Numerous

Chanika, mssn., r.h.	T3	Handeni	5 25 S 38 01 E	
Chanjaani, popl.	P	Pemba	5 17 S 39 45 E	Greenway
Chanlers Falls, falls	K1	Northern Frontier	0 47 N 38 05 E	
Chao = Kapchok, mt.	K2	Turkana	1 50 N 35 07 E	
Chapani, I.	Z	Zanzibar	6 07.5 S 39 11.5 E	Vaughan
Chapota, popl., sw.	T4	Ufipa	8 09 S 31 14.5 E	Bullock, Richards
Chaputuka (near Kalambo Falls)	T4	Ufipa	near 8 36 S 31 14 E	Richards
Charawe, popl., area	Z	Zanzibar	6 11 S 39 26 E	Vaughan
Charity Farm	K4	South Nyeri	0 18 S 36 46 E	van Someren
Chariwindi, Mt.	T7	Iringa	7 49 S 35 00 E	
Chasafa, hill	T5	Kondoa	4 52.5 S 35 30.5 E	Dale, Templer, Drummond & Hemsley
Cha Shimba, area, for.	K7	Kwale	c.4 14 S 39 27 E	
Cha Shimba, str.	K7	Kwale	c.4 09 S 39 18 E–4 04.5 S 39 32 E	
[Cha Simba] see Cha Shimba	K7	Kwale	c.4 14 S 39 27 E	Drummond & Hemsley
Chasimba, fort	K7	Kilifi	3 44 S 39 42 E	Musyoki & Hansen, B.R. Adams, Faden
Chawia, popl.	K7	Teita	3 29 S 38 21.5 E	Drummond & Hemsley
[Chawka] see Chwaka	Z	Zanzibar	6 10 S 39 26 E	
Chaya, L.	T5	Dodoma	5 37 S 34 03.5 E	Numerous
Chaya, rsta.	T5	Dodoma	5 34.5 S 34 06 E	Numerous
Chebarbar, mkt.	K3	Nandi	0 12 N 35 08 E	Brunt
[Chebele] see Chewele	K1	Northern Frontier	1 10.5 S 39 59 E	J. Adamson
[Chebereria] see Chepareria	K2	West Suk	1 19 N 35 12 E	Tweedie
Chebiemit, area	K3	Trans-Nzoia	1 09.5 N 34 43 E	Tweedie
Chebloch, popl., t.c.	K3	Baringo	0 27.5 N 35 41 E	Tweedie, Bally
Chebroa, R. = Kaboroa	K3	Trans-Nzoia	1 09 N 34 42.5 E–1 11 N 34 44 E	Tweedie
Chebuswa, hill	K3	Naivasha	0 14 S 36 35 E	Mabberley
Chehe, popl.	K4	South Nyeri	0 25 S 37 11 E	Elliot
Chei, hill	U1	West Nile	3 44 N 31 12 E	Eggeling
Chejoni ? see Cheju	Z	Zanzibar	?6 14 S 39 20.5 E	R.M. Davies
Cheju, popl., area	Z	Zanzibar	6 14 S 39 20.5 E	Vaughan

41

Name	Code	Region	Coordinates	Collector
[Cheptoket] see Chepkotet	K3	Elgeyo	1 14.5 N 35 26 E	Townsend
Cheptongei, popl.	K3	Elgeyo	0 56.5 N 35 31 E	Tothill
Cheptui, popl.	U3	Mbale	1 24 N 34 20.5 E	Kerfoot
Cheptuiyet, r.h.	K5	Kericho	0 33 S 35 11 E	
Cheptuiyet, sch.	K5	Kericho	0 29 S 35 03.5 E	
Cheptulon, str., popl.	K3	Elgeyo	c.0 58 N 35 30 E	Townsend
[Chepukupuinoi] see Chepkuloi	K2	Turkana	2 20 N 35 05 E–2 35 N 35 07 E	J. Wilson
Cherangani, for. stn.	K3	Trans-Nzoia	1 02 N 35 19 E	Irwin
Cherangani, hills	K2/3	West Suk/Elgeyo	c.1 15 N 35 27 E	Numerous
Cherangany Hills = Cherangani Hills	K2/3	West Suk/Elgeyo	c.1 15 N 35 27 E	
[Cherongonu] see Cherangani	K2/3	West Suk/Elgeyo	c.1 15 N 35 27 E	Bogdan
Chesegon, popl., mssn.	K3	Elgeyo	1 17.5 N 35 37 E	Tweedie
Chesoi, popl.	K3	Elgeyo	1 04 N 35 35 E	
Chesongoch, popl.	K3	Elgeyo	1 08 N 35 39 E	A.G. & L.C. Miller
Chesoweri, area, popl.	U3	Mbale	1 20 N 34 40 E	Tweedie
Chessera, ln, str.	K2	West Suk	1 27 N 35 08 E	Bogdan
Chewele, popl.	K1	Northern Frontier	1 10.5 S 39 59 E	
[Chiagaia] see Chiegea	T5	Mpwapwa	6 10 S 36 15 E	B.D. Burtt
[Chiawante] see Cawente	U1	Lango	1 47 N 32 39.5 E	H.B. Johnston
Chicot ? see Chiko	U3	Busoga	?0 34 N 33 16 E	Dummer
Chidya, popl.	T8	Masasi	10 37 S 39 04 E	Leonhardt
Chiegea = Chiyegea, hills	T5	Mpwapwa	6 10 S 36 15 E	
Chieni, for.	K4	South Nyeri	0 23 S 37 02 E	
Chifumbo (Maisome I.)	T1	Mwanza	c.2 18 S 32 02 E	Carmichael
Chigara (Kome I.)	T1	Mwanza	c.2 22 S 32 28 E	Carmichael
Chigunga = Chizungu	T4	Tabora	Not traced	Leonard
[Chigurufumi, f.r.] see Shikurufumi	T6	Morogoro	7 10 S 37 31 E	Semsei
Chiko, popl.	U3	Busoga	0 34 N 33 16 E	
Chikuku, f.r. (Kome I.)	T1	Mwanza	c.2 22 S 32 28 E	Forest Herb.
Chikulwe, popl.	T4	Ufipa	8 09 S 32 28 E	Bullock

Name	Code	Region	Coordinates	Collector
Chikumbi, near Sakalilo	T4	Ufipa	near 8 12 S 31 58 E	Michelmore
Chikunda, hill	U2	Ankole	0 48.5 S 30 33 E	Eggeling
[Chikundu] see Chikunda	U2	Ankole	0 48.5 S 30 33 E	Polhill & Paulo
[Chikuyu] see Kikuyu	T5	Dodoma	6 52 S 35 05 E	Michelmore
Chilambwa, ln.	T4	Ufipa	prob. 7 40 S 31 34 E	Gillman
Chilangala, popl.	T8	Newala	10 33 S 39 08 E	Greenway & Eggeling
[Chilema] see Chelima	U2	Kigezi	c.1 05 S 29 55 E	Carmichael
Chilibata (Rubondo I.)	T1	Mwanza	c.2 20 S 31 52 E	Numerous
Chimala, popl., sch.	T7	Mbeya	8 51 S 34 01 E	Numerous
Chimala, R.	T7	Njombe/Mbeya	c.9 03 S 33 51 E-8 43 S 34 02 E	Procter
Chimala Scarp, f.r.	T7	Mbeya/Njombe	c.8 52 S 33 52 E	Braun
Chiminda	T8	Lindi	near 10 08 S 39 15 E	Richards
[Chinaputa] see Kinyantupa	T7	Iringa		Richards
[Chinatupa] see Kinyantupa	T7	Iringa		
Chingaza – vernacular name of plant				
Chingoma, Rukwa Central	T4	Ufipa	c.7 50 S 31 40 E	Vesey-FitzGerald
Chingwede, popl.	K7	Kwale	4 29.5 S 39 26.5 E	Bally
Chiope, area	U2	Bunyoro	c.2 10 N 32 10 E	Dawe
Chipangati, camp	T4	Mpanda	7 29 S 31 53 E	
Chipogolo, popl., hill	T5	Mpwapwa	6 52 S 36 02 E	Disney, Polhill & Paulo
[Chipogoro] see Chipogolo	T5	Mpwapwa	6 52 S 36 02 E	
Chipoli, mssn.	T8	Songea	c.10 50 S 35 15 E	Milne-Redhead & Taylor
[Chipongolo] see Chipogolo	T5	Mpwapwa	6 52 S 36 02 E	Verdcourt
Chiromo, suburb, former est.	K4	Nairobi	c.1 16 S 36 48 E	Napier, Tweedie
Chisokwe, mssn., sch.	T5	Mpwapwa	6 19 S 36 25 E	
Chisungu, popl.	T4	Ufipa	near 7 47 S 31 29.5 E	Richards
Chita, popl. sch.	T6	Ulanga	8 30.5 S 35 55.5 E	Carmichael
Chitara (L. Burigi)	T1	Bukoba	c.2 07 S 31 16 E	Braun
Chito = Kito, mt.	T4	Ufipa	8 33.5 S 31 29 E	Sanane
Chito, popl.	T4	Ufipa	prob. 8 32 S 31 29 E	Richards

Chitukutu, popl.	T4	Ufipa	8 08 S 32 14 E	Bullock, Siame
Chitwe, popl.	U2	Ankole	0 58 S 30 28 E	Eggeling
Chiumo, popl.	T8	Newala	10 55 S 39 44 E	Hay
Chivanjee, tea est.	T7	Rungwe	9 22.5 S 33 39.5 E	Cribb & Grey-Wilson, Leedal
[Chiwanje] see Chivanjee	T7	Rungwe	9 22.5 S 33 39.5 E	Greenway
Chiyegea = Chiegea, hills	T5	Mpwapwa	6 10 S 36 15 E	
Chizungu	T4	Tabora	Not traced	Leonard
Choba, Grof, crater	K1	Northern Frontier	2 24 N 38 03 E	Bally & Smith
Chobi, safari lodge	U1	Acholi	2 15 N 32 08 E	G. Jackson, Angus
Chogni, hills	K7	Kilifi	c.3 43 S 39 42 E	Battiscombe
Chogo, popl.	T7	Iringa	8 18 S 35 43 E	Carmichael
Chogoria, popl.	K4	Embu	0 14 S 37 37.5 E	W.H. Lewis
Chogoria Track	K4	Embu	c.0 15 S 37 35 E	
Chogwe, old area	T3	Pangani	c.5 23 S 38 49 E	
[Chokwe] ? see Chogwe	T3	Pangani	? c.5 23 S 38 49 E	Zimmermann
Chole = Chore, hill, area	T1	Mwanza	2 45.5 S 32 55.5 E	Tanner
Chole, I.	T6	Rufiji	7 58.5 S 39 45.5 E	Greenway
Cholol, hill	U1	Karamoja	1 56 N 34 50 E	J. Wilson, Tweedie
Chombe, mssn.	T6	Ulanga	8 18.5 S 36 07 E	
Chome, f.r.	T3	Pare	c.4 18 S 37 57 E	
Chome, mt.	T3	Pare	4 18 S 37 53 E	Greenway
Chonga, popl.	P	Pemba	5 18.5 S 39 45 E	Greenway
Chonwe, mts.	T6	Kilosa	7 36 S 36 57 E	Akeroyd & Mayuga
Chonyi, popl.	K7	Kilifi	3 47 S 39 41 E	Bally & Smith, Faden
Chopeh, old area	U2	Bunyoro	c.1 53 N 32 15 E	Grant
Chop Plain	K1	Northern Frontier	3 25 N 38 45 E	Bally
Chosan, str.	U1	Karamoja	c.1 58 N 34 48 E	Symes
[Cho-sen] see Chosan	U1	Karamoja	c.1 58 N 34 48 E	Tweedie
[Chowea] see Chawia	K7	Teita	3 29 S 38 21.5 E	Gardner
[Chowia] see Chawia	K7	Teita	3 29 S 38 21.5 E	

Name		Region	Coordinates	Authority
Chua, co.	U1	Acholi	c.3 15 N 33 15 E	Numerous
Chuaka, popl., area	P	Pemba	5 23 S 39 47 E	Burtt Davy, Vaughan
[Chuaka] see Chwaka				
[Chuca] see Chuka	Z	Zanzibar	6 10 S 39 26 E	
Chuele = Chwele, str.	K4	Embu	0 20 S 37 38.5 E	Tweedie
Chuere = Chuele, str.	K5	North Kavirondo	0 44 N 34 35 E	
Chuini, popl., area	K5	North Kavirondo	0 44 N 34 35 E	
Chuini, old palace	Z	Zanzibar	6 03 S 39 13.5 E	Greenway, Vaughan
Chuka, area	Z	Zanzibar	6 04.5 S 39 12.5 E	
Chuka, popl.	K4	Embu	c.0 21 S 37 45 E	Fries
Chukwani, palace, area	K4	Embu	0 20 S 37 38.5 E	Bally, M.D. Graham
Chukwani, Ras, pt.	Z	Zanzibar	6 15 S 39 12.5 E	Faulkner, Vaughan, R.O. Williams
Chumbati, hill	Z	Zanzibar	6 15 S 39 12.5 E	
Chumbuni, popl.	T8	Lindi	10 26 S 38 50 E	
Chumo, popl., old area	Z	Zanzibar	6 09 S 39 13.5 E	Vaughan, R.M. Davies
Chumo Pass (in Ugogo)	T8	Kilwa	8 33 S 39 02 E	von Trotha
Chundani, popl.	T5		Not traced	
Chungai, popl., r.h.	P	Pemba	5 16 S 39 49 E	Greenway
[Chungarumo] see Chunguruma	T5	Kondoa	4 40.5 S 35 52 E	Polhill & Paulo
Chunguruma, area	T6	Rufiji	7 50 S 39 44 E	Greenway
Chungururu, L.	T6	Rufiji	7 50 S 39 44 E	
Chunya, distr.	T7	Rungwe	9 18 S 33 52 E	
Chunya, escarp.	T7	Chunya	c.7 50 S 33 05 E	
Chunya, popl.	T7	Mbeya	c.8 45 S 33 35 E	Richards
Chunyu, popl.	T7	Chunya	8 32 S 33 25 E	Boaler, Richards
Churi, R.	T5	Mpwapwa	6 18 S 36 20 E	
Chuvwi, for., mt.	K4	Meru	prob. c.0 05 S 37 45 E	Fries
[Chuwaka] see Chwaka	T7	Mbeya	8 58 S 33 39 E	Cribb & Grey-Wilson
Chwaka, popl., bay	Z	Zanzibar	6 10 S 39 26 E	Numerous
Chwaka, popl., area	Z	Zanzibar	6 10 S 39 26 E	Vaughan
	P	Pemba	4 57.5 S 39 47 E	

Name	Code	Region	Coordinates	Collector
Chwele, sch., str., foothills	K5	North Kavirondo	c.0 44 N 34 35 E	Tweedie
Chyulu, hills, foothills	K4/6	Machakos/Masai	c.2 18–2 50 S 37 40–38 00 E	Bally, Gibbons
Cimambwe = Simambwe, sw.	T7	Mbeya	8 58.5 S 33 36.5 E	Cribb & Grey-Wilson
Cis Mara Masai = area of E. of R. Mara	K6	Masai		Glover et al.
Closeburn, est.	K4	Kiambu	1 12 S 36 47 E	Bell
Cobb's Farm	K3	Nakuru	c.0 40 S 35 55 E	Glover et al.
Cochran's Farm	K3	Trans-Nzoia	1 07 N 34 49 E	Tweedie
Cole estate	K3	Naivasha	c.0 30 S 36 17 E	Cole
Cole's Estate	K4	North Nyeri	c.0 11 S 36 56 E	Cole
Cole's Mill	K4	North Nyeri	0 11 S 37 03 E	Fries
Condcona	K7	Kwale	4 32 S 39 04 E	Fries
Condcuna = Condcona	K7	Kwale	4 32 S 39 04 E	Kassner
[Congano] see Mkongani				Kassner
[Corbessa] see Malka Gorbesa	K7	Kwale	4 17 S 39 16 E	
Cottar's Camp	K1	Northern Frontier	0 13 N 38 20 E	J. Adamson
Cox's Farm	K6	Masai	1 29 S 35 22 E	Gillett
Crampton's Inn	K3	Trans-Nzoia	1 05 N 35 06 E	Tweedie
Crater Highlands, mts.	K3	Trans-Nzoia	0 57 N 34 58.5 E	Townsend
Crater Lake, L.	T2	Masai/Mbulu	c.2 40–3 20 S 35 07–36 00 E	Newbould, Peter
Crater Lakes (Ndali)	K3	Naivasha	0 47 S 36 15.5 E	McCallum Webster, E. Polhill, J. Thomson
Crescent, I.	U2	Toro	0 29 N 30 17 E	Eggeling
Cross Hill	K3	Naivasha	0 46 S 36 24 E	
	K3	Trans-Nzoia	1 09.5 N 34 49 E	Tweedie
Dabaga, f.r.	T7	Iringa	c.8 05 S 35 56 E	Numerous
Dabaga, highlands	T7	Iringa	c.8 05 S 35 56 E	Numerous
Dabaga, popl., t.c.	T7	Iringa	8 07 S 35 55 E	Tanner
Dabajira (Buhoro Flats)	T7	Mbeya	c.8 30 S 34 30 E	Gillett & Gachathi
Dadaab, w.h.	K1	Northern Frontier	0 04 N 40 19 E	Robertson
Dagamra, r.h.	K7	Kilifi	3 11 S 39 56 E	
[Dagoretti] see Dagoretti	K4	Kiambu	c.1 18 S 36 40 E	

Name	Region	District	Coordinates	Authority
Dagoretti, for.	K4	Kiambu	c.1 18 S 36 40 E	
Dagoretti, popl.	K4	Kiambu	1 17 S 36 41 E	Tweedie
Dagoretti Corner, ln.	K4	Nairobi	1 18 S 36 46 E	Numerous
Dagusi, I.	U3	Busoga	0 08 N 33 34 E	G.H. Wood
Dahali (near Madanga)	T3	Pangani	near 5 21 S 38 59 E	Tanner
Dahari (S. Hanang)	T2	Mbulu	c.4 29 S 35 25 E	Carmichael
Daje, popl.	U4	Masaka	0 27.5 S 32 16.5 E	A.S. Thomas
Dakabuka, hill	K7	Kilifi	c.2 54 S 39 38.5 E	C. Field, Dale
Dakacha, popl.	K1	Northern Frontier	c.0 05 S 38 55 E	Sampson
Dakadima, hill, airstrip	K4	Kitui	2 26 S 39 24 E	Parker
Dakatcha, popl.	K7	Kilifi	3 01 S 39 48 E	Dale
Dakawa, popl., t.c., area	T6	Morogoro	6 27 S 37 32 E	Drummond & Hemsley, Peter
Dakawa, popl.	T6	Morogoro	7 25 S 37 42.5 E	
Dakawachu, ln.	K7	Kilifi	2 40.5 S 39 39 E	Koss
Dalai, R.	T5	Kondoa	c.4 58 S 35 58.5 E–5 02 S 35 57 E	J.L. Newman
Daluni, area	T3	Lushoto	4 49 S 38 46.5 E–4 43 S 38 45 E	Numerous
Daluni, popl., r.h., t.c.	T3	Lushoto	4 46.5 S 38 46 E	
Daluni, str.	T3	Lushoto	4 43 S 38 45 E–4 48 S 38 45 E	
Damasa, b.p., w.p., airstrip	K1/Som. boundary	Northern Frontier	3 09 N 41 20 E	Gillett, Bally & Smith
[Damassa] see Damasa	K1/Som. boundary	Northern Frontier		
Damba, I.	U4	Mengo	c.0 00 32 48 E	Dawkins, Maitland
[Dambala Faichana] see Dambala Fulchana	K1	Northern Frontier	3 30.5 N 38 45 E	Bally & Smith
Dambala Fulchana, w.h.	K1	Northern Frontier	3 30.5 N 38 45 E	
[Dambelo] see Gombelo	T3	Lushoto	4 41 S 38 37 E	Gilbert
Danacre, ln.	T7	Iringa	c.8 40 S 35 15 E	Carmichael
Dandu, mt.	K1	Northern Frontier	3 26 N 39 52 E	Delamere, Gillett
Danicha, popl.	K7	Kilifi	3 30 S 39 44.5 E	
[Danisa] ? see Danicha	K7	Kilifi	?3 30 S 39 44.5 E	Dale

Name	Code	District	Coordinates	Collector
Derema, tea est.	T3	Lushoto	5 05 S 38 39 E	Numerous
Derisa ("Tanaland")	K1	Northern Frontier	Not traced	J. Adamson
Derobo or Dorobo (tribe)	K1	Northern Frontier		J. Bally
Deschi, L. (near Mbirira)				
Diabohika, popl.	T4	Buha	near 4 21 S 30 10 E	Peter
Diani Beach, popl.	T4	Kahama	3 18 S 31 37 E	Procter
Dick's Head = Ras Chiambone	K7	Kwale	4 18 S 39 35 E	Moomaw, Napier, Napper
	Som.		1 39 S 41 36 E	J. Adamson
Dida, popl, for.	K7	Kilifi	3 25.5 S 39 48 E	Jeffery
Dida Galgalu, area	K1	Northern Frontier	2 57 N 38 12 E	Bally & Smith
Dida Wachuf, area	K1	Northern Frontier	4 05.5 N 40 23 E	Bally & Smith
[Digidigo] see Digodigo	T2	Masai	2 08 S 35 43 E	Bally
Digo, t.a.	K7	Kwale	c.4 05–4 35 S 39 25 E	Dale, Elliot
Digodigo, popl.	T2	Masai	2 08 S 35 43 E	Bally
[Djruina] see Duruma	K7	Kwale	4 33.5 S 39 05 E	Kassner
Dilangilo, popl.	T6	Uzaramo	c.6 50 S 38 49 E	Stuhlmann
Dimani (Ngulakula f.r.)	T6	Rufiji	c.7 50 S 38 54 E	Ngoundai
Dimbani, popl.	Z	Zanzibar	6 26 S 39 28 E	
[Dinda] see Diuda	T7	Njombe	c.9 12 S 34 02 E	
[Dindila] see Dindira	T3	Lushoto	5 00.5 S 38 26 E	
Dindini, popl.	T5	Singida	5 12 S 34 09 E	
Dindira, tea est.	T3	Lushoto	5 00.5 S 38 26 E	Drummond & Hemsley, Mwamba
[Diobahika] see Diabohika	T4	Kahama	3 18 S 31 37 E	Procter
Diroro = Liroro, R.	T7	Njombe	c.9 15 S 34 04 E	Goetze
Ditima, mt.	T7	Njombe	9 25 S 35 17 E	Schlieben
Diuda, mt.	T7	Njombe	c.9 12 S 34 02 E	Goetze
Diwali = Diwale, str.	T6	Morogoro	6 01 S 37 34 E –6 07.5 S 37 35 E	Carmichael
[Djalla] see Chala	K7/T2	Teita/Moshi	3 19 S 37 42 E	Geilinger
Djamimbi = Yamimbi	T7	Njombe	9 40 S 34 18 E	
[Djenye] see Njenje	T8	Kilwa	c.10 20 S 36 52 E –9 05 S 37 26 E	
Djilulu, mt.	T7	Njombe	9 20 S 34 09 E	Goetze

49

Name		Region	Coordinates	Collector
Dobel, popl.	K1	Northern Frontier	3 06 N 39 16 E	Bally
Doda, popl., hill	T3	Tanga	4 55 S 39 05.5 E	Holst
Dodola, ln..	K1	Northern Frontier	3 15 N 39 40 E	Gillett
Dodoma, distr.	T5	Dodoma	c.6 30 S 35 00 E	Staples
Dodoma, popl., rsta., mssn.	T5	Dodoma	6 11 S 35 45 E	Numerous
Dodoma, popl.	T8	Songea	10 38 S 35 49 E	Milne-Redhead & Taylor
[Dodosi] see Dodoth	U1	Karamoja	c.3 30 N 34 00 E	
Dodoth, co.	U1	Karamoja	c.3 30 N 34 00 E	
[Dodotho] see Dodoth	U1	Karamoja	c.3 30 N 34 00 E	
Dodwe, str.	T3	Lushoto	c.5 05 S 38 36—38 39 E	
Dogogicha, hill	K1	Northern Frontier	3 25 N 39 13.5 E	Numerous
[Doinyo Sapuk] see Ol Dane Sapuk	K3	Uasin Gishu	0 28 N 35 15 E–0 32 N 35 08.5 E	Bally & Smith
Dokolo, co.	U1	Lango	c.1 56 N 33 03 E	Greenway, G.R. Williams
Dol Dol = Don Dol, Escarpment	K4	North Nyeri	c.0 23 N 37 09 E	Powys
[Dole] see Dule	T3	Lushoto	4 34 S 38 19 E	Greenway
Dole, popl.	Z	Zanzibar	6 06 S 39 15 E	Faulkner, R.O. Williams
Dolelo, area	T3	Lushoto	c.4 43 S 38 19 E	Drummond & Hemsley
Domoni, Ras, point	P	Pemba	5 27 S 39 42.5 E	
Donde, area	T8	Kilwa	c.9 00 S 37 45 E	Busse
[Dongabesh] see Dongobesh	T2	Mbulu	4 03 S 35 23 E	Carmichael
Dongobesh, popl.	T2	Mbulu	4 03 S 35 23 E	B.D. Burtt, Procter
Dongobesh, str.	T2	Mbulu	c.4 05 S 35 28 E–4 22 S 35 05 E	
Dongwe, well	Z	Zanzibar	6 11 S 39 32 E	Vaughan
[Donje Sambu] see Ol Donyo Sambu	T2	Arusha	3 10 S 36 39 E	
[Donyo Sabouk] see Donyo Sabuk	K4	Machakos	c.1 08 S 37 12.5 E	
Donyo Sabuk, est.	K4	Machakos	c.1 08 S 37 12.5 E	Tweedie
Donyo Sabuk, road	K4	Machakos	c.1 04 S 37 13 E	
Donyo Sabuk North, road	K4	Machakos	c.1 08 S 37 20 E	
Dors Rocks	K3	Trans-Nzoia	1 08 N 34 45.5 E	Tweedie
Dors Turning, ln.	K3	Trans-Nzoia	1 09 N 34 46.5 E	Tweedie

Name	Code	Region	Coordinates	Collector(s)
Dorwa = Wasa, R.	U2	Toro	0 48 N 30 15 E–1 02 N 30 30 E	Geilinger, Bally, Volkens
[Dschalla] see Chala	K7/T2	Teita/Moshi	3 19 S 37 42 E	Winkler
[Dschindscha] see Jinja	U3	Busoga	0 26 N 33 13 E	Schlieben
[Dualle] see Diwale	T6	Morogoro	6 01 S 37 34 E–6 07.5 S 37 35 E	Eggeling
[Duamsikizi] see Nyamunuka	U2	Toro	0 05 S 29 59 E	Bally
Duduntu (near Lamu)	K7	Lamu	near 2 16 S 40 54 E	Greenway & Eggeling
Dufile, popl., ft.	U1	West Nile	3 35 N 31 57 E	Eggeling
[Dufili] see Dufile	U1	West Nile	3 35 N 31 57 E	Holst, Drummond & Hemsley
Duga, popl.	T3	Tanga	5 07 S 39 06 E	
[Dugoretti] see Dagoretti	K4	Kiambu	c.1 18 S 36 40 E	Carmichael
Dulang, R. (S. Hanang)	T2	Mbulu	c.4 29 S 35 25 E	Moomaw
Duldul, popl.	K7	Lamu	1 55 S 40 46 E	
Duldul, str.	K1/7	Northern Frontier/Lamu	c.1 43 S 40 35 E–1 56 S 40 46 E	
Dule, mt.	T3	Lushoto	4 34 S 38 19 E	Uhlig, Geilinger, Carmichael
Duluti, L.	T2	Arusha	3 23 S 36 47 E	Greenway
Duma, R.	T1	Maswa/Mwanza	c.2 50 S 34 40 E–2 32 S 33 25 E	Greenway & Turner
Duma River airstrip	T1	Maswa	2 40 S 34 24 E	Carmichael
Dumanang, str.	T2	Mbulu	4 26.5 S 35 23.5 E–4 24 S 35 19 E	Maitland
Dumba, fy.	U3	Busoga	prob. 0 44 N 33 53 E	Gregory
Dumi, popl., L.	K7	Tana River	2 16 S 40 08 E	Eggeling
Dumu, hill, pt., popl.	U4	Masaka	0 39 S 31 48 E	Dawe, Eggeling
Dumu, popl.	U4	Masaka	0 39.5 S 31 48 E	
Dumu, pt.	U4	Masaka	0 39 S 31 48.5 E	Carmichael
Dunacheri = Isumacheri, I.	T1	Mwanza	c.2 24 S 31 57 E	
Dunda, hill	T5	Dodoma	6 11 S 36 04 E	Stuhlmann
Dunda, mssn.	T6	Bagamoyo	6 34 S 38 50 E	
Dunda, mt.	T7	Chunya	8 32 S 33 34 E	
Dunda, popl.	T7	Mbeya	8 10 S 34 15 E	
Dundani, popl.	T6	Rufiji	7 57.5 S 39 37.5 E	Greenway, Wingfield
Dundori, store	K3	Nakuru	0 11.5 S 36 14 E	Baxendall

51

Eastleigh, popl., airport	K4	Nairobi	1 16 S 36 52 E
East Madi, co.	U1	West Nile	3 15 N 31 50 E
East Usambaras, mts.	T3	Lushoto/Tanga	c.4 50–5 15 S 35 39–35 50 E
Ebenda (prob. near L. Lutamba)	T8	Lindi	prob. near 10 02 S 39 28 E
[Ebuguziwa] see Ibuguziwa	T7	Iringa	7 42 S 34 53 E	Schlieben
Eburru, ln., rsta. (closed)	K3	Nakuru	0 35 S 36 15.5 E	Richards
Eburu, for., mts.	K3	Naivasha/Nakuru	c.0 40 S 36 12 E	Bogdan
[Echuja] see Echuya	U2	Kigezi	1 17 S 29 49 E	Heriz-Smith, E. Polhill, Pegler
Echuya, f.r.	U2	Kigezi	1 17 S 29 49 E	Tothill
Edith, Bay	T4	Mpanda	6 30 S 29 55 E	Paulo
Edward, L. = Lake Idi Amin Dada	U2/Zaire	Toro/Ankole/Kigezi	c.0 25 S 29 30 E	B.D. Burtt, Van Meel
[Egalok] see Keekorok	K6	Masai	1 36 S 35 14.5 E	Purseglove, Eggeling
[Egelok] see Keekorok	K6	Masai	1 36 S 35 14.5 E	Bally, Verdcourt
[Egerok] see Keekorok	K6	Masai	1 36 S 35 14.5 E
Egerton, Agric. Coll.	K3	Nakuru	0 22 S 35 56 E	Bally
[Egolok] see Keekorok	K6	Masai	1 36 S 35 14.5 E
[Eil Lass] see El Das	K1	Northern Frontier	c.2 32 N 39 33 E	Richards
Eisero, popl., sch.	K3	Nandi	0 26 N 35 08 E	Dale
Ela Nairobi, volcano	T2	Masai	2 54 S 35 50 E
[Eland Hill] see Chebuswa	K3	Naivasha	0 14 S 36 35 E	B.D. Burtt, Jaeger
El Barta Plains, area	K1	Northern Frontier	c.1 43 N 36 55 E	Mabberley
[El Bata] see El Barta	K1	Northern Frontier	c.1 43 N 36 55 E	Carter & Stannard, Tweedie
Elburgon, rsta., popl.	K3	Nakuru	0 18 S 35 48 E	Jex-Blake
El Bururi, popl.	K1	Northern Frontier	3 56 N 39 57 E	Numerous
Eldama, str.	K3	Uasin Gishu	c.0 02 N 35 43 E	Gillett
Eldama Ravine, popl., mssn.	K3	Ravine	0 03 N 35 43.5 E
El Das, hills, w.h.'s	K1	Northern Frontier	c.2 32 N 39 33 E	Numerous
El Dera, ln.	K1	Northern Frontier	0 36 N 38 50.5 E	Gillett
[Eldoma] see Eldama	K3	Uasin Gishu	c.0 02 N 35 43 E	Whyte
[Eldonumara] see Ol Doinyo Mara	K1	Northern Frontier	c.2 15 N 37 02 E	Cockburn

53

Eldoret, popl., rsta.	K3	Uasin Gishu	0 31 N 35 16 E	Numerous
[Eleanata] see Eluanata	T2	Masai	3 23 S 36 17 E	
Elemo, hills	K2	Turkana	c.2 35 N 35 30 E	
Elephant Path Swamp, Kanaba	U2	Kigezi	c.1 15 S 29 47.5 E	Purseglove
Elephant's Cave = Kiboi, ln.	K3	Trans-Nzoia	1 07 N 34 45 E	Tweedie
Elephant's Wallow, ln.	K3	Trans-Nzoia	1 08.5 N 34 41.5 E	Tweedie
[El Gerre] ? see Il Gerai	K1	Northern Frontier	?1 36 N 37 20 E	J. Bally
Elgese Gonyek (8 km. from Mara bdg.)	K6	Masai	Not traced	Glover et al.
Elgeyo, distr.	K3	Elgeyo	c.0 45 N 35 35 E	
Elgeyo, escarp.	K3	Elgeyo	c.0 18—1 20 N 35 33—35 38 E	Brodhurst-Hill, Bally
Elgeyo-Marakwet, distr.	K3	Elgeyo	c.0 45 N 35 35 E	
[Elgijada] see El Kajarta	K1	Northern Frontier	2 43 N 36 57 E	Bally, J. Bally
Elgon, mt.	U3/K3/5	Mbale/Trans-Nzoia/ N. Kavirondo	c.1 08 N 34 33 E	Numerous
Elgon Club	K3	Trans-Nzoia	0 53.5 N 34 55.5 E	Tweedie
Elgon Nyanza, distr.	K5	North Kavirondo	c.0 45 N 34 30 E	
		(considered as part of North Kavirondo Distr. for F.T.E.A.)		
[Elionata Dam] see Eluanata	T2	Masai	3 23 S 36 17 E	
El Kajarta, gorge	K1	Northern Frontier	2 43 N 36 57 E	Greenway
El Kekhotoito, L.	T2	Arusha	3 14 S 36 52.5 E	Bally
[El Lass] see El Das	K1	Northern Frontier	c.2 32 N 39 33 E	Richards, Greenway
Ellis, L.	K4	Meru	0 07.5 S 37 24 E	Townsend
Elmenteita, L.	K3	Nakuru	c.0 26.5 S 36 15 E	Numerous
Elmenteita, p.p.	K3	Nakuru	0 29 S 36 09.5 E	
El Mole, popl., w.h.	K1	Northern Frontier	4 02 N 40 12 E	Gillett
El molo, bay	K1	Northern Frontier	c.2 50 N 36 42 E	
Elmolo Island prob. = Loyeni	K1	Northern Frontier	prob.2 47 N 36 41.5 E	J. Adamson
Elphon's Pass	T6	Kilosa	7 22 S 36 42 E	Eggeling, Troll
[Elpon's Pass] see Elphon's Pass	T6	Kilosa	7 22 S 36 42 E	Eggeling
Elton Plateau	T7	Njombe/Mbeya	c.9 00 S 33 50 E	Numerous

Name	Code	Region	Coordinates	Collector
El Tulli, wells	K1	Northern Frontier	1 38 N 40 21 E	Hornby, Leippert, Greenway
Eluanata, L., fm.	T2	Masai	3 23 S 36 17 E	Gillett, West
El Wak, popl., t.c., wells	K1	Northern Frontier	2 49 N 40 56 E	Bally, Carter & Stannard
Emali, rsta.	K4	Machakos	2 05 S 37 28.5 E	Archer, Bally, V.G. van Someren
Emali, hill	K6	Masai	2 03 S 37 22 E	Birch
Ematundu, sch.	K5	North Kavirondo	0 10 N 34 33 E	Numerous
Embagai, crater lake	T2	Masai	2 56 S 35 49 E	
Embakasi, for.	K6	Masai	c.1 19 S 36 39 E	
Embakasi, popl., rsta.	K4	Nairobi	1 21 S 36 53.5 E	M.D. Graham, Orde Browne
Emberre, t.a.	K4	Embu	c.0 36 S 37 45 E	Mabberley, Tweedie
Embobut, R., f.r.	K3	Elgeyo	1 15 N 35 35 E	
Embu, distr.	K4	Embu	c.0 35 S 37 40 E	Numerous
Embu, popl., airstrip	K4	Embu	0 32 S 37 27 E	Peter
Embulbul, crater	T2	Masai	2 29 S 35 45 E	Vessey
Eninit, for.	K2	Turkana	c.3 08 N 35 03 E	Morgan
Emin Pasha Gulf, bay	T1	Biharamulo/Mwanza	2 32 S 31 52 E	Batty
Emmanuel, est.	T2	Moshi	near 3 11 S 37 13 E	Peter
Emmau, mssn. (near Mashewa)	T3	Lushoto	prob. near 4 46 S 33 38 E	Dyson-Hudson
[Emoruagaberru] see Moruangaberu	U1	Karamoja	1 58 N 34 39 E	Dyson-Hudson
[Emoruangaberru] see Moruangaberu	U1	Karamoja	1 58 N 34 39 E	Frame
[Empakaai] see Embagai	T2	Masai	2 56 S 35 49 E	Richards
[Empeta] see Ipeta	T4	Ufipa	8 08 S 31 17 E	Glover & Samwell
Empurputia	K6	Masai	prob. c.1 35 S 36 25 E	Greenway
[Emsambulai] see Nasampolai	K6	Masai	c.0 48.5 S 36 07 E	Vincens
Emsos, springs	K3	Baringo	0 09 N 36 06 E	
Emugur, R.	T2	Masai	2 40 S 35 20 E–2 50 S 35 37 E	
Emugur Belek, str.	T2	Masai	c.3 05 S 36 08 E	Peter
[Emugur Belekj] see Emugur Belek	T2	Masai	c.3 05 S 36 08 E	
Emugur Berek = Emugur Belek	T2	Masai	c.3 05 S 36 08 E	
Emurgur Ojine	T2	Masai	c.2 45 S 35 25 E	Oteke

55

Engare Len = Rombo, R.	K6,7/T2	Masai/Teita/Moshi	c.3 03 S 37 38 E – 3 03 S 37 55 E	St. Clair-Thompson
Engare Longischo, old camp	T2	Masai	2 51 S 36 16 E	Richards
Engare Naibor, popl.	T2	Masai	2 24 S 36 24 E	Numerous
Engare Nairobi, popl.	T2	Moshi	3 03 S 37 00 E	
[Engare Nairobi, airstrip] see West Kilimanjaro	T2	Moshi	3 03 S 37 00 E	
Engare Nairobi (North), str.	T2	Moshi	c.3 02 S 37 19 E – 2 59 S 36 59 E	Newbould, Greenway, Paulo
Engare Nairobi (South), str.	T2	Moshi	c.3 01 S 37 09 E – 3 01 S 36 57 E	Numerous
Engare Nanyuki, air strip, spr.	T2	Masai	2 37 S 35 13 E	
Engare Nanyuki = Ngare Nanyuki, popl.	T2	Arusha	3 09 S 36 51.5 E	
Engare Nanyuki, t.a.	T2	Arusha	3 05 S 36 53 E	
Engare Nanyuki, R.	T2	Arusha/Masai	3 15 S 36 48 E – 2 48 S 36 55 E	Peter, Richards
Engare Narok, str.	K1	Northern Frontier	1 13 N 36 33 E – 0 52 N 36 47 E	Newbould
Engare Narok Game Sanctuary, g.r.	K1	Northern Frontier	c.0 56.5 N 36 43 E	
Engare Narok, str./s	K1	Northern Frontier	1 20 N 37 15 E – 1 25 N 37 08 E	J. Bally
Engare Narok = Ngare Narok, R.	K6	Masai	0 42.5 S 35 44 E – 1 12.5 S 35 53 E	
Engare Narok, str.	T2	Arusha	c.3 15 S 36 45 E – 3 25 S 36 40 E	
[Engare Nyanuki Springs] see Engare Nanyuki	T2	Masai	c.2 37 S 35 13 E	Greenway
Engare Olduroto = Stony Athi, R.	K4/6	Machakos/Masai	1 46 S 37 02 E 1 25.5 S 37 00 E	Bogdan
Engare Olmotoni, R.	T2	Arusha/Masai	c.3 15 S 36 42 E – 3 35 S 36 41 E	Numerous
Engare Olmotoni, t.a.	T2	Arusha	c.3 19 S 36 36 E	Numerous
Engare Olmotoni Chini, popl.	T2	Arusha	3 21.5 S 36 36 E	
Engare Olmotoni Juu, popl.	T2	Arusha	3 18.5 S 36 38.5 E	
[Engare Olmotonj] see Engare Olmotoni	T2	Arusha	c.3 19 S 36 36 E	
[Engare Olmotonj] see Engare Olmotoni	T2	Arusha/Masai	c.3 15 S 36 42 E – 3 35 S 36 41 E	
Engare Rongai, str./s	K6/T2	Masai/Moshi	c.2 58.5 S 37 27 E – 2 53 S 37 31 E	Schlieben, Vesey-FitzGerald
Engare Siapei = Seyabei, R.	K6	Masai	c.0 37 S 35 59 E – 1 20 S 35 59.5 E	Pfennig
[Engari Rongi] see Engare Rongai	K6/T2	Masai/Moshi	2 58.5 S 37 27 E – 2 53 S 37 31 E	Greenway
Engaruka, popl.	T2	Masai	2 59 S 35 57 E	Numerous

[Engili See] see Longil				
[Engitai] see Engitati				
Engitati, hill	T2	Arusha	3 15.5 S 36 53 E	Uhlig
English Point = Ras Kidomoni	T2	Masai	3 08 S 35 33 E	Greenway
Engomeni valley	T2	Masai	3 08 S 35 33 E	
	K7	Mombasa	4 03 S 39 41 E	Napier
[Engongo-Engare] see Ngongongare	T2	Masai	c.2 41 S 36 42 E	Carmichael
[Engulia] see Ngulia	T2	Arusha	c.3 18 S 36 54 E	Uhlig
Engurdoto, area	K6	Masai	c.3 00 S 38 10 E	Bally
[Engurdoto] see Ngurdoto	T2	Masai	2 41 S 36 43 E	
Engushai, str.	T2	Arusha		
Engwaki, peak	T2	Moshi	2 58 S 37 12 E–2 50 S 37 07 E	Procter
Enkare Narok = Engare Narok, str.	K4	North Nyeri	0 16 N 37 06 E	Moreau
Enkorika, r.h., borehole	K1	Northern Frontier	1 13 N 36 33 E–0 52 N 36 47 E	
[Ensembe] see Msembe	K6	Masai	1 58 S 36 55.5 E	Kuchar
Entasekera, ln.	T7	Iringa	7 44 S 34 57 E	Greenway
Entebbe, popl., pier	K6	Masai	1 51 S 35 50 E	Glover et al.
Enunki	U4	Mengo	0 03 N 32 29 E	Numerous
Enziu, popl.	K6	Masai	prob. c.0 45 S 36 05 E	Greenway & Kanuri
Eorengitok, ln.	K4	Kitui	0 51 S 38 15 E	
Epanko, popl.	K6	Masai	0 55 S 35 54 E	
Equator, rsta., popl.	T6	Ulanga	8 43 S 36 40 E	Schlieben
Equator, str.	K3/5	Ravine/Kericho	0 00 35 34 E	Bogdan, Tweedie
Era, f.r.	K3	Nakuru	c.0 02 N 36 21 E	Lacey, Barney
Eruba, popl.	U1	West Nile	3 33 N 31 40 E	Eggeling
Erusi, popl.	U1	West Nile	2 58 N 30 55 E	
Erusi, mt.	U1	West Nile	2 20 N 31 06 E	Eggeling, Beaton
Erusi East, mt.	U1	West Nile	2 20 N 31 05.5 E	
Erute, co.	U1	West Nile	2 20 N 31 07 E	
Erute, popl.	U1	Lango	c.2 10 N 32 55 E	
[Eruti] see Erute	U1	Lango	2 14 N 32 55 E	
	U1	Lango	2 14 N 32 55 E	

Name	Code	Region	Coordinates	Collector
Esageri, rsta.	K3	Ravine	0 04 S 35 46.5 E	Bogdan
Esaigeri = Asagari, area, w.h.	K6	Masai	1 22 S 36 32 E	
Escarpment, popl.	K4	Kiambu	1 01 S 36 37 E	
[Esero] see Eisero	K3	Nandi	0 26 N 35 08 E	Brunt
Esimingor = Essimingor	T2	Masai	c.4 05 S 36 06 E	
Esoit ol Origa, hill	T2	Masai	c.2 20 S 36 10 E	Bally
Esokota, str.	K6	Masai	c.2 19 S 36 46 E–2 32 S 36 56 E	
Essimingor, mt., f.r.	T2	Masai	c.3 24 S 36 06 E	Foster, Greenway
Essimingor, area	T2	Masai	c.4 05 S 36 30 E	
[Eti] see Wati	U1	West Nile	3 13 N 31 02 E	Chancellor
[Euaso Narok] see Ewaso Narok	K3	Laikipia	0 00 36 22 E–0 32 N 36 52 E	
[Eusso Nyiro] see Uaso Nyiro (S.)	K6	Masai	c.0 39 S 35 44 E–2 04 S 36 07 E	Verdcourt et al.
Ewaso Narok, R.	K3	Laikipia	0 00 36 22 E–0 32 N 36 52 E	
Ewaso Narok Swamp	K3	Laikipia	c.0 19 N 36 36 E	
Ewaso Ngiro, popl.	K6	Masai	1 09 S 35 46 E	Glover et al.
Ewaso Ngiro = Uaso Nyiro (N.), R.	K1/3/4	several	0 19 S 36 39 E–0 27 N 39 55 E	D.C. Edwards, G. Dalton, J. Adamson, Pratt
[Ewaso Ngiro] see Uaso Nyiro (S.)	K6	Masai	c.0 39 S 35 44 E–2 04 S 36 07 E	
Ewaso Rongai, R.	K3	Trans-Nzoia	1 02 N 34 46.5 E–0 45 N 34 56 E	
Eyasi, L.	T1/2	Maswa/Masai/Mbulu	c.3 40 S 35 05 E	Numerous
[Faio] see Faiyu, Gara	K1	Northern Frontier (now in Ethiopia but considered as K1 for F.T.E.A.)	3 28 N 39 33 E	Gillett
Faiyu, Gara, mt.	K1	Northern Frontier (now in Ethiopia but considered as K1 for F.T.E.A.)	3 28 N 39 33 E	
Fajao, popl.	U2	Bunyoro	2 16 N 31 42 E	Eggeling
Falck's Farm	K3	Trans-Nzoia	1 09.5 N 34 49 E	Tweedie
Fanusi = Panusi, popl., str.	T3	Lushoto	5 07 S 38 40 E	Greenway
Farhani, popl.	T6	Kilosa	6 47 S 37 08 E	Stuhlmann
Farkwa, popl., t.c.	T5	Kondoa	5 25 S 35 36 E	Newman

59

60

Fort Jesus, museum	K7	Mombasa	4 04 S 39 41 E	Bally
Fort Portal, popl.	U2	Toro	0 40 N 30 17 E	Numerous
Fort Ternan, popl., rsta.	K5	Kisumu-Londiani	0 12 S 35 20.5 E	Drummond & Hemsley, Verdcourt
Fourteen Falls, falls	K4	Kiambu/Machakos	1 04.5 S 37 15 E	Numerous
Foweira, popl.	U2	Bunyoro	2 10 N 32 19 E	Bagshawe
French Island	Z	Zanzibar	prob. c.6 07 S 39 10 E	Kirk
French Mission, Nairobi	K4	Nairobi	1 17 S 36 49 E	Napier, Bally
Freretown, popl.	K7	Mombasa	4 01 S 39 42 E	Napier, W.E. Taylor
Freshfield, Pass	U2	Toro	0 20.5 N 29 54 E	Osmaston
Fufu, escarp.	T5	Dodoma/Mpwapwa	c.6 45 S 35 50 E	Milne-Redhead & Taylor
Fufu, popl.	T5	Mpwapwa	6 42 S 35 58 E	Polhill & Paulo
Fufu, str.	T5	Dodoma/Mpwapwa	c.6 56 S 35 50 E	Polhill & Paulo, Wigg
Fufuni, area, popl.	P	Pemba	5 52.5 S 39 41 E	Greenway
[Fugogo] see Fugugo				
Fugugo, mt., w.h.	K1	Northern Frontier	3 19 N 39 35.5 E	Bally & Smith
Fullekullesat, ln.	K1	Northern Frontier	3 19 N 39 35.5 E	
[Fulua] see Furua	K7	Tana River	c.0 29 S 39 32 E	Paulo, F. Thomas
Fumba, popl., area	T6	Ulanga	c.9 03 S 36 28 E–8 51 S 36 05 E	
Fumbini, beach, popl.	Z	Zanzibar	6 19 S 39 17 E	Numerous
[Fumbwa] see Jumba	K7	Kilifi	3 36 S 39 49 E	Swynnerton
Fumve = Funve, I.	U4	Mengo	0 32 N 32 52 E	Dummer
[Funda] see Ifunda	U4	Masaka	c.0 30 S 32 19 E	G.H. Wood, A.S. Thomas
Fundi Isa = Fundisa, Is.	T7	Iringa	8 02 S 35 28 E	Goetze
Fundi Isa = Fundisa, popl.	K7	Kilifi	c.2 57 S 40 10 E	C.W. Elliot, R.M.Graham
Fundisa, Is.	K7	Kilifi	2 56 S 40 05.5 E	
Fundisa, popl.	K7	Kilifi	c.2 57 S 40 10 E	
Fundo, I.	K7	Kilifi	2 56 S 40 05.5 E	
Funga, popl.	P	Pemba	c.5 03 S 39 39 E	Vaughan
[Funge] see Funga	T6	Ulanga	8 11 S 36 45 E	Haerdi
Fungoni, popl., for.	T6	Ulanga	8 11 S 36 45 E	Haerdi
	T6	Uzaramo	7 01 S 39 25 E	Mgaza, Procter

Fungos (L. Manyara–Mbulu)	T2	Mbulu	Not traced	Haarer
Funve, I.	U4	Masaka	c.0 30 S 32 19 E	
Funzi, beach	K7	Kwale	4 35 S 39 26 E	
Furrole, mt.	K1	Northern Frontier	c.3 42 N 38 01 E	Delamere, Gillett
Furroli = Furrole, mt.	K1	Northern Frontier	c.3 42 N 38 01 E	
Furrow, ln.	K3	Trans-Nzoia	1 10 N 34 43.5 E	Tweedie
Furua, R.	T6	Ulanga	c.9 03 S 36 28 E–8 51 S 36 05 E	
Gaba, l.s.	U4	Masaka	0 28 S 32 28 E	
Gaba, for., est.	U4	Mengo	c.0 07.5 N 32 56.5 E	Dummer
Gaba, popl., mssn., t.c.	U4	Mengo	0 16 N 32 38 E	? Dummer
[Gabaralome] ? see Gablaron	U3	Mbale	1 15.5 N 34 29.5 E	A.S. Thomas
Gablaron, popl.	U3	Mbale	1 15.5 N 34 29.5 E	Eggeling
Gabr Bori, ln.	K1	Northern Frontier	3 27 N 37 46 E	Bally
Gadaduma, mt., well	K1	Northern Frontier	3 27 N 39 33 E	Gillett
		(now in Ethiopia but considered as K1 for F.T.E.A.)		
[Gaddaduma] Gadaduma	K1	Northern Frontier	3 27 N 39 33 E	
		(now in Ethiopia but considered as K1 for F.T.E.A.)		
Gaditu	K4	Machakos/Kitui	c.1 53 S 37 59 E	Kassner
Gadu, ln.	K7	Kwale	c.4 00 S 39 17 E	Kassner
Gadumire, L.	U3	Busoga	c.1 15 N 33 26 E	
Gadumire, ln.	U3	Busoga	1 07 N 33 30 E	Maitland
[Gahinga] see Mgahinga	U2/Rwanda	Kigezi	1 23 S 29 39 E	
Gaikuyu, for.	K4	Kitui	0 35 S 38 00 E	
Gairo, hill, r.h.	T6	Kilosa	6 09 S 36 53 E	
Gakoe = Gakai, for. stn.	K4	Kiambu	0 53 S 36 48 E	Dyson, Archer
Gala, popl.	T5	Kondoa	5 04 S 36 08 E	
Galana, R.	K4/7	Kitui/Teita/Kilifi	2 59.5 S 38 31 E–3 10 S 40 08 E	Numerous
Galana Ranch	K4/7	Kitui/Kilifi	c.2 40 S 39 30 E	Bally
Galana River Scheme Airstrip = Dakadima	K4	Kitui	2 26 S 39 24 E	Van Praet

Name		Region	Coordinates	Collector
[Gale] see Gare				
Galiraya, popl.	U4	Mengo	1 21 N 32 49 E	Wakefield
Galla, old t.a.	K7	Kilifi/Tana River/Lamu	c.2 00—3 00 S 39 20—40 30 E	J. Adamson
Galma Galla, area, hill, w.h.	K1	Northern Frontier	c.1 11 S 40 48 E	Sampson, Makin, Bally & Smith
Galole, t.a.	K4	Kitui	c.2 20 S 39 10 E	
Galole, popl., irrigation scheme	K7	Tana River	1 30 S 40 02 E	Geilinger, Michelmore, Wingfield
Galole = Thowa, R.	K4/7	Kitui/Tana River	1 21 S 38 08 E—1 30 S 40 03 E	
Galula, mssn.	T7	Chunya	8 36.5 S 33 01 E	Kassner
Galunka	K4	Kitui	1 51 S 37 55 E	
[Gamagalla] see Galma Galla				
Gamura, w.h.	K1	Northern Frontier	1 11 S 40 48 E	J. Adamson
Ganda, popl.	K1	Northern Frontier	2 55 N 37 34 E	Bally
Ganda, hill	K7	Kilifi	3 13 S 40 04 E	Dale
Gandajega, R.	K7	Kwale	4 31 S 39 16 E	B.D. Burtt
Gandi, str./s	T5	Singida	4 48 S 34 18 E—4 51 S 33 58 E	Dale
Gangu, f.r.	K7	Kilifi	c.2 59 S 39 56 E	Eggeling
Gani, area	U4	Mengo	c.0 10 N 32 12 E	Grant*
Ganze, popl.	U1	Acholi*/Lango	c.1 30—3 30 N 32 15 E	L.B. Evans
[Gapus] see Gapuss	K7	Kilifi	3 32 S 39 43 E	
Gapuss, area	K2	Turkana	c.3 13 N 35 27 E	
Gapuss – Kaapus, str./s.	K2	Turkana	c.3 13 N 35 27 E	
Garabani = Karibani, hill	K2	Turkana	c.3 11 N 35 20 E	Dale
Gara Faiyu, mt.	K5	Masai	2 01.5 S 37 21.5 E	van Someren
	K1	Northern Frontier	3 28 N 39 33 E (now in Ethiopia but considered as K1 for F.T.E.A.)	
[Garagua] see Gararagua	T2	Moshi	3 07.5 S 37 00.5 E	Greenway
[Garanga] see Karanga	T2	Moshi	3 06 S 37 22 E—3 30 S 37 18 E	Uhlig
Gararagua, area	T2	Moshi	3 07.5 S 37 00.5 E	
Gararagua, est.	T2	Moshi	3 08 S 37 02.5 E	Greenway
[Garasari] see Ngaserai	T2	Masai	2 48 S 36 54 E	
Garashi, popl., t.c., D.C.'s camp.	K7	Kilifi	3 08 S 39 53 E	Moomaw

Name	Code	Province	Coordinates	Collector
Garaya, str.	T3	Lushoto	c.5 00–5 02 S 38 26 E	Peter
[Gara ya Niuki] see Engare Nanyuki				
Garba Tula, popl., air strip	T2	Arusha/Masai	3 15 S 36 48 E–2 48 S 36 55 E	Richards
Gare, popl., mssn., pl., r.h.	K1	Northern Frontier	0 32 N 38 31 E	Bally, Davidson, J. & T. Adamson
[Garguez] see Uaraguess	T3	Lushoto	4 47 S 38 20 E	Numerous
Garissa, distr.	K1	Northern Frontier	0 57 N 37 23.5 E	Heller
	K1/7	Northern Frontier/ Tana River	c.0 25 S 39 50 E	
Garissa, popl., r.h.	K7	Tana River	0 28 S 39 38 E	Numerous
[Garissia] see Karisia	K1	Northern Frontier	1 05 N 36 50 E	Kerfoot
Gar Jirimi, hill	K1	Northern Frontier	2 21 N 37 58 E	Carter & Stannard
Garolola, w.h.	K1	Northern Frontier	3 52 N 41 39 E	Ellenbeck
Garri, hills	K1	Northern Frontier	3 25 N 40 57 E	J. Adamson
Garsen, popl., p.p., air strip	K7	Tana River	2 16 S 40 07 E	Numerous
Garuweni	K4	South Nyeri	prob.0 35 S 37 08 E	Brunt
[Gasi] see Gazi	K7	Kwale	4 25 S 39 30 E	
Gatab, popl.	K1	Northern Frontier	2 37 N 36 55 E	Hepper & Jaeger
[Gatamaiyo] see Gatamayu				
Gatamayu, popl., t.c.	K4	Kiambu	0 59 S 36 42 E	
Gatamayu, for., str.	K4	Kiambu	0 59 S 36 42 E	
Gatarurumo Glade, ln.	K4	Kiambu	0 53 S 36 37 E–0 59 S 36 42 E	C. van Someren, Verdcourt
Gathiba, R.	K4	Kiambu	c.0 51 S 36 41 E	Dyson
Gathigiriri, popl.	K4	South Nyeri	0 17.5 S 37 17 E–0 26 S 37 20.5 E	Ossent
[Gatindiri] see Gatundiri	K4	South Nyeri	0 41 S 37 24.5 E	Kabuye
Gatondo, sch.	K4	Kiambu	c.1 07 S 36 58 E	Kirrika
[Gattuba] see Gathiba	K4	South Nyeri	0 26 S 37 10 E	Balbo
Gatundiri, popl.	K4	South Nyeri	0 17.5 S 37 17 E–0 26 S 37 20.5 E	Robertson
Gayaza, popl., r.h., t.c.	K4	Kiambu	c.1 07 S 36 58 E	Kirrika
Gayaza, popl.	U2	Ankole	0 47 S 30 48 E	Numerous
Gayaza, popl.	U4	Masaka	0 21 S 31 43 E	
Gayaza, popl.	U4	Masaka	0 27 S 31 11 E	
Gayaza, popl.	U4	Masaka	0 37.5 S 31 29 E	

Name	Code	District	Coordinates	Collector
Gayaza, popl.	U4	Masaka	0 37 S 31 45 E	
Gayaza, popl., mssn.	U4	Mengo	0 27 N 32 36 E	Michelmore
Gayaza, popl., mssn.	U4	Mengo	0 06 N 32 24 E	
[Gaze] see Gazi	K7	Kwale	4 25 S 39 30 E	
Gazi, popl.	K7	Kwale	4 25 S 39 30 E	Dale, Drummond & Hemsley, Moomaw
[Gazi Bay] see Maftaha Bay	K7	Kwale	4 25 S 39 31 E	
[Gazita] see Kazita	K4	Meru	0 04 S 37 25 E – 0 14 S 38 00 E	
Gebogo = Jebogo, hill	T2	Mbulu	4 38.5 S 35 32.5 E	Copley
Gede = Gedi, for., popl., area	K7	Kilifi	c.3 19 S 40 01 E	B.D. Burtt
Gede = Gedi, Royal Nat. Pk.	K7	Kilifi	3 18.5 S 40 01 E	Numerous
Gedi, for., popl., area	K7	Kilifi	c.3 19 S 40 01 E	Numerous
Gedi, Royal Nat. Pk.	K7	Kilifi	3 18.5 S 40 01 E	
Gehandu, popl., area	T2	Mbulu	3 50 S 35 34.5 E	Richmond
[Geikara] see Geikwara	K2	Turkana	4 34 N 35 27 E – 4 27 N 35 18 E	Poultney
Geikwara, valley	K2	Turkana	4 34 N 35 27 E – 4 27 N 35 18 E	
Geita, distr.	T1	Mwanza	c.2 40 S 32 15 E	Eggeling
(considered as part of Mwanza district for F.T.E.A.)				
Geita, f.r.	T1	Mwanza	2 50 S 32 12 E	Procter
Geita, gold mine	T1	Mwanza	c.2 52 S 32 10 E	B.D. Burtt
Geita, hill, popl.	T1	Mwanza	2 52 S 32 12 E	Numerous
Geitokitok = Leitokitok, spr.	T2	Masai	c.3 12 S 35 37 E	Eggeling, Verdcourt
Gelai, hill	K1	Northern Frontier	1 44 N 36 51 E	Carter & Stannard
Gelai, mt., f.r.	T2	Masai	2 36 S 36 06 E	Richards, Bally, Fischer
Gelai, popl.	T2	Masai	2 38 S 36 12 E	
Gelai Meru-goi, boma	T2	Masai	2 39 S 36 06 E	
Gele (Ijassi)	T5	Kondoa	c.4 40 S 35 50 E	Peter
Gem, ln.	K5	Central Kavirondo	c.0 02 N 34 28 E	Davidson, Butler
Gendabi = Kendabi, popl.	T2	Mbulu	4 27 S 35 20 E	Carmichael
Gendagenda North, f.r.	T3	Handeni	c.5 30 S 38 39 E	Procter
Gendagenda South, f.r.	T3	Handeni	c.5 34 S 38 38 E	Procter

Gendoya, hill	T7	Iringa	7 58 S 35 31 E	Rae
[Gengamo] see Gingama	T8	Songea	10 32 S 34 45 E	Busse
George, L.	U2	Toro/Ankole	c.0 00 30 12 E	Numerous
Gera, popl., sw.	T1	Bukoba	1 14 S 31 44.5 E	Gillman
[Gerengera] see Ngerengere	T6	Morogoro/Uzaramo	6 52 S 37 36.5 E–7 03 S 38 31 E	Stuhlmann
Gereza, popl., sch.	T3	Tanga	5 11 S 38 01 E	Semsei
Gerezani, popl.	T6	Uzaramo	6 50 S 39 16 E	
[Gerezanikreek] see Gerezani	T6	Uzaramo	6 50 S 39 16 E	B.D. Burtt
[Gerima] see Jirma	K1	Northern Frontier	3 58 N 41 46 E	Ruspoli & Riva
Geta, for. stn.	K3	Naivasha	0 28 S 36 36 E	
Gethagul = Getaqul, area	T2	Mbulu	c.4 26 S 35 29 E	Carmichael
Ghaloghungul (Hanang Forest)	T2	Mbulu	c.4 29 S 35 26 E	Carmichael
Ghost Mountain Forest	T5	Kondoa	Not traced	B.D. Burtt
Gidas, popl.	T2	Mbulu	4 25 S 35 41 E	Polhill & Paulo, Richards
Gidoghandajek (W. Mt. Hanang)	T2	Mbulu	c.4 26 S 35 23 E	Carmichael
Gikohi Ridge	K4	Kiambu	c.0 54 S 36 39 E	Forest Dept.
[Gikuyu Escarpment] see Kikuyu Escarp.	K4	Kiambu	c.0 55 S 36 40 E	
Gilgil, popl., rsta.	K3	Naivasha	0 30 S 36 19 E	Numerous
Gilgil, R.	K3	Naivasha	0 11 S 36 14 E–0 43.5 S 36 20 E	Scott Elliot
[Gilgit] see Gilgil	K3	Naivasha	0 30 S 36 19 E	
Gimba, hill, area	K7	Teita	3 27 S 38 33 E	Gardner, Uhlig
Gingama, popl.	T8	Songea	10 32 S 34 45 E	Busse
[Giombene] see Nyambeni	K4	Meru	c.0 15 N 37 55 E	Balbo
Giriama, popl., mssn.	K7	Kilifi	3 48 S 39 35 E	Dale
Giriama, t.a.	K7	Kilifi	c.3 00 S 39 45 E	Dale
Giriama Point, ln.	K7	Kwale	4 15.5 S 39 27 E	Moomaw, Magogo & Glover
[Girma] see Jirma	K1	Northern Frontier	3 58 N 41 46 E	Ruspoli & Riva
[Giro] see Gairo	T6	Kilosa	6 09 S 36 53 E	Carmichael
Girtalo, ln.	T2	Masai	2 02 S 35 35 E	Moreau
[Giryama] see Giriama	K7	Kilifi		W.E. Taylor

66

Githi, ln.	K4	South Nyeri	0 35 S 37 07 E	Kibui
Giyedabash (Marang Forest)	T2	Mbulu	c.3 42 S 35 40 E	Carmichael
Glades, The, ln.	T2	Arusha	3 17 S 36 55 E	Greenway & Kanuri
Glanville Halt, rsta.	K3	Trans-Nzoia	0 57.5 N 35 04 E	Irwin
Gobat, hill	K3	Baringo	0 10.5 N 35 55.5 E	Pratt
Gochot = Gochow, area	T2	Mbulu	c.4 24 S 35 25 E	Carmichael
Godegode, popl., rsta., t.c.	T5	Mpwapwa	6 32 S 36 33.5 E	B.D. Burtt, Busse
Godoni, for.	K7	Kwale	4 09 S 39 26 E	
Gof Bongole, crater	K1	Northern Frontier	2 12 N 37 55.5 E	J. Adamson, Magogo
Gof Choba, hill, crater	K1	Northern Frontier	2 24 N 38 03 E	Bally & Smith, Carter & Stannard
Gof Redo, hill	K1	Northern Frontier	2 23 N 38 01 E	van Swinderen
Gof Sokorte Guda = Lake Paradise	K1	Northern Frontier	2 16 N 37 56 E	
[Gof Sokota Guda] see Gof Sokorte Guda	K1	Northern Frontier	2 16 N 37 56 E	Bally & Smith
Gogo Falls, falls	K5	South Kavirondo	0 54 S 34 21 E	
Gogoi (upper Wami R.)	T6	Kilosa	prob. c.6 40 S 37 05 E	Busse
[Gogoni] see Gongoni	K7	Kwale	c.4 25 S 39 28 E	Dale
Gogonyo, popl.	U3	Mbale	1 15 N 33 35 E	
[Gohingo] see Ngohingo	T6	Uzaramo	6 47 S 38 50 E	Peter
[Goitokitok] see Ngoitokitok	T2	Masai	3 12 S 35 37 E	Greenway
Goka (Shagayu f.r.)	T3	Lushoto	c.4 30 S 38 18 E	Carmichael, Mgaza
Golana Gof, str.	K1	Northern Frontier	c.0 32 N 38 31 E–0 46 N 39 24 E	Bally, J. Adamson
Golbanti	K7	Tana River	2 27 S 40 11.5 E	Sampson
Goli, popl. mssn.	U1	West Nile	2 23 N 31 02 E	
Goli, hill, mssn.	U1	West Nile	2 42 N 31 06 E	
Gol Kopjes = Ol Doinyo Gol, hills	T2	Masai	c.2 42 S 35 26 E	Greenway
Golo, ln.	K4	Meru	0 10 N 38 16 E	P.H. Hamilton
Gologolo, area	T3	Lushoto	4 42 S 38 13 E	Drummond & Hemsley, Eggeling, Mgaza
Gologolo, mts.	T7	Iringa	7 40 S 36 53 E	Thulin
Gologolo East, f.r.	T3	Lushoto	4 42 S 38 14 E	
Gologolo West, f.r.	T3	Lushoto	4 42 S 38 14 E	

67

Name	District	Grid	Coordinates	Collector
Gomba, co.	Mengo	U4	c.0 10 N 31 45 E	Maitland, Eggeling
Gomba, hill	Lushoto	T3	5 01 S 38 17 E	
Gomba, pl.	Lushoto	T3	5 00 S 38 18 E	
Gomba, sisal est.	Lushoto	T3	5 02 S 38 15 E	Peter, Rykebusch
Gombe, str.	Kigoma	T4	4 42 S 29 36 E	Numerous
Gombelo, hill	Lushoto	T3	4 41 S 38 37 E	Holst
Gombelo, popl.	Lushoto	T3	4 31.5 S 38 22 E	
Gombelo, popl.	Lushoto	T3	4 50 S 38 48.5 E	
Gombelo, popl.	Lushoto	T3	4 59 S 38 21 E	
Gombelo, popl.	Lushoto	T3	5 10.5 S 38 43 E	
Gombelo = Gombero	Tanga	T3	4 57 S 38 57 E	Peter, Holst
[Gombero] see Gombelo	Lushoto	T3	4 50 S 38 48.5 E	Peter
Gombero, f.r.	Tanga	T3	4 58 S 39 00 E	Mgaza
Gombero, popl., r.h.	Tanga	T3	4 57 S 38 57 E	Peter 22945, 39543, 48159
Gombero, popl.	Handeni	T3	5 30 S 37 53 E	
Gombe Stream, g.r.	Buha/Kigoma	T4	c.4 35—49 S 29 35—40 E	Numerous
Gombo, L.	Mpwapwa	T5	6 38 S 36 42 E	B.D. Burtt
[Gombopo] see Gombero	Tanga	T3	4 58 S 39 00 E	Cribb & Grey-Wilson
[Gombora] see Nkombola	Lushoto	T3	4 56.5 S 38 41 E	Moreau
[Gonda] see Igonda	Tabora	T4	5 33 S 32 40 E	Boehm
[Gonga Berg] see Gonja Mt.	Lushoto	T3	4 47 S 38 33 E	Busse
Gongolamboto, cemetery copse	Uzaramo	T6	6 53 S 39 09 E	Wingfield
Gongolo, str.	Tanga	T3	5 12 S 39 01 E—5 13 S 39 03.5 E	
Gongoni, for.	Kwale	K7	c.4 25 S 39 28 E	Gardner
Gongoni, popl., for.	Kilifi	K7	3 49.5 S 39 46 E	Bally (Apr. 1953)
Gongoni, for.	Lamu	K7	c.2 23 S 40 28 E	Dale, Mohammed Abdullah
Gongoni, popl., for.	Kilifi	K7	3 02 S 40 08 E	Bally, Polhill & Paulo
Gongoni, popl.	Tana River	K7	2 08 S 40 11 E	Dale
Gonja, popl.	Kwale	K7	4 33 S 39 04 E	Dale
Gonja, mt.	Lushoto	T3	4 47 S 38 33 E	Holst, Engler

Gonja, popl.	T3	Pare	4 16.5 S 38 02.5 E	Greenway, Engler, Hughes, Zimmerman, Eggeling, Semsei, Mgaza, Sangiwa
Gonja, mt.	T3	Lushoto	5 11 S 38 36 E	
Gonja = Ngonja, old area	T3	Lushoto	c.4 59 S 38 45 E	Peter
Gonja, old area	T3	Lushoto	c.5 03 S 38 38 E	Holst
[Gonya Berg] see Gonja Mt.	T3	Lushoto	5 11 S 38 36 E	Peter
Gopolal Mwaru, ln.	K3	Laikipia	c.0 13 N 36 18 E	Gregory
Goregore = Ngore Ngore	K6	Masai	1 02 S 35 30 E	Glover
[Gori] see Migori	K5/6	S. Kavirondo/Masai	c.0 54 S 35 07 E–0 57 S 34 08 E	
Gori Cheboa, ln.	K1	Northern Frontier	3 34 N 41 32 E	
Gortheg, mt.	U3	Mbale	1 18.5 N 34 41.5 E	Eggeling
Goshi = Voi, R.	K7	Teita/Kilifi	3 22 S 38 23 E–3 21 S 39 40 E	
Gotani, popl.	K7	Kilifi	3 47 S 39 32 E	Moomaw
Goweko, rsta.	T4	Tabora	5 20 S 33 08 E	Holtz, Peter
Grand Falls, falls	K4	Meru/Kitui	0 16 S 38 00 E	
Grassland Research Station	K3	Trans-Nzoia	0 21 S 35 39 E	
[Grave] see Chapani	Z	Zanzibar	6 07.5 S 39 11.5 E	
Great Ardai Plain = Ardai Plains	T2	Masai	c.3 21–3 26 S 36 20–36 25 E	Greenway
Greater Kiboko = Kiboko (part), R.	K4/5	Machakos/Masai	c.2 30 S 37 15 E–2 09 S 37 54 E	
Great North Road	U1/3, K3/4/6, T2/5/7			
Great Ruaha gorge	T6/7	Kilosa/Iringa	c.7 31 S 36 37 E	Cribb & Grey-Wilson
		(road is in Kilosa District)		
Great Ruaha, R.	T5/5/7	several	9 10 S 34 10 E–7 56 S 37 52 E	Numerous
[Greek, R.] see Kelim	U3	Mbale	c.1 22 N 34 47 E–1 34 N 34 16 E	Bryce
Greek River, popl.	U3	Mbale	1 27 N 34 43 E	
Grewal's sawmills	T3	Lushoto	c.4 43 S 38 18 E	Greenway
[Grumchem] see Grumechen	T1	Musoma	c.2 18 S 35 15 E	Greenway
Grumechen, hill	T1/2	Musoma/Masai	2 18 S 35 15 E	Greenway
Grumechen Gate, ln.	T1	Musoma	c.2 18 S 35 15 E	Greenway

Grumeti, R.	T1	Musoma	c.1 54 S 34 57 E−2 05 S 33 57 E	Greenway
[Guaso Narok] see Ewaso Narok				
[Guaso Nyiro] see Uaso Nyiro (N.)	K3	Laikipia	0 00 36 22 E−0 32 N 36 52 E	Dalton
[Guaso Nyiro] see Uaso Nyiro (N.)	K1/3/4	several	0 19 S 36 39 E−0 27 N 39 55 E	
[Guaso Nyiro] see Uaso Nyiro (S.)	K6	Masai	c.0 39 S 35 44 E−2 04 S 36 07 E	
Gucha, R.	K5	South Kavirondo	c.0 38 S 35 00 E−0 55 S 34 08 E	
Gulmarg Farm	K3	Trans-Nzoia	0 57 N 34 49 E	Tweedie
[Gulmi] see Ngurunit	K1	Northern Frontier	1 45 N 37 18 E	Sato
[Gulni] see Ngurunit	K1	Northern Frontier	1 45 N 37 18 E	Sato
Gulu, popl., mssn., r.h.	U1	Acholi	2 47 N 32 17 E	Numerous
Gulwe, hill	T5	Mpwapwa	6 30 S 36 25 E	
Gulwe, popl., rsta., t.c.	T5	Mpwapwa	6 26.5 S 36 24.5 E	Numerous
[Gumba] see Gomba	U4	Mengo	c.0 10 N 31 45 E	Eggeling
Gumbira, popl., area	T7	Iringa	c.7 46 S 35 45 E	Goetze
Gumbiro, popl.	T8	Songea	10 16 S 35 39 E	Milne-Redhead & Taylor
Gunda Mkali = Mgunda Mkali, area	T5	Dodoma	5 50 S 33 45 E−6 15 S 34 45 E	Stuhlmann
Gungu = Ngungu, popl.	T4	Kigoma	4 51.5 S 29 38.5 E	
Gura, R.	K4	South Nyeri	0 37 S 36 43 E−0 31.5 S 37 05 E	Gardner
Gura Falls, falls	K4	South Nyeri	0 31 S 36 46 E	
Gurar, p.p.	K1	Northern Frontier	3 22 N 39 34 E	Bally & Smith
Gurika, hill	K1	Northern Frontier	1 48.5 N 37 00 E	Carter & Stannard
[Guru] see Hanang	T2	Mbulu	4 26 S 35 24 E	Jaeger
[Gurumbi] see Ngurumbi	T7	Rungwe/Mbeya	c.9 02 S 33 29 E	Goetze
[Gurunet] see Ngorinit	K1	Northern Frontier	1 45 N 37 18 E	J. Adamson
Gurungu, well	K2	Turkana	4 26 N 34 32 E	
[Guruwe] see Hanang	T2	Mbulu	4 26 S 35 24 E	
Gwagalo (Bufumira I.)	U4	Masaka	c.0 21 S 32 24 E	A.S. Thomas
Gwahata = Kwahata, ln.	T2	Mbulu	3 56 S 35 39 E	
Gwanaya (SE. Hanang)	T2	Mbulu	c.4 29 S 35 26 E	Carmichael
[Gwari] see Kware	T2	Moshi	3 17 S 37 09 E	Haarer
[Gwashi] see Gwasi	K5	South Kavirondo	0 37 S 34 09 E	Napier

Name		Region	Coordinates	Collector
Gwasi, hills	K5	South Kavirondo	0 37 S 34 09 E	Napier, Vuyk
[Gwaso Nyiro] see Ewaso Ngiro				
Gwen's Corner, ln.	K6	Masai	1 09 S 35 46 E	Tweedie
Gweri, popl.	K3	Trans-Nzoia	1 10 N 34 50 E	
Gwongongween (Near Lumbwa)	U3	Teso	1 43 N 33 44 E	Fries
Gyengo	K5	Kisumu-Londiani	near 0 12 S 35 28 E	A.S. Thomas
	U3	Mbale	Not traced	
Habaswein, t.c.	K1	Northern Frontier	1 01 N 39 29 E	Bally & Smith, J. Adamson
Hadado	K1	Northern Frontier	prob. c.2 45 N 36 45 E	J. Adamson
Hadu, popl.	K7	Kilifi	2 51 S 39 58 E	Dale
Hagadera, w.h.	K1	Northern Frontier	0 01 N 40 22 E	Hale, Brenan et al.
Hagafilo = Hagafiro, falls	T7	Njombe	9 23 S 34 50 E	Milne-Redhead & Taylor
Hagafilo = Hagafiro, R.	T7	Njombe	c.9 31 S 34 37 E–9 22 S 35 02 E	Milne-Redhead & Taylor, Schlieben
Hagafiro = Hagafilo, falls	T7	Njombe	9 23 S 34 50 E	
Hagafiro = Hagafilo, R.	T7	Njombe	c.9 31 S 34 37 E–9 22 S 35 02 E	
Haidedonga, hill	T2	Mbulu	4 08 S 35 55 E	Schlieben
Haitajwa, hill	Z	Zanzibar	6 16 S 39 16 E	Richards
[Haiwani] see Hewani				
Hakitengya, popl.	K7	Tana River	2 14 S 40 11 E	Greenway, Vaughan
Halau, R.	U2	Toro	0 44 N 30 03 E	Katende
Hale, popl, rsta.	T8	Songea	c.11 00 S 34 55 E	Maitland, Eggeling
Halembe, hill, popl, r.h.	T3	Pangani	5 19 S 38 37 E	Milne-Redhead & Taylor
Hallowes Farm	T4	Kigoma	5 44 S 29 55 E	Numerous
Hall Tarns, lakes	K3	Naivasha	0 15 S 36 20 E	Bally
[Hamayi] see Hameyi				
Hambalai, popl. escarp.	K4	Meru	0 08.5 S 37 20.5 E	Le Pelley
Hameyi, ln.	K7	Tana River	c.0 04 S 39 01 E	Sampson
Hamit, str.	T3	Lushoto	4 52.5 S 38 20 E	Moreau
	K7	Tana River	c.0 04 S 39 01 E	
	T2	Mbulu	4 27 S 35 25 E–4 35 S 35 24 E	
Hamugoma (Ruwenzori)	U2	Toro	prob. c.0 19 N 29 54 E	Eggeling
Hanang, mt.	T2	Mbulu	4 26 S 35 24 E	Numerous

Handajega, airstrip, spr.	T1	Maswa	2 15 S 34 06 E	Greenway
Handei, old area	T3	Lushoto	c.5 10 S 38 40 E	Peter, Holst
Handei, popl.	T3	Lushoto	4 48 S 38 22 E	Holst
Handei, popl.	T3	Lushoto	5 05 S 38 21 E	
Handeni, distr.	T3	Handeni	c.5 30 S 38 00 E	
Handeni, hill	T3	Handeni	5 27 S 38 03 E	Numerous
Handeni, popl., mssn., r.h.	T3	Handeni	5 26 S 38 01.5 E	Numerous
Hanga, popl., fm., str.	T8	Songea	10 13 S 35 30 E	Milne-Redhead & Taylor
Hangi, popl.	T7	Njombe	9 10 S 35 14 E	Schlieben
Hangula, mssn.	T6	Morogoro	7 02 S 37 38.5 E	
[Hannington] see Bogoria	K3	Baringo	c.0 15 N 36 06 E	Bally, Faden, Napper
Haraa, mt., f.r.	T2	Mbulu	prob. c.4 19 S 35 46 E	Carmichael
Hara mbuga, sw.	T2	Mbulu	4 43 S 35 30 E	B.D. Burtt
Hara Daua, area	K1/Eth	Northern Frontier	4 07 N 40 23 E	Bally & Smith
Harnett's Farm	K3	Trans-Nzoia	1 10 N 34 48 E	Tweedie
Hassama, hill, f.r.	T2	Mbulu	3 55 S 35 40 E	Carmichael
[Hassiga] see Kasiga	T3	Lushoto	4 50 S 38 14 E	Holst
Hatati, ln.	K1	Northern Frontier	c.1 58 N 40 04 E	Bally
Häuptlings Boma, Mwika	T2	Moshi	c.3 17 S 37 34.5 E	Volkens
Hausburg Valley	K4	North Nyeri	c.0 06.5 S 37 13 E–0 08.5 S 37 18 E	Hedberg
Hayton Falls	K6	Masai	1 51 S 36 07 E	Bally
Heartbreak Camp, ln.	K7	Teita	3 14 S 39 05 E	
Heathcote's Farm = Ol Lalang	K3	Trans-Nzoia	1 02.5 N 34 57 E	Tweedie
Heboma, popl.	T3	Lushoto	4 50.5 S 38 17 E	Holst
Hedaru, popl., mssn., rsta.	T3	Pare	4 30 S 37 54.5 E	Greenway, Peter
Hekulo, area, popl.	T3	Lushoto	4 49.5 S 38 26.5 E	
Hell's Gate, gorge	K3	Naivasha	c.0 52–0 57 S 36 18.5 E	Numerous
Helu, p.p.	K1	Northern Frontier	3 32 N 39 05 E	Gillett
[Hema] see Hima	U2	Toro		Eggeling
Hemagoma, popl., old pl.	T3	Lushoto	4 53 S 38 35 E	Peter

Hemai, popl.	K7	Kwale	4 29 S 39 16 E	
Hemashai, popl.	T3	Lushoto	4 46.5 S 38 17 E	
[Hemsteads Bridge] see Moi's Bridge	K3	Trans-Nzoia/Uasin Gishu	0 45.5 N 35 04 E	Greenway
[Hemsted's Bridge] see Moi's Bridge	K3	Trans-Nzoia/Uasin Gishu	0 45.5 N 35 04 E	
[Hemvera] see Hemwera	T3	Pare	4 23 S 37 55 E	Leechman
Hemwera, mt.	T3	Pare	4 23 S 37 55 E	
[Herere] see Iherehe	T1	Mwanza	3 02 S 32 47 E	
[Herkulu] see Hekulo	T3	Lushoto	4 49.5 S 38 26.5 E	Lovett
Hertahöhe (Lutindi Mt.)	T3	Lushoto	c.4 54 S 38 37 E	Peter
Heru, area	T4	Buha	c.4 35 S 30 05 E	
Heru Chini, popl.	T4	Buha	? c.4 34 S 30 06 E	Rounce
Heru Juu, popl.	T4	Buha	4 32 S 30 04 E	Eggeling, Procter, Carmichael
Hewani, popl.	K7	Tana River	2 14 S 40 11 E	
Higulu Plateau (N. Uzungwa Mts.)	T7	Iringa	Not traced	Goetze
Hika, popl, area	T5	Dodoma	5 37 S 34 59 E	B.D. Burtt
Hima, popl.	U2	Toro	0 17.5 N 30 10.5 E	
Hima, R.	U2	Toro	0 23 N 30 02 E–0 15 N 30 13 E	
[Himit] see Hamit	T2	Mbulu	4 27 S 35 25 E–4 35 S 35 24 E	Carmichael
Himo, popl, sisal est.	T2	Moshi	3 23 S 37 33 E	Numerous
Himo, rsta.	T2/3	Moshi/Pare	3 26.5 S 37 33.5 E	
Himo, str.	T2/3	Moshi/Pare	c.3 15 S 37 30 E–3 32 S 37 31 E	Greenway, Volkens
Hinde Valley	K4	Meru	c.0 07 S 37 22 E	J. Bally
Hindi, ln.	K7	Lamu	2 11 S 40 49 E	Gillett
Hippo Point, pt.	K3	Naivasha	0 48 S 36 18 E	Pegler
Hippo Pool	K4	Nairobi	1 25 S 36 54 E	
Hippo Pool, ln.	T2	Masai	c.3 15 S 35 35 E	Greenway & Kanuri
Hippo Pools	T6	Kilosa	7 19 S 37 06 E	Greenway & Kanuri
[Hiru Chini] see Heru Chini	T4	Buha	? c.4 34 S 30 06 E	Rounce
Hissey's Rocks, ln.	K3	Trans-Nzoia	1 09 N 34 50 E	Tweedie
Hiwaga	T6	Kilosa	Not traced	Swynnerton

Name		District	Coordinates	Collector
[Hobley's Volcano] see Orgaria	K3	Naivasha	0 53 S 36 16 E	Verdcourt
Hoddinott's Rocks, ln.	K3	Trans-Nzoia	1 10 N 34 49.5 E	Tweedie
Hodi, R., Buhindi F.R.	T1	Mwanza	c.2 25 S 32 15 E	Carmichael
[Hoey's Bridge] see Moi's Bridge	K3	Trans-Nzoia	0 53 N 35 07 E	Numerous
Hoey's Turning, ln.	K3	Trans-Nzoia	c.1 10 N 34 47.5 E	Tweedie
Höhenfriedeberg, mssn.	T3	Lushoto	4 34.5 S 38 21 E	Geilinger
[Höhnel Island] see South I.	K1	Northern Frontier	c.2 38 N 36 36 E	
Höhnel-Schmellen, rapids	T3	Lushoto	c.4 52 S 38 02 E	
Höhnel Valley	K4	North Nyeri/South Nyeri	c.0 11 S 37 14–37 17.5 E	Fries
Hoima, popl., mssn.	U2	Bunyoro	1 26 N 31 21 E	?Bagshawe, Fyffe, Eggeling
[Hoina] see Hoima	U2	Bunyoro	1 26 N 31 21 E	Bagshawe
Hola, popl.	K7	Tana River	1 29 S 40 02 E	Bally, Hooper & Townsend, Robertson
Holoba, ln.	T7	Iringa	c.8 07 S 35 56 E	Carmichael
Hololo = Holola, popl.	K7	Tana River	0 10 S 39 20 E	F. Thomas
Homa, bay	K5	South Kavirondo	c.0 28 S 34 28 E	
Homa, mt.	K5	South Kavirondo	0 23 S 34 30 E	
Homa, pt.	K5	South Kavirondo	0 21 S 34 28 E	
Homa Bay, distr.	K5	South Kavirondo	c.0 45 S 34 15 E	
		(considered as part of South Kavirondo for F.T.E.A.)		
Homa Bay, popl., p.p.	K5	South Kavirondo	0 31 S 34 27 E	Napier, Allen Turner
Hombe, popl.	K4	South Nyeri	0 21 S 37 07 E	Rayner
[Homo Bay] see Homa Bay	K5	South Kavirondo	0 31 S 34 27 E	Napier
Honi = Amboni (in part), R.	K4	North/South Nyeri	0 18.5 S 36 37.5 E–0 24 S 36 59.5 E	
Horombo Hut = Peters Hut	T2	Moshi	3 08 S 37 26 E	
Horr Valley	K1	Northern Frontier	c.2 10 N 36 55 E	Jex-Blake
Hoshingo, hill, w.h.	K4/7	Kitui/Kilifi	2 38.5 S 39 35.5 E	Bally
[Hosiga] see Kasiga	T3	Lushoto	4 50 S 38 14 E	Holst
[Hosigo] see Kasiga	T3	Lushoto	4 50 S 38 14 E	Holst
Hotulwe, peak	T3	Pare	prob. c.4 25 S 37 55 E	Leechman
Howbury Farm	K3	Naivasha	0 41 S 36 37.5 E	Polhill

Name		Province	Coordinates	Authority
[Huala Falls] see Hululu Falls	T6	Morogoro	c.7 02 S 37 38 E	
Hujuwi, popl.	T7	Mbeya	8 48 S 32 39 E	Goetze
Hululu Falls (on Mgeta R.)	T6	Morogoro	c.7 02 S 37 38 E	Drummond & Hemsley, Greenway & Eggeling
[Humwa] see Ihumwa	T5	Dodoma	6 10 S 35 53 E	Sperling
Hundogo, f.r.	T6	Uzaramo	c.6 47 S 39 00 E	Paulo, Procter
[Hundwi] see Itundwi	T5	Kondoa	4 39 S 35 51 E	B.D. Burtt
Hunters' Lodge [Inn]	K4	Machakos	2 12.5 S 37 43 E	Tweedie
Huri, hills	K1	Northern Frontier	c.3 30 N 37 50 E	Bally, T. Adamson
Huruhuru Mbuga	T1	Shinyanga	3 27 S 33 10 E	B.D. Burtt, Staples
Huyuwi = Hujuwi, popl.	T7	Mbeya	8 48 S 32 39 E	Goetze
Hyrax Hill	K3	Nakuru	0 17 S 36 07 E	Mwangangi
[Iara] see Itara	T1	Bukoba	1 09 S 31 28 E	
Ibaba, popl.	T7	Rungwe	9 24.5 S 33 22 E	Wigg
Ibambalo, f.r.	U2	Toro	c.0 36 N 30 52 E	Cribb & Grey-Wilson
[Ibambaro] see Ibambalo	U2	Toro	c.0 36 N 30 52 E	
Ibanda, hill	U2	Ankole	0 07 S 30 29 E	Osmaston
Ibanda, popl., r.h., t.c.	U2	Ankole	0 08 S 30 29 E	Fishlock & Hancock
[Ibanda] see Banda	U4	Mengo	0 17 N 32 49.5 E	
Ibanda, ln., for.	T1	Mwanza	2 47 S 32 30 E	Welch, Snowden, Maitland
Ibanda, sw.	T4	Buha	4 12 S 30 52 E	Pool
Ibando, f.r., ? = Ibanda	T1	Mwanza	?2 47 S 32 30 E	Bullock, Trapnell
Ibaya, lodge	T3	Pare	3 58 S 37 48 E	Carmichael
Ibonde, popl., r.h.	U2	Toro	0 41 N 30 12 E	Richards, Harris
Ibondo, popl.	T1	Mwanza	2 38 S 32 40.5 E	A.S. Thomas
Ibuga, popl.	T1	Bukoba	1 36 S 31 36 E	B.D. Burtt, Tanner
Ibuguziwa, fy.	T7	Iringa	7 42 S 34 53 E	Ford
Ibuje, popl., r.h., t.c.	U1	Lango	1 54 N 32 24 E	Richards, Greenway, Bjornstad
Ibumu, popl.	T7	Iringa	7 27 S 36 08 E	Richards, Carmichael

Name	Code	Region	Coordinates	Authority
Ibwera, hill, popl.	T1	Bukoba	c.1 29 S 31 38 E	Herbert & Partridge
Ichachiri, popl.	K4	Kiambu	0 59 S 36 53 E	Kirrika
[Ichaweri] ? see Ichachiri	K4	Kiambu	?0 59 S 36 53 E	
Ichemba, area	T4	Tabora	c.4 45 S 32 20 E	Wigg, Procter
Idakho, ln.	K5	North Kavirondo	0 13 N 34 41 E	Greenway
Idege, popl.	T7	Iringa	8 41 S 35 12 E	Polhill & Paulo
Iderero	T5	?Mpwapwa	Not traced	Hornby
Idete, str.	T7	Iringa	7 29.5 S 36 12.5 E–7 27 S 36 14.5 E	Carmichael
Idewa, f.r.	T7	Iringa	c.8 17 S 35 47 E	
[Idewe] see Idewa	T7	Iringa	c.8 17 S 35 47 E	Polhill & Paulo
Idi Amin Dada = Edward, L.	U2/Zaire	Toro/Ankole/Kigezi	c.0 25 S 29 30 E	
Idodi, popl.	T7	Iringa	7 47 S 35 11 E	Ward, Procter
Idudumo, popl.	T4	Nzega	4 21 S 33 13 E	Ruffo
Idumu (Image Mt., NW)	T7	Iringa	c.7 26 S 36 09 E	Carmichael
Idunduge (Image Mt., E.)	T7	Iringa	c.7 27 S 36 11 E	Carmichael
Iduo, area	T5	Mpwapwa	c.6 18 S 36 37 E	Busse
Iduo, hill	T5	Mpwapwa	6 16 S 36 40 E	
Iduo, mssn.	T5	Mpwapwa	6 19.5 S 36 40 E	
Idweli, popl.	T7	Rungwe	9 02 S 33 36 E	Cribb & Grey-Wilson
Idya, popl., area	P	Pemba	5 6.5 S 39 43 E	
Ifakara, popl., mssn.	T6	Ulanga	8 08 S 36 41 E	Anderson, Haerdi, Eggeling
[Ifigulu] see Ifuguru	T7	Iringa	7 36 S 35 03 E	Greenway
[Ifiguru] see Ifuguru	T7	Iringa	7 36 S 35 03 E	Greenway
Ifinga, popl.	T6	Ulanga	9 30 S 35 30 E	Schlieben
Iflaga = Uflaga, plain	T4	Kigoma	4 42 S 31 35 E	Vesey-FitzGerald
Ifua, popl.	T7	Iringa	8 05 S 35 53 E	Carmichael
Ifuguru, popl.	T7	Iringa	7 36 S 35 03 E	
Ifumba, Kopje (or Ipumba)	T1	Maswa	c.3 15 S 34 50 E	Greenway
Ifunda, popl.	T7	Iringa	8 02 S 35 28 E	Numerous
Igagala, resettlement area	T4	Tabora	c.4 56 S 31 35 E	Hooper & Townsend

Name		Region	Coordinates	Collectors
[Igalala] see Igagala	T4	Tabora	c.4 56 S 31 35 E	Hooper & Townsend
[Igale] see Igali	T7	Mbeya	9 03 S 33 24 E	Cribb & Grey-Wilson
Igali, pass	T7	Mbeya	9 01 S 33 24 E	Goetze, Richards, Stolz
Igali, popl.	T7	Mbeya	9 03 S 33 24 E	Milne-Redhead & Taylor
Igalukiro, popl.	T1	Maswa	3 08 S 33 31 E	
Igalukiro, popl.	T1	Mwanza	2 25 S 33 43 E	Tanner
Igalula, popl., rsta.	T4	Tabora	5 14.5 S 33 00 E	C.H. Jackson, Lunan, Peter
Igamba, popl.	T7	Rungwe	9 34 S 33 27 E	
Iganga, popl.	U3	Busoga	1 10 N 33 03 E	
Iganga, popl.	U3	Busoga	0 43 N 33 14.5 E	
Iganga, popl., rsta.	U3	Busoga	0 37 N 33 29 E	
Iganjo, ln.	K4	South Nyeri	0 35 S 37 07 E	Tweedie, G.H. Wood
Igara, co.	U2	Ankole	c.0 30 S 30 10 E	Kibui
Igarama, hill (Buhindi f.r.)	T1	Mwanza	c.2 25 S 32 15 E	Cree, Purseglove, Snowden
Igawa, popl.	T7	Mbeya	8 46 S 34 23 E	Carmichael
Igawira, hill	T7	Iringa	7 47 S 34 50 E	Pole Evans & Erens, Polhill & Paulo, Richards, Eggeling, Carmichael
Igeri, area	T7	Njombe	9 40 S 34 40 E	Greenway
[Igezi] see Kigezi	U2	Kigezi	c.0 50 S 29 45 E	S.A. Robertson, Hounsell
Igigwa, popl.	T4	Tabora	5 26 S 32 52 E	Norman
Igila, hill	T7	Chunya	c.7 38 S 33 24 E	
Igisi, hill	U2	Bunyoro	1 59.5 N 31 44 E	Richards
Igitschu ? see Gucha R.	K5	South Kavirondo	9 10 S 33 32 E	Eggeling
Igogwe, str., popl.	T7	Rungwe	0 11.5 S 37 40 E	Fischer
[Igoje] see Igoji	K4	Meru	0 11.5 S 37 40 E	
Igoji, mkt., mssn.	K4	Meru	8 59.5 S 33 38.5 E	Cribb et al.
Igoma, popl.	T7	Mbeya	4 18 S 33 04 E –4 25 S 31 58 E	
Igombe, R.	T4	Tabora/Nzega	4 50 S 32 42 E	
Igombe Dam, f.r.	T4	Tabora	c.4 35 S 32 25 E	Cribb & Grey-Wilson
Igombe River, f.r.	T4	Tabora		Lindeman, Procter

[Igomena] see Ngomeni	K4	Kitui	0 39 S 38 24 E	Rauh
Igonda, popl.	T4	Tabora	5 33 S 32 40 E	Boehm
Igongo, popl.	T5	Dodoma	6 15 S 35 12 E	Busse
Igongo, str.	T7	Iringa	c.8 30 S 35 45 E	Carmichael
Igosi, popl.	T7	Njombe	9 19 S 34 32 E	Wingfield
Igunga, popl.	T4	Nzega	4 17 S 33 53 E	Doggett
Igurusi, popl.	T7	Mbeya	8 50.5 S 33 51 E	Procter, R.A. Nicholson
Igwe, f.r.	U3	Busoga	c.0 29 N 33 51 E	
Igwe, popl.	U3	Busoga	0 31 N 33 49 E	G.H. Wood
Ihale, popl.	T1	Mwanza	2 36 S 33 31 E	Tanner
Ihangana, f.r.	T7	Iringa	c.8 18 S 35 43 E	Polhill & Paulo
Ihangiro, sultanate	T1	Bukoba	c.1 50 S 31 20 E	Stuhlmann, Gillman
Ihango, popl.	T7	Mbeya	8 55 S 33 40 E	Cribb & Grey-Wilson
[Ihelele] see Iherehe	T1	Mwanza	3 02 S 32 47 E	Tanner, Procter
Iheme, popl.	T7	Iringa	8 27 S 35 02 E	Numerous
Iheme, popl., agric. stn.	T7	Iringa	8 01 S 35 31 E	Polhill & Paulo
Iherehe, hill	T1	Mwanza	3 02 S 32 47 E	Meyer, Burgess
Ihihizo, str.	U2	Kigezi	1 05 S 29 42 E–0 57 S 29 43 E	
[Ihimba] see Ikimba	T1	Bukoba	1 28 S 31 30 E	Carmichael
Ihimbo, ln.	T7	Iringa	c.8 15 S 35 50 E	Carmichael, Procter
Ihoho, f.r.	T7	Mbeya	8 57 S 33 38 E	Carmichael
Ihongali, popl.	T7	Iringa	c.7 30 S 36 10 E	Tanner
Ihumba, area	T1	Mwanza	3 00 S 32 50 E	Sperling
Ihumwa, rsta.	T5	Dodoma	6 10 S 35 53 E	
Ihunga, hill	U2	Bunyoro	1 47 N 31 23 E	
[Ihunga] see Ihungu	U2	Bunyoro	1 40 N 31 42 E	
Ihungu, popl., mssn., t.c.	U2	Bunyoro	1 40 N 31 42 E	Bally, Greenway, J. Adamson
Ijara, p.p., w.p., r.h.	K1	Northern Frontier	1 36 S 40 31 E	Tanner
Ijijawenda, hill	T1	Mwanza	2 45 S 33 01 E	Goetze
Ijunga, popl.	T7	Mbeya	9 04 S 33 11 E	

Place	Grid	District	Coordinates	Collector
Ijunga = Iyungu, popl.	T7	Chunya	8 36 S 33 02 E	Goetze
Ikanga, mkt.	K4	Kitui	1 42 S 38 04 E	Kuchar, Hildebrandt
Ikanga, ln.	T7	Iringa	c.8 30 S 35 31 E	Carmichael
Ikapa, L.	T7	Rungwe	9 22 S 33 48.5 E	Wingfield
Ikapo = Ikapa, L.	T7	Rungwe	9 22 S 33 48.5 E	
Iki Iki, popl.	U3	Mbale	1 06 N 34 01 E	
Ikikota, popl.	T7	Rungwe	9 29 S 33 33 E	Cribb & Grey-Wilson
Ikimba, L.	T1	Bukoba	1 28 S 31 30 E	Stuhlmann, Procter
Ikizu, popl., mssn., t.c.	T1	Musoma	1 55 S 34 03 E	Tanner, Willan
Ikoga, popl., hill	T7	Mbeya	8 24 S 34 37 E	B.D. Burtt, Zimmermann
Ikola, popl.	T4	Mpanda	6 44 S 30 24 E	Richards
Ikoma, popl.	T1	Musoma	2 04 S 34 37 E	Moore, Tanner, Procter
Ikomba, old popl.	T7	Mbeya	9 07 S 32 16 E	
Ikombe, mssn.	T7	Njombe	9 31 S 34 04 E	
Ikonge, popl.	K5	South Kavirondo	0 32 S 35 01 E	Goetze
Ikowa, hill	T5	Dodoma	6 16 S 36 12 E	Perdue & Kibuwa
Ikowa, popl., mssn.	T5	Dodoma	6 13 S 36 13 E	Bally
Ikoyo, rsta.	K4	Machakos	2 14 S 37 47 E	
[Iku] see Ikuu	T4	Mpanda	6 56 S 31 11 E	
Ikuiwa = Kuywa, R.	K5	North Kavirondo	0 48.5 N 34 36 E–0 28 N 34 38 E	Bullock, Sanane
Ikuka, R.	T7	Iringa	7 28 S 34 58 E–7 38 S 34 50 E	Tweedie
Ikula, R.	T7	Iringa	c.7 45 S 36 40 E	Greenway
Ikulwe, popl.	U3	Busoga	0 27 N 33 29 E	Carmichael
Ikundu, str.	K4	Kitui	c.1 23 S 38 01 E	
Ikungi, popl., mssn., r.h.	T5	Singida	5 08 S 34 46 E	
Ikungu, popl.	T4	Tabora	5 30 S 33 51 E	Polhill & Paulo
Ikuru, I.	T1	Mwanza	c.2 15 S 32 33 E	Holtz
[Ikusa] see Ikuza	T1	Biharamulo	2 08 S 31 46 E	
Ikutha, popl.	K4	Kitui	2 04 S 38 11 E	Bally, Edwards, F. Thomas
Ikutwe = Itikony	T2	Arusha	c.3 15 S 36 50 E	Greenway

79

80

Name	Code	Region	Coordinates	Collector
Il Gerai, w.h.	K1	Northern Frontier	1 36 N 37 20 E	Kuchar
Ilgeri, hill, area	K6	Masai	1 47 S 35 42 E	Carmichael
Iliangelo, R., ridge	T7	Iringa	prob. c.7 38 S 36 33 E	Carmichael
Ilinde = Ilindi, ridge (Image Mt.)	T7	Iringa	c.7 24 S 36 13 E	Busse, Stuhlmann
Ilindi, popl.	T5	Dodoma	c.5 58 S 35 29 E	Carmichael
Ilindi, mt.	T7	Mbeya	8 53 S 33 07 E	Haerdi
Ilingera, popl.	T6	Ulanga	7 57 S 36 31 E	J. Wilson
Ilipath (near Mt. Moroto)	U1	Karamoja	near 2 32 N 34 46 E	Gillett
[Ilkamasya] see Kamasia	K3	Baringo	c.0 23 N 35 48 E	Tweedie
Ilkek, rsta.	K3	Naivasha	0 36 S 36 22 E	
[Ilkinangop] see Kinangop	K3'4	Naivasha/Fort Hall/S. Nyeri	0 37.5 S 36 42.5 E	
[Ilkinopop] see Kinangop	K3'4	Naivasha/Fort Hall/S. Nyeri	0 37.5 S 36 42.5 E	
[Ilkinopop Plateau] see Kinangop	K3	Naivasha	0 30—0 53 S 36 30—36 37 E	
Illaut, w.h.	K1	Northern Frontier	1 53 N 37 16 E	Carter & Stannard
Illembo, popl., plain	T4	Mpanda	6 24 S 31 20 E	Sanane
[Iloma] see Ilomo	T7	Chunya	8 02 S 32 35 E	Goetze
Ilomba, popl., mssn.	T7	Mbeya	8 52 S 33 29 E	Harwood
Ilomo, hills	T4	Mpanda	c.6 50—6 57 S 30 30 E	C.H.B. Grant
Ilomo, mt.	T7	Chunya	8 02 S 32 35 E	Goetze
Ilonga, res. stn., r.h., str., dam	T6	Kilosa	6 45 S 37 02 E	Numerous
Ilongelo	T7	Iringa	near 8 00 S 36 00 E	Carmichael
Ilongero, popl., r.h., t.c., p.p.	T5	Singida	4 40 S 34 52.5 E	Geilinger
Ilongo, area	T7	Mbeya	8 52 S 33 29 E	Cribb & Grey-Wilson
Ilongo, popl.	T7	Mbeya	8 47 S 33 43 E	
Ilsenstein, ln.	T3	Lushoto	c.5 06 S 38 49 E	Peter
Itoroto, hill	K6	Masai	2 06 S 37 25 E	Napper
Ilunda, r.p.	T7	Iringa	6 53 S 34 54 E	Pegler, Renvoize, Greenway
Ilunga, popl.	T4	Tabora	5 51 S 33 22 E	Ukiriguru Res. Stn.
Ilunga, mts.	T7	Chunya	c.8 20 S 33 00 E	Goetze
Iwilo (Mbarika)	T1	Mwanza	c.2 55 S 32 51 E	Tanner

Name	Region	District	Coordinates	Authorities
Image, mt. f.r.	T7	Iringa	7 30 S 36 10 E	Polhill & Paulo, Procter, Carmichael
Imagi, hill	T5	Dodoma	6 13 S 35 46 E	Polhill & Paulo
Imagit Peak, mt.	U1	Karamoja	2 34 N 34 43 E	Myers, J. Wilson
Imatong, mts.	U1/Sudan	Acholi	c.4 00 N 32 50 E	Eggeling
(formerly all in Uganda; transferred to Sudan in 1926 leaving only most southerly part in Uganda)				
Imega, popl.	T5	Singida	c.5 17 S 34 47 E	Polhill & Paulo
Imenti, Lower, for.	K4	Meru	0 07 N 37 43 E	Balbo, Verdcourt
Imenti, Upper, for.	K4	Meru	0 03 N 37 32 E	Balbo, Verdcourt
Immini, f.r.	T4	Buha	c.4 35 S 30 03 E	Procter
Impenetrable Forest, f.r.	U2	Kigezi	c.1 05 S 29 35 E	Butt, Eggeling, Purseglove
Inde, popl.	U1	West Nile	2 53 N 31 14 E	
Inepanga	K7	Kwale	3 56 S 39 31 E	Kassner
[Ingomani] see Ngomeni				
Inkipikoni, area, for.	T3	Tanga	5 09 S 38 54 E	Greenway
Inshala, str.	K6	Masai	c.1 21 S 36 38 E	
Intona, area	T2	Mbulu	c.3 50 S 35 05 E	Richards
[Intoto wa Andei] see Mtito Andei	K6	Masai	1 09 S 34 50 E	Kuchar, Msafiri
Inyala, hill, popl.	K4	Machakos	2 41 S 38 10 E	Gregory
Inyonga, popl., hill	T7	Mbeya	8 52 S 33 38 E	Hooper & Townsend
Ipafu, hill	T4	Mpanda	6 43 S 32 04 E	Richards, Ruffo
Ipala, popl.	T7	Iringa	8 33 S 35 31 E	Perdue & Kibuwa, Paget-Wilkes
[Ipana] see Ipyana	T5	Dodoma	6 01 S 35 57 E	Busse, Stuhlmann
Ipata, popl.	T7	Rungwe	?9 36 S 33 52 E	Greenway
Ipeta, sw.	T7	Mbeya	9 02 S 32 58 E	R.M. Davies
[Ipiana] see Ipyana	T4	Ufipa	8 08 S 31 17 E	Richards
Ipin, hill ? see Ipiri	T7	Rungwe	9 36 S 33 52 E	
Ipinda (N. Usafwa f.r.)	T2	Mbulu	?4 19 S 35 46 E	Carmichael
	T7	Mbeya	near 8 50 S 33 30 E	Procter, Eggeling, Moreau, Vesey-FitzGerald
Ipinda, mssn.	T7	Rungwe	9 29 S 33 54 E	Richards
[Ipindi] see Ipinda	T7	Rungwe	9 29 S 33 54 E	Richards

Name	Region	District	Coordinates	Collector
Ipiri, hill	T2	Mbulu	4 19 S 35 46 E	Richards
Ipogoro = Mapogoro, popl.	T7	Mbeya	8 19 S 34 42 E	
[Ipugulu] see Ifuguru	T7	Iringa	7 36 S 35 03 E	
Ipumba, kopje (or Ifumba)	T1	Maswa	c.3 15 S 34 50 E	Greenway
Ipumbuli, popl.	T4	Nzega	4 19 S 33 04 E	Greenway
Ipume = Ipumi, str.	T7	Njombe	c.9 08 S 34 00 E	
Ipumi = Ipume, str.	T7	Njombe	c.9 08 S 34 00 E	Richards
Ipunguli, popl.	T5	Dodoma	5 51 S 34 23 E	B.D. Burtt
[Ipusulu] see Ipusuro	T5	Singida	5 20 S 34 33 E	
Ipusuro, popl.	T5	Singida	5 20 S 34 33 E	
Ipyana, mssn.	T7	Rungwe	9 36 S 33 52 E	R.M. Davies, Stolz
Irago, hill (near Ufiome Mt.)	T2	Mbulu	near 4 14 S 35 48 E	B.D. Burtt
Iragua, t.c.	T6	Ulanga	8 33 S 36 31 E	Rees
[Iragwa] see Iragua	T6	Ulanga	8 33 S 36 31 E	
Iraku, t.a.	T2	Mbulu	c.3 55 S 35 33 E	Merker
Iramba, mt.	T1	Mwanza	2 00 S 33 25 E	Carmichael
Iramba, popl.	T1	Mwanza	1 59 S 33 23 E	Numerous
Iramba, popl.	T1	Musoma	1 42 S 34 17 E	
Iramba, plateau	T5	Singida	4 25 S 34 25 E	Cribb & Grey-Wilson, Lewis
Irambo, mssn.	T7	Mbeya	8 57.5 S 33 40 E	Schelpe, Zogg
Irangi, for. stn.	K4	Embu	0 21 S 37 29 E	Fischer, Stuhlmann
Irangi, t.a.	T5	Kondoa	c.4 45 S 36 00 E	Procter
Irangi, popl.	T7	Iringa	8 49 S 35 39 E	Fries
Iraru, str.	K4	Meru	c.0 09 S 37 45 E	Geilinger
[Irente] see Kwamweleti	T3	Lushoto	4 47.5 S 38 16 E	
Iriani, ln.	K4	South Nyeri	0 35 S 37 07 E	Kibui
Irima, hill	K7	Teita	3 17 S 38 32 E	Bally, Greenway
Irima, rsta.	K7	Teita	3 21 S 38 32 E	Carter & Stannard
Iringa, province	T7	various		
Iringa, distr.	T7	Iringa	c.8 00 S 35 35 E	Greenway, Schlieben, McLoughlin

Iringa, popl., airstrip	T7	Iringa	7 46 S 35 42 E	Numerous
Iringa, escarp., f.r.	T7	Iringa	c.7 45 S 35 42 E	Numerous
Iriri, popl., w.h.	U1	Karamoja	2 05 N 34 12 E	Tweedie, J. Wilson, W.H. Lewis
[Irma] see Jirma	K1	Northern Frontier	3 58 N 41 46 E	Ruspoli & Riva
[Iroma] see Ilomo	T7	Chunya	8 02 S 32 35 E	J. Bally
Irunda, for.	T7	Iringa	8 32 S 35 16 E	Procter
Irunde, ln.	T4	Mpanda	5 57 S 31 22 E	Boehm
Irundi, hill	T7	Iringa	8 27 S 35 15 E	Childs, Procter, Benedicto
Irunga, hill	U2	Ankole	near 1 00 S 30 30 E	Bagshawe
Iruru, f.r.	K3	Nandi	0 11 N 35 05 E	
Irwin Farm (Mpui-Zambia rd.)	T4	Ufipa	Not traced	McCallum Webster
Isalala, f.r.	T7	Mbeya	8 54 S 32 38 E	
Isalala, popl.	T7	Mbeya	9 02 S 32 45 E	
[Isalolo] see Isalala	T7	Mbeya	9 02 S 32 45 E	Knight
Isanga, popl.	T7	Iringa	c.7 45 S 36 38 E	Carmichael
Isangati, mssn.	T7	Mbeya	9 04 S 33 26 E	
Isangati, mts.	T7	Mbeya/Rungwe	c.9 04 S 33 26 E	Cribb et al.
Isansu, popl.	T5	Singida	4 04 S 34 45 E	Kohl-Larsen
Isata, ln.	T7	Iringa	c.8 07 S 35 55 E	Carmichael
Iseke, str./s.	U2	Bunyoro	c.1 28—1 35 N 31 54 E	B. J. Turner
Isessa	T1	Mwanza	Not traced	Tanner
Ishasha, gorge	U2	Kigezi	c.0 54 S 29 42 E	Numerous
Ishasha, R.	U2/Zaire	Kigezi	1 07 S 29 54 E—0 27 S 29 39 E	
Ishenta, popl.	T7	Rungwe	9 21.5 S 33 13.5 E	Leedal
Ishiara, popl.	K4	Embu	0 27 S 37 47 E	S.A. Robertson, Beentje, Scholte, G.R. van Someren
Ishinga = Salala, mt.	T7	Mbeya	8 55 S 33 49 E	Cribb & Grey-Wilson
[Ishingiro] see Isingiro	U2	Ankole	c.0 50 S 30 55 E	Purseglove
Ishozi, mt.	T1	Bukoba	1 35 S 30 52 E	Gillman
Ishozi, popl.	T1	Bukoba	1 08.5 S 31 46 E	Gillman

Isiki, str.	T7	Mbeya/Chunya/Iringa	7 28 S 34 11.5 E–7 46.5 S 34 10 E	Bjornstad
[Isimbila] see Isimbira				
Isimbira, popl.	T4	Tabora	6 02 S 32 09 E	Richards
Isimila, prehistoric site	T4	Tabora	6 02 S 32 09 E	Richards
Isingiro, co.	T7	Iringa	c.7 57 S 35 35 E	Bidgood
Isiolo, distr.	U2	Ankole	c.0 50 S 30 55 E	Harker
Isiolo, popl., airstrip	K1/4	Northern Frontier/Meru	c.0 55 N 38 50 E	Numerous
Isoko, mssn.	K4	Meru	0 21 N 37 35 E	Numerous
Isongo = Issongo, mssn., popl.	T7	Rungwe	9 29 S 33 30 E	Stolz, Cribb & Grey-Wilson
Isongole = Idweli	T6	Ulanga	8 45 S 36 43 E	Cribb & Grey-Wilson
Isonso, popl.	T7	Rungwe	9 02 S 33 36 E	Cribb & Grey-Wilson
Isowi, area	T7	Mbeya	9 15 S 33 23 E	Leedal
Issambil	T7	Njombe	c.9 05 S 34 50 E	Schlieben
Issanga, popl.	T6	Ulanga	prob. c.7 45 S 36 50 E	Carmichael
[Issansu] see Isansu	T5	Dodoma	c.6 06 S 35 40 E	Polhill & Paulo
Issaras, popl.	T5	Singida	4 04 S 34 45 E	Kohl-Larsen
Issaua, str.	T2	Mbulu	3 56 S 35 35 E	Haarer
Issongo, mssn.	T5	Dodoma	6 26 S 33 57 E–6 38 S 34 08 E	Yohannes
Issuna, popl., mssn.	T6	Ulanga	8 45 S 36 43 E	Schlieben
Issuri, popl.	T5	Singida	5 23 S 34 46 E	Greenway & Paulo
[Isujana] see Ipyana	T4	Tabora	5 18 S 33 44 E	Ukiriguru Res. Stn.
Isukha, ln.	T7	Rungwe	9 36 S 33 52 E	Stolz
Isumacheri = Dunacheri, I.	K5	North Kavirondo	0 15 N 34 45 E	Paulo
Isunga, area	T1	Mwanza	c.2 24 S 31 57 E	
Isunkaviola, mt., plateau	U2	Toro	0 30 N 30 25 E	Dawe
Isunura, popl.	T7	Mbeya	c.7 48 S 34 00 E	Bjornstad
Isuria = Soit Ololol, escarp.	T7	Mbeya	8 37 S 34 25 E	B.D. Burtt, Thulin & Mhoro
Isyeseye, popl.	K6	Masai	c.1 00 S 35 10 E–1 25 S 34 47 E	Glover et al.
Isyonje = Izyonje, popl.	T7	Mbeya	c.8 53 S 33 30 E	Harwood, Hooper & Townsend
Itabua, str./s.	T7	Mbeya	8 59 S 33 37.5 E	Hooper & Townsend
	K4	Embu	c.0 29 S 37 30 E–0 44 S 37 33 E	M.D. Graham

Name		District	Coordinates	Authority
Itaka, popl.	T7	Mbeya	8 52 S 32 48 E	Greenway
Itala, hills	T4	Ufipa	c.7 46 S 31 30 E	Richards
Italavanda	T7	Iringa	prob. c.7 40 S 36 33 E	Carmichael
Itamba, L.	T7	Rungwe	9 21 S 33 51 E	R.M. Davies
Itanga	T7	Iringa	8 07.5 S 35 54 E	Carmichael
Itara, mt.	T1	Bukoba	1 08.5 S 31 27 E	Procter
Itara, popl.	T1	Bukoba	1 09 S 31 28 E	Holtz, Mildbraed, Procter
Itare, popl.	K5	Kericho	0 35 S 35 15 E	Bally, Copley, Honoré
Itare, R.	K5	Kericho	c.0 25 S 35 30 E–0 35 S 35 15 E	Copley, Maas Geesteranus
Itarige ? = Uthiri	?T1	?Mara	?1 20 S 33 50 E	Fischer
[Itatie] see Ititie	T4	Kigoma	c.5 27 S 29 48 E	Azuma
[Ite] see Wati	U1	West Nile	3 13 N 31 02 E	Eggeling
Itemba (Mahali Mts.)	T4	Mpanda	Not traced	Jefford & Newbould
Iten, ln.	K3	Elgeyo	0 40 N 35 30 E	Tweedie
Itend, R.	T7	Mbeya	Not traced	Mgaza
Itende, popl.	T7	Mbeya	8 54 S 33 25 E	Mgaza
Itende, L.	T7	Rungwe	9 19 S 33 47 E	Goetze
Itete, mssn., r.h.	T6	Ulanga	8 40 S 36 25 E	Rees
[Itete] see Idete	T7	Iringa	7 29.5 S 36 12.5 E–7 27 S 36 14.5 E	Carmichael
Ithaba, sw.	K4	Machakos	c.2 32 S 37 50 E	Bally
Ithanga, hills	K4	Fort Hall	c.0 54–1 02 S 37 19–37 25 E	C. van Someren
[Ithange] see Ithanga	K4	Fort Hall	c.0 54–1 02 S 37 19–37 25 E	Birch
[Ithapa] see Ithaba	K4	Machakos	c.2 32 S 37 50 E	Bally
Ithumbi, hill	K4	Kitui	0 53 S 38 06 E	Gatheri et al.
[Itiani] see Itieni	K4	Meru	0 14 N 37 53 E	Mabberley
Itibo, popl.	K5	Kericho	0 33 S 35 00 E	
Itibol, popl.	Sudan *not* U1		4 01 N 32 52 E	A.S. Thomas
Itiene, peak	K4	Meru	0 14 N 37 53 E	
Itigi, popl., rsta.	T5	Dodoma	5 42 S 34 29 E	Numerous
Itigi Thicket	T5	Dodoma	c.5 20–6 20 S 34–35 E	Numerous

Itikony, ln.	T2	Arusha	c.3 15 S 36 50 E	Greenway
Itimbua, popl.	T4	Tabora	5 47 S 32 11 E	Boehm
[Itimbury] see Itimbua	T4	Tabora	5 47 S 32 11 E	Boehm
Itiso, popl., area	T5	Dodoma	5 38 S 36 01 E	Bally
Ititie, area (Kabogo Mts.)	T4	Kigoma	c.5 27 S 29 48 E	Kyoto Univ. Exped.
[Ititoe] see Ititie	T4	Kigoma	c.5 27 S 29 48 E	Azuma
[Itilye] see Ititie	T4	Kigoma	c.5 27 S 29 48 E	Kyoto Univ. Expd.
Itizi, for.	T7	Mbeya	9 07 S 33 27 E	Cribb et al.
[Ito] see Itor	U1	West Nile	3 21 N 31 00 E	Eggeling
Ito, I.	T1	Mwanza	2 15 S 32 32 E	Carmichael
Itogoro – soil type – not a locality				
Itolio, popl.	T1	Bukoba	c.1 37 S 31 44 E	Stuhlmann
Itonjo, hills	T1	Maswa	2 40 S 34 45 E	Greenway
Itor, mt.	U1	West Nile	3 21 N 31 00 E	Eggeling
Itula, popl.	T6	Ulanga	8 00 S 36 33 E	Haerdi
Itulo, fm.	T7	Njombe	9 03 S 33 53 E	Gillett
Itulu, hill	T4	Tabora	6 18 S 33 45 E	Groome
[Itumba] see Ukaguru	T6	Kilosa	6 20 S 37 10 E	Last
[Itumbwa] see Tumba	T4	Mpanda	7 31 S 31 39.5 E	Richards
Itundufula, popl.	T7	Iringa	7 57 S 36 45 E	Haerdi
Itundwi, popl., scarp	T5	Kondoa	4 39 S 35 51 E	B.D. Burtt
[Itura] see Tura	T4	Tabora	5 31 S 33 50.5 E	Hannington
Iturce, I. ? see Ikuru	T1	Mwanza	?2 15 S 32 33 E	Carmichael
Ituri, Forest	U2/Zaire	Toro	c.0 50 N 29 59 E	

(N.B. Most collections are from Zaire)

Itwara, f.r.	U2	Toro	c.0 48 N 30 28 E	Eggeling, Greenway & Eggeling, St. Clair-Thompson
Iveti, hills	K4	Machakos	1 31 S 31 18 E	Bogdan, Clayton
Ivigi (Bundali Hills)	T7	Rungwe	c.9 30 S 33 27 E	Fuller
Ivuna, L.	T7	Mbeya	8 26 S 32 29 E	Bullock, Michelmore

87

Name	Region	District	Coordinates	Collector
Ivuna, popl.	T7	Mbeya	8 24 S 32 32 E	Procter
Ivuna, f.r.	T7	Mbeya	8 29 S 32 26 E	B.D. Burtt
Iwanga	T5	Kondoa	Not traced	Polhill & Paulo
Iwerewere, hill	T5	Singida	5 35 S 34 40 E	Power
Iwezo (? in Boni For.)	K7	Lamu	Not traced	
[Iwondo] see Ibondo	T1	Mwanza	2 38 S 32 40.5 E	Goetze
Iwugu, hills	T7	Rungwe	9 28 S 33 37 E	B.D. Burtt
[Iwumbu] see Wamba	T5	Singida	c.5 20 S 34 37 E–5 04 S 34 01 E	
Iwungu, popl.	T7	Chunya	8 26 S 32 54 E	Polhill & Paulo, Procter
Iyayi, popl.	T7	Njombe	8 53 S 34 33 E	Richards
[Iyonga] see Inyonga	T4	Mpanda	6 43 S 32 04 E	Goetze (Nos. 1115–1120)
Iyunga = Iyungu, popl.	T7	Chunya	8 36 S 33 02 E	Goetze (Nos. 1369–1372)
Iyunga = Ijunga, popl.	T7	Mbeya	9 04 S 33 11 E	Carmichael
Iyungi, hill	T7	Iringa	near 8 00 S 36 00 E	Goetze
Iyungu = Iyunga, popl.	T7	Chunya	8 36 S 33 02 E	Polhill & Paulo
Izazi, popl.	T7	Iringa	7 12 S 35 44 E	Leedal
Izumbwe, popl., Natural Bridge	T7	Mbeya	8 58 S 33 22.5 E	
Izyonje, popl.	T7	Mbeya	8 59 S 33 37.5 E	
[Jaba] see Gaba	U4	Mengo		Dummer
Jacaranda, est., coffee res. stn.	K4	Kiambu	c.1 05.5 S 36 53 E	
Jackson's Summit, peak	U3	Mbale	1 08.5 N 34 31 E	Tothill, Eggeling
Jadini, beach	K7	Kwale	4 19 S 39 34 E	Numerous
Jaegertal, popl., old pl.	T3	Lushoto	4 47 S 38 18 E	Drummond & Hemsley, Peter, Archbold
[Jakumia] see Jekuhama	T2	Arusha	c.3 15 S 36 49 E	Richards
Jamara, str., gorge	T2	Arusha	3 16 S 36 52 E–3 19 S 36 53 E	Richards
Jambangome, popl.	P	Pemba	5 21 S 39 43 E	Vaughan, Greenway
[Jambe, Little] see Yambe, Little	T3	Tanga	5 05 S 39 10 E	
[Jambiana] see Jambiani, area	Z	Zanzibar	6 20 S 39 30 E	Stuhlmann
[James Corner] see Nyololo	T7	Iringa	8 31 S 35 03 E	Gillett

Name	Code	Region	Coordinates	Collector
[Jamora] see Jamara				
Jamvini, popl.	T2	Arusha	3 16 S 36 52E–3 19 S 36 53 E	Richards
Japata, est.	P	Pemba	5 12.5 S 39 42 E	Vaughan
Jaracuma, R. (near Meru)	K3	Trans-Nzoia	1 13 N 34 46 E	Hedberg
[Jardini] see Jadini	K4	Meru	near 0 03 N 37 39 E	Fries
Jaribuni, popl.	K7	Kwale	4 19 S 39 34 E	Drummond & Hemsley, Greenway
Jasini, f.r., popl.	K7	Kilifi	3 38 S 39 44.5 E	Musyoki & Hansen
Jasini (? Kinyasini)	T3	Pangani	5 23 S 38 56 E	Faulkner, Milne-Redhead & Taylor, Tanner
[Jassini] see Jasini	Z	Zanzibar	?5 58 S 39 18 E	T.C. Vaughan
[Jebago] see Jebogo	T3	Tanga	4 40 S 39 11 E	Kassner
Jebel Kei = Kei Mt.	T2	Mbulu	4 38.5 S 35 32.5 E	B.D. Burtt
Jebel Midigo = Midigo Mt.	U1	West Nile	3 36 N 31 06 E	
Jebogo = Geboga, hill	U1	West Nile	3 37 N 31 14 E	
[Jekuhama] see Jekukumia	T2	Mbulu	near 4 35 S 35 30 E	Eggeling, A.S. Thomas
Jekukumia, str., camp	T2	Arusha	3 14 S 36 45.5 S–3 15 S 36 49 E	B.D. Burtt
Jensen's Farm	T2	Arusha	3 14 S 36 45.5 E–3 15 S 36 49 E	Greenway
Jeroko, w.h.	K3	Trans-Nzoia	1 12.5 N 34 48.5 E	Pegler, Renvoize
[Jezani] see Jozani	K1	Northern Frontier	3 24 N 41 18 E	Tweedie
Jibana, area, Kaya	Z	Zanzibar	6 16 S 39 25.5 E	Ellenbeck
Jibondo, I.	K7	Kilifi	3 50 S 39 41 E	Vaughan
Jie, co.	T6	Rufiji	8 03 S 39 43.5 E	Gilbert & May
Jilore, for. stn.	U1	Karamoja	c.3 00 N 34 00 E	Greenway
Jilore Mwisho, ln.	K7	Kilifi	3 12 S 39 55 E	J. Wilson
Jilori = Jilore, Lake, sw.	K7	Kilifi	3 11 S 39 52 E	Spjut
[Jimba] see Shimba Hills	K7	Kilifi	3 12 S 39 54 E	Polhill & Paulo, Tweedie, Verdcourt
Jimba, mssn.	K7	Kwale	c.4 07–4 20 S 39 25 E	Kassner
Jindani Beach	K7	Kilifi	3 54 S 39 33 E	Kassner
Jinja, popl., mssn., pier	Z	Zanzibar	6 19 S 39 17 E	R.M. Davies
Jinjo, area	U3	Busoga	0 26 N 33 13 E	Numerous
Jipe, L.	T5	Kondoa	5 02.5 S 36 10 E	B.D. Burtt
	K7/T3	Teita/Pare	3 35 S 37 45 E	Numerous

90

Judyland, fm.	T7	Mbeya	c.9 02 S 32 56 E	Richards
Juja, est.	K4	Kiambu/Nairobi	1 11 S 37 07 E	Mearns
Jujuliet, hill	K5	Kericho	0 44 S 35 06 E	Dale
Juma, I.	T1	Mwanza	2 26 S 32 45 E	
[Jumam] see Jonam	U1	West Nile	c.2 28 N 31 25 E	
Jumba, str./s.	U2	Bunyoro	c.1 42 N 31 53 E–1 26.5 N 31 55 E	
Jumba, hill	U4	Mengo	0 32 N 32 52 E	Dummer
Jumbe Salim's, popl.	T8	Songea	10 35 S 36 30 E	Hay, Muir
[Juna] see Juma	T1	Mwanza	2 26 S 32 45 E	Procter
Jungo, popl.	U4	Mengo	0 10 N 32 26 E	Lye
Jungu, hill	T2	Mbulu	3 36 S 35 24 E	Williams, Bally
Juniper, hill	T2	Arusha	c.3 16.5 S 36 52 E	Richards
Kaabong, popl., r.c.	U1	Karamoja	3 31 N 34 08 E	Dale, A.S. Thomas, Eggeling
Kaagya, popl.	T1	Bukoba	1 09.5 S 31 51 E	Gillman, Haarer
Kaapus, str./s.	K2	Turkana	c.3 11 N 35 20 E	
Kaazi, popl.	U4	Mengo	0 12.5 N 32 37 E	Lye
Kabage, ln.	K4	South Nyeri	0 24 S 36 50 E	Dale
Kabagole, rsta.	U2	Ankole	0 12 N 30 54 E	
Kabaka's Lake, L.	U4	Mengo	0 18 N 32 33.5 E	
[Kabaku] see Kabuku	T3	Handeni	Not traced	Faulkner
Kabale, popl.	U2	Kigezi	1 15 S 29 59 E	Numerous
Kabale, popl., area	T1	Bukoba	1 55 S 31 05 E	Gillman
Kabale, popl.	T1	Bukoba	1 20.5 S 31 46 E	Reakes Williams
Kabale Forest	T7	Rungwe	9 12 S 33 43 E	Cribb & Grey-Wilson
Kabalega, Falls	U1/2	Acholi/Bunyoro	2 17 N 31 41 E	Katende
Kabalega, Nat. Park	U1/2	Acholi/Bunyoro	c.2 15 N 31 50 E	
Kabaleka, L.	U2	Toro	0 15.5 N 30 15 E	Osmaston
Kabamba, r.c.	U2	Toro	0 21 N 29 55 E	Eggeling
Kabanda, popl.	U2	Bunyoro	1 32 N 30 59 E	Purseglove

Kabanga, mssn., hills, str.	T4	Buha	c. 4 30 S 30 08 E	Procter
Kabango, r.h.	U2	Toro	0 47 N 30 08 E	Hazel, A.S. Thomas, Eggeling
Kabango-Muntandi, f.r.	U2	Toro	0 47 N 30 08 E	Eggeling, Mukasa
[Kabangu] see Kabungu	T4	Mpanda	6 18 S 30 59 E	Boaler
[Kabangya] see Kibangya	U2	Bunyoro	1 33 N 32 01 E	Purseglove
Kabarak, for.	K3	Baringo	0 23 N 35 49 E	
Kabarak Range, hills	K3	Baringo	c. 0 23 N 35 49 E	Fischer, Tweedie
Kabaras = Kabras, t.a.	K5	North Kavirondo	c. 0 25 N 34 50 E	Ukiriguru Res. Centre.
Kabare, ln.	T1	Bukoba	c. 1 55 S 31 40 E	Numerous
Kabarnet, popl.	K3	Baringo	0 29.5 N 35 44.5 E	Snowden
[Kabaroni] see Kaburoni	U3	Mbale	c. 1 25 N 34 37 E	Bonnefille & Riollett
Kabarsero, ln.	K3	Baringo	0 53 N 35 51 E	Stolz
Kabasa, prob. = Kipassa, popl.	T7	Rungwe	prob. 9 15.5 S 33 41.5 E	
Kabasanda, popl.	U4	Mengo	0 16 N 32 13 E	Eggeling
Kabatanagi, hill, popl. (near Rwenyaga)	U2	Ankole	near 0 50.5 S 30 39 E	
Kabatoro, popl.	U2	Toro	0 08 S 29 55 E	Toyoshima
Kabeko (Kabogo Mts.)	T4	Kigoma	Not traced	
Kaberamaido, co.	U3	Teso	c. 1 48 N 33 15 E	Shabani
[Kabere] see Kaberi	T4	Buha	3 40 S 31 05 E	Bullock, Shabani
Kaberi, sw.	T4	Buha	3 50 S 31 05 E	Newbould & Jefford, Procter
Kabesi, str.	T4	Mpanda	6 12 S 29 53 E–6 00 S 29 47 E	Numerous
Kabete, popl., t.c.	K4	Kiambu	1 15.5 S 36 44 E	
Kabete, Lower, popl., t.c.	K4	Kiambu	1 14.5 S 36 43 E	
Kabewyan, bdg.	K3	Trans-Nzoia	0 59 N 34 50 E	
Kabiok, f.r.	K3	Baringo	0 40 N 35 50 E	Snowden
Kabira, popl., r.h.	U2	Ankole	0 41 S 30 02 E	Dyson Hudson
Kabiramorok, str.	U1	Karamoja	2 28 N 34 42 E–2 25 N 34 30 E	Haarer, Watkins
Kabirizi, popl.	T1	Bukoba	1 34 S 31 29 E	Richards
[Kabisama] see Kambisama	T4	Mpanda	c. 6 26 S 31 22 E	Watkins, Sangiwa
[Kabobwa] see Kabwoba	T1	Bukoba	1 05 S 31 25 E	

Name	Code	Region	Coordinates	Collector
Kaboen, str.	K3	Nandi	c.0 12 N 35 09 E	Brunt, Hooper & Townsend
Kabogo (near Shanga)	T1	Ngara	near 2 32.5 S 30 34 E	Tanner
Kabogo, mts.	T4	Kigoma	c.5 27 S 29 48 E	Numerous
Kabogo Head, pt.	T4	Kigoma	5 28 S 29 45 E	Kyoto Univ.
[Kaboko] see Koboko	U1	West Nile		Hazel
[Kabolet] see Kapolet	K3	Trans-Nzoia	c.1 09 N 35 10 E	Tweedie
[Kabolet] see Kapolet	K3	Trans-Nzoia/Elgeyo	1 03 N 35 09 E – 1 09 N 35 11 E	Tweedie
Kabondo, popl.	K5	South Kavirondo	0 27 S 34 53 E	Bogdan
Kabonge, for.	K4	Kitui	1 20 S 38 03 E	
Kaboroa, str.	K3	Trans-Nzoia	1 09 N 34 42.5 E – 1 11 N 34 44 E	
Kaboya, popl.	T1	Bukoba	1 35.5 S 31 45 E	Gillman
Kaboyo, popl., t.c.	U4	Masaka	0 21 S 31 34 E	
Kabras = Kabaras, t.a.	K5	North Kavirondo	c.0 25 N 34 50 E	Tweedie
Kabuga, hill	U2	Toro	0 13.5 N 30 29 E	Bagshawe
Kabuku (Korogwe area), for.	T3	Handeni	Not traced	Faulkner
Kabula, popl., r.h.	U4	Masaka	0 21 S 31 09 E	Eggeling, Purseglove
Kabula, co.	U4	Masaka	c.0 15 S 31 12 E	Michelmore, Purseglove
Kabulasoke, popl.	U4	Mengo	0 09 N 31 48 E	
Kabungu, hill	T4	Mpanda	6 23 S 31 03.5 E	Eggeling, Semsei, F.G. Smith
Kabungu, popl.	T4	Mpanda	6 18 S 30 59 E	Groome, Carmichael, Procter
Kabungu, t.a.	T4	Mpanda	c.6 13 S 31 16 E	Boaler, Friend, Semsei
[Kabuoch] see Kabwoch	K5	South Kavirondo	0 45 S 34 28 E	Jarrett
Kabura	T?4	?Tabora	Not traced	Swynnerton
[Kaburon] see Kaburoron	U3	Mbale	1 25 N 34 37 E	Eggeling, J. Wilson, Tweedie
Kaburoni = Kaburon area	U3	Mbale	c.1 25 N 34 37 E	Tweedie
Kaburoron, popl., p.p.	U3	Mbale	1 25 N 34 37 E	G.H. Wood, Eggeling, Tweedie
[Kabururoni] see Kaburoron	U3	Mbale	1 25 N 34 37 E	G.H. Wood
Kabwe, R.	T4	Mpanda	c.6 05 S 29 45 E	Harley
Kabwoba, popl.	T1	Bukoba	1 05 S 31 25 E	Carmichael, Procter
Kabwoch, area, for.	K5	South Kavirondo	0 45 S 34 28 E	

94

Name	Region	District	Coordinates	Reference
Kafu, R.	U2/4	Bunyoro/Mengo	1 12 N 31 17 E–1 39 N 32 05 E	Langdale-Brown
Kafu, [fy.] bdg.	U2/4	Bunyoro/Mengo	1 33 N 32 02.5 E	Eggeling, A.S. Thomas
[Kafu] see Kafufu	T4	Ufipa/Mpanda	c.7 00 S 31 17 E–7 40 S 31 46 E	Richards
Kafufu, R.	T4	Ufipa/Mpanda	c.7 00 S 31 17 E–7 40 S 31 46 E	

(Sometimes the R. Katuma which flows into the N. end of L. Chada is given this name. The lower reaches of the Kafufu R. are also known as the Kavu or Kavuu R.)

Name	Region	District	Coordinates	Reference
Kafukola, popl.	T4	Ufipa	8 05 S 31 57 E	Richards, Siame, McCallum Webster
Kafulu, hill	T4	Kigoma	5 32 S 30 32 E	Hoyle, Eggeling
Kafunzo, popl.	U2	Ankole	1 01 S 30 25 E	
[Kafuro Lake] see Kazinga Channel	U2	Toro/Ankole	0 04–0 12 S 29 53–30 09 E	Dawe
Kafuro, spr.	T1	Bukoba	1 43 S 31 00 E	Stuhlmann
[Kafuru Lake] see Kazinga Channel	U2	Toro/Ankole	0 04–0 12 S 29 53–30 09 E	
Kagaa, hill	U3	Teso	1 41 N 33 01 E	Eggeling
Kagali	T7	Rungwe	Not traced	Stolz
[Kagama] see Kagoma	U3	Busoga	0 38 N 33 11.5 E	G.H. Wood
Kagara prob. = Kagwara	U3	Teso	prob. 1 30 N 33 05.5 E	A.S. Thomas
[Kagata, est.] see Kayata	K4	Machakos	1 09 S 37 21 E	Gardner
[Kagatende] see Kogatende	T1	Musoma	1 35 S 34 56 E	Greenway
Kagehe, mt.	T1	Bukoba	1 06 S 30 29 E	Stuhlmann
Kagehi, popl.	T1	Mwanza	2 44 S 31 54 E	Stuhlmann
[Kagehi] see Kayenzi	T1	Mwanza	2 23 S 33 06 E	Fischer
[Kagei] see Kayenzi	T1	Mwanza	2 23 S 33 06 E	Hannington
[Kageli] see Kayenzi	T1	Mwanza	2 23 S 33 06 E	Fischer
Kagera, R.	U2, 4/T1/Rw./Bur.	several	2 21 S 30 22 E–0 57 S 31 47 E	Bagshawe, Eggeling, Mildbraed, Snowden
Kagera Port = Nyakanyasi, popl.	T1	Bukoba	1 11 S 31 13 E	Haarer
[Kageyi] see Kayenzi	T1	Mwanza	2 23 S 33 06 E	
[Kaghara] see Kajara	U2	Ankole	c.1 00 S 30 10 E	
Kagochi, fm.	K4	South Nyeri	0 23 S 37 08.5 E	Kerfoot
[Kagoli] see Kayenzi	T1	Mwanza	2 23 S 33 06 E	Fischer
Kagoma, popl.	U3	Busoga	0 38 N 33 11.5 E	Osmaston, G.H. Wood

[Kagome] see Kakoma			
Kagulu, popl.	T4	Tabora	5 47 S 32 26 E	Boehm
Kagunguli, mssn.	U3	Busoga	1 15 N 33 18 E	G.H. Wood
Kaguru = Ukaguru, mts.	T1	Mwanza	2 00 S 33 04 E	
Kagurukeru (Bukwaya)	T6	Kilosa	c.6 25 S 36 50 E	Swynnerton
	T1	Musoma	c.1 34 S 33 46 E	Tanner
Kagwalas = Kauwalathe, str./s.	K2	Turkana	c.3 35 N 35 04 E–3 07 N 35 38 E	Parry
[Kagwalathe] see Kagwalas	K2	Turkana	c.3 35 N 35 04 E–3 07 N 35 38 E	
Kahama, distr.	T4	Kahama	c.3 40 S 32 00 E	
Kahama, popl.	T4	Kahama	3 50 S 32 36 E	Bullock, B.D. Burtt, Glover
Kahama, str.	T4	Kigoma	c.4 41.5 S 29 37 E	Clutton-Brock
Kaharati, popl.	K4	Fort Hall	0 50.5 S 37 08 E	Bally & Smith
Kahe, rsta.	T2	Moshi	3 30 S 37 26.5 E	Volkens, Winkler, Cribb & Grey-Wilson
Kahendero = Kahenderu, popl.	U2	Toro	0 04 N 30 03 E	Eggeling
Kahoko, ridge	T4	Mpanda	6 10.5 S 29 46 E	Newbould & Harley
Kahorongole, ln.	K2	Turkana	2 55 N 35 23 E	Newbould
Kahororo, popl.	T1	Bukoba	1 15 S 31 50 E	Haarer
Kahunge, popl., mssn.	U2	Toro	0 20 N 30 27 E	
Kahuroini, area	K4	South Nyeri	0 28 S 37 03 E	Kibui
[Kaiango] see Kayango				
Kaibibich = Kaibwibich, peak, r.c.	U3	Busoga	0 35 N 34 51 E	G.H.S. Wood
[Kaibichbich] see Kaibibich	K3	Elgeyo	1 12 N 35 17 E	Mabberley & McCall, Tweedie
Kaibos Farm	K3	Elgeyo	1 12 N 35 17 E	Tweedie
Kaibwibich, peak	K3	Trans-Nzoia	1 09.5 N 35 08.5 E	Mabberley & McCall, Tweedie
Kaiemothia, p.p.	K3	Elgeyo	1 12 N 35 17 E	Thulin & Tidigs
Kaigi (Rubare, f.r.)	K2	Turkana	4 54 N 35 21 E	Carter & Stannard
Kaikamosi, popl.	T1	Bukoba	c.1 23 S 31 48.5 E	Gillman
Kailekong	U3	Teso	1 51 N 34 08 E	
Kaili, mt.	U1	Karamoja	Not traced	J. Wilson
Kailongol. mt.	T3	Pare	3 46 S 37 34.5 E	Peter
Kaimat, Escarp.	K1	Northern Frontier	1 53 N 35 46 E	Champion, Mathew
	K2	West Suk	1 36 N 35 26 E	Carter & Stannard

Name	Code	Region	Coordinates	Collector
Kaimosi, mssn.	K3	Nandi	0 08 N 34 56 E	Numerous
Kaimosi, popl.	K3	Nandi	0 07.5 N 34 51 E	Malaki, Gillett
Kaimosi, tea est.	K3	Nandi	c.0 09 N 34 56 E	Tweedie
Kainam, popl. mssn.	T2	Mbulu	3 55 S 35 35 E	Moreau
[Kainook] see Kainyunyok	K1	Northern Frontier	2 33 N 35 21 E	Coppock
Kainyunyok, popl., area	K1	Northern Frontier	2 33 N 35 21 E	
[Kaioleni] see Kaloleni	K7	Kilifi	3 49 S 39 38 E	Hooper & Townsend
Kaipos, peak	K2/3	West Suk/Trans-Nzoia/Elgeyo	1 12 N 35 09 E	Mabberley & McCall
Kaiser Wilhelm Spitze = Uhuru Point, peak	T2	Moshi	3 04.5 S 37 21 E	Maitland
[Kaisi] see Kaiso	U4	Masaka	c.0 58 S 31 32 E	
Kaiso, f.r.	U4	Masaka	c.0 58 S 31 32 E	
[Kaisungor] see Kaisungur	K3	Elgeyo	1 03.5 N 35 24.5 E	Numerous
Kaisungur, for.	K3	Elgeyo	c.1 03 N 35 26 E	
Kaisungur, peak, graves	K3	Elgeyo	1 03.5 N 35 24.5 E	Tweedie
Kaisut Desert	K1	Northern Frontier	1 50 N 37 45 E	Carter & Stannard
Kaiti = Keite, R.	K4	Machakos	c.1 44 S 37 21 E–1 45 S 37 42 E	Katumani Staff
Kaitokoi Pools	T2	Masai	5 03 S 37 17.5 E	Vesey-FitzGerald, Procter
Kaizi River Bridge	U2	Ankole/Kigezi	0 25.5 S 29 52 E	Lock & Synnott
Kaja, R.	K4	Meru	0 01 S 37 44 E–0 00.5 S 37 49 E	
Kajala	T7	Rungwe	c.9 22 S 33 49 E	Stolz
[Kajamoro] see Kijomoro	U1	West Nile	3 10 N 30 54 E	A.S. Thomas
Kajansi, f.r., popl.	U4	Mengo	0 12 N 32 33 E	Chandler, Dawkins
Kajara, co.	U2	Ankole	c.1 00 S 30 10 E	
Kajado, distr.	K6	Masai	c.2 10 S 37 00 E	
		(considered as part of Masai Distr. for F.T.E.A.)		
Kajiado, popl.	K6	Masai	1 51 S 36 47 E	Numerous
Kajiado, R.	K6	Masai	1 44 S 36 44 E–2 27 S 37 17 E	
[Kajomoro] see Kijomoro	U1	West Nile	3 10 N 30 54 E	
[Kajumjumere] see Kanjumjumere	T7	Rungwe	9 36 S 33 55 E	

98

Name	Code	Region	Coordinates	Source
Kakabara, popl.	U2	Toro	0 32 N 30 57 E	Tateoka
[Kakage] see Kakoge	U4	Mengo	1 03 N 32 28 E	
Kakakala, popl.	U4	Mengo	0 36 N 32 49 E	
Kakamari, popl.	U1	Karamoja	3 29 N 34 08 E	
Kakamega, distr.	K3/5	Uasin Gishu/N. Kavirondo	c.0 25 N 34 45 E	Liebenberg
Kakamega, for.	K5	North Kavirondo	c.0 15 N 34 52 E	Numerous
Kakamega, popl., exper. fm.	K5	North Kavirondo	0 17 N 34 45 E	Numerous
[Kakamongole] see Kokumongole	U1	Karamoja	1 54 N 34 38 E	
[Kakamori] see Kakamari	U1	Karamoja	3 29 N 34 08 E	Liebenberg
[Kakamoyok] see Kochemaluk	U1	Karamoja	2 11 N 34 21 E	Dyson-Hudson
[Kakamuri] see Kakamari	U1	Karamoja	3 29 N 34 08 E	
Kakeani, popl.	K4	Kitui	1 09 S 37 57 E	Napper
Kakep, pass	K2	Turkana	2 33 N 35 10 E	J. Wilson, Newbould
[Kakesio] see Kakessio	T2	Masai	3 19 S 35 02 E	Paulo
Kakessio, popl.	T2	Masai	3 19 S 35 02 E	Newbould
[Kakesyo] see Kakessio	T2	Masai	3 19 S 35 02 E	
Kakiani = Kakeani, popl.	K4	Kitui	1 09 S 37 57 E	Napper
Kakindo, mssn.	U2	Bunyoro	1 28 N 31 26 E	Purseglove
[Kakindu] see Kakindo	U2	Bunyoro	1 28 N 31 26 E	
Kakindu, popl.	U4	Mengo	0 19 N 32 09 E	
Kakindu, popl.	T1	Bukoba	1 10 S 31 29 E	Haarer, Holtz, Procter
Kakinzi, popl.	U4	Mengo	0 56 N 32 28 E	Lye
Kakira, popl., t.c. rsta.	U3	Busoga	c.0 30 N 33 17 E	Jameson
Kakira, sugar est.	U3	Busoga	0 31 N 33 18 E	
Kakira, popl.	U3	Busoga	0 45 N 33 33 E	
Kakiri, popl.	U4	Mengo	0 25 N 32 23 E	
Kakitumba, popl., customs post	U2/Rwanda	Ankole	1 03 S 30 28 E	B.D. Burtt
Kakitumba, str.	U2/Rwanda	Ankole	c.1 04 S 30 21—30 28 E	
Kakoge, popl., r.h., t.c.	U4	Mengo	1 03 N 32 28 E	Langdale-Brown, Eggeling, Tateoka
Kakoge, springs, popl.	U4	Mengo	0 56 N 32 57.5 E	

Kakoma, popl.	T1	Bukoba	1 54 S 31 26.5 E	Hornby
Kakoma, popl.	T4	Tabora	5 47 S 32 26 E	Numerous
Kakombe, str.	T4	Buha	c.4 38 S 29 36 E	Newbould & Harley, Pirozynski
[Kakomoyok] see Kochemaluk	U1	Karamoja	2 11 N 34 21 E	Dyson-Hudson
Kakoneni, popl.	K7	Kilifi	3 10 S 39 52 E	Battiscombe, Moomaw, Tweedie
[Kakoneri] see Kakoneni	K7	Kilifi	3 10 S 39 52 E	Moomaw
Kakonji, w.h.	T4	Mpanda	6 59.5 S 31 04 E	Richards
Kakoro, popl.	U3	Mbale	1 11 N 34 03 E	
Kakuma, p.p., airstrip	K2	Turkana	3 43 N 34 52 E	Buxton, Itani
Kakumiro, popl., t.c., exp. farm	U4	Mubende	0 47 N 31 19.5 E	Numerous
[Kakumongole] see Kokumongole	U1	Karamoja	1 54 N 34 38 E	A.S. Thomas
Kakuto, popl.	U4	Masaka	0 51 S 31 28 E	Johnston
Kakuyuni, hill	K7	Kilifi	3 13 S 39 59 E	
Kakuyuni, popl.	K7	Kilifi	3 13 S 40 00 E	Tweedie
Kakuzi, hills	K4	Fort Hall	c.1 01 S 37 19 E	Vesey-FitzGerald
Kala, ln.	T4	Mpanda	prob. 6 43 S 31 17 E	Carmichael
Kala, bay	T4	Ufipa	8 08 S 30 57 E	Richards, Van Meel
Kalabata = Kalapata, str.	K1	Northern Frontier	1 58 N 35 49 E–2 33 N 36 15 E	Mwangangi & Gwynne
Kalabwe, R.	T7	Rungwe	c.9 10 S 33 40 E	Stolz
Kalachwa Wells	K1	Northern Frontier	3 07 N 37 25 E	Bally
Kalagwe = Kalabwe R.	T7	Rungwe	c.9 10 S 33 40 E	Stolz
Kalait, popl.	K5	North Kavirondo	0 44 N 34 19 E	Lyne Watt
Kalaki, popl.	U3	Teso	1 49 N 33 20 E	
Kalakol, R.	K2	Turkana	3 34 N 35 24 E–3 33 N 35 54.5 E	
Kalalamuka, mssn.	T7	Rungwe	9 19 S 33 35 E	Stolz
[Kalalokoil] see Kalakol	K2	Turkana	3 34 N 35 24 E–3 33 N 35 54.5 E	Mortimer
Kalama, hill, area	K4	Machakos	1 37 S 37 21 E	Fliervoet
Kalambo, falls	T4/Zambia	Ufipa	8 36 S 31 14 E	Numerous
Kalambo, popl.	T4	Ufipa	8 11 S 31 23 E	Richards
Kalambo, R.	T4/Zambia	Ufipa	c.7 59 S 31 17 E–8 36 S 31 11 E	Numerous

Kalimbili	T7	Iringa	near 8 00 S 36 00 E	Carmichael
Kalimbo	T7	Rungwe	near 9 17 S 33 39 E	Stolz
Kalinga, popl.	T7	Iringa	8 29 S 35 21 E	Ward
Kalinzi = Mkalinzi, popl.	T4	Buha	4 37 S 29 44 E	Carmichael, Procter, Verdcourt
Kalinzu, f.r.	U2	Ankole	c.0 25 S 30 00 E	Numerous
[Kaliokoil] see Kalakol	K2	Turkana	3 34 N 35 24 E–3 33 N 35 54.5 E	Mortimer
[Kaliolokwel] see Kalakol	K2	Turkana	3 34 N 35 24 E–3 33 N 35 54.5 E	
Kaliro, popl.	U3	Busoga	0 54 N 33 30 E	
Kaliro, rsta.	U3	Busoga	0 52.5 N 33 30 E	G.H. Wood
[Kalisama] see Kambisama	T4	Mpanda	c.6 26 S 31 22 E	Richards
[Kalish] see Katesh	T2	Mbulu	4 31 S 35 23 E	Richards
Kalisizo, popl.	U4	Masaka	0 32 S 31 37.5 E	Purseglove
Kaliua, popl., rsta.	T4	Tabora	5 04 S 31 48 E	Bullock, Ramazani, Shabani
[Kaliuwa] see Kaliua	T4	Tabora	5 04 S 31 48 E	Numerous
Kaloleni, popl.	K4	Nairobi	1 18 S 36 51 E	H.J. Taylor
Kaloleni, popl.	K7	Kilifi	3 49 S 39 38 E	Verdcourt
Kalossia, ex govt. stn.	K2	West Suk	1 38.5 N 35 46 E	
Kalubi, pass	T7	Rungwe	prob. c.9 29 S 33 30 E	Stolz
Kaluguza, popl., mssn.	U4	Mubende	0 46 N 31 05 E	
Kalumbalesa, popl.	T4	Ufipa	7 36 S 31 32.5 E	Michelmore
Kalungi, popl.	U4	Mengo	1 18 N 32 40 E	Langdale-Brown
Kalungu, popl., r.h., mssn.	U4	Masaka	0 10 S 31 45 E	Lye
Kalya, popl.	T4	Mpanda	6 28 S 30 00 E	Harley et al.
Kama, popl., sw.	Z	Zanzibar	6 03 S 39 12.5 E	Faulkner
Kamachumu, popl., mssn. r.h.	T1	Bukoba	1 35 S 31 37 E	Haarer
[Kamakia] see Kimakia	K4	Fort Hall	c.0 46 S 36 44 E	Napper
[Kamakoia] see Kamakoiwa	K3/5	Tranz-Nzoia/N. Kavirondo	0 56 N 34 45 E–0 44.5 N 34 49 E	Tweedie
Kamakoiwa, popl.	K5	North Kavirondo	0 47 N 34 47 E	Mainwaring
Kamakoiwa, str.	K3/5	Trans-Nzoia/N. Kavirondo	0 56 N 34 45 E–0 44.5 N 34 49 E	Tweedie
Kamakuywa, popl., sch.	K5	North Kavirondo	0 48 N 34 47.5 E	

101

Name	Code	Region	Coordinates	Authority
Kami, str.	U3	Mbale	c.0 29 N 34 13 E–0 35 N 33 59.5 E	G.H. Wood
Kamiana ? see Kaniamwia	K5	South Kavirondo	?0 40–0 45 S 34 10–34 20 E	Fischer
Kamigo, ln.	U3	Busoga	0 38 N 33 16 E	G.H. Wood
Kamilat = Kamelet, hill	K3	Nandi/Uasir Gishu	0 31 N 35 06.5 E	Dale
Kamion, popl.	U1	Karamoja	3 43 N 34 12 E	Eggeling, Dale, A.S. Thomas
Kamiti, R.	K4	Kiambu	1 04.5 S 36 39 E–1 13 S 37 02 E	Kirrika
Kamlambwe, popl.	T4	Mpanda	c.6 25 S 31 10 E	Richards
Kammengo, popl., r.h.	U4	Mengo	0 09 N 32 14 E	
[Kamoda] see Kamuda	U3	Teso	1 45 N 33 31 E	Fiennes
[Kamolo] see Komolo	U1	Karamoja	2 43 N 34 15 E	
Kamori, area	T3	Pare	3 50 S 37 42 E	
Kamorin, ln.	K3	Elgeyo	0 40.5 N 35 30 E	Tweedie
Kamor Jila, area, w.h.	K7	Tana River	1 21 S 39 55 E	Gillett
Kamothing, R.	U1	Karamoja	2 12 N 34 29 E–2 02 N 34 29 E	
Kampala, popl.	U4	Mengo	0 19 N 32 35 E	Numerous
Kampi ya Bibi, ln.	K6	Masai	c.1 33 S 36 32 E	Harding, Greenway
Kampi ya Farasi, ln.	K4	North Nyeri	0 08 S 37 16 E	Hedberg
Kampi ya Faru, ln.	T2	Mbulu	c.3 37 S 35 36 E	Peter
Kampi ya Fisi, ln.	T2	Arusha	c.3 15 S 36 52 E	
Kampi ya Mawe, hill	K4	Machakos	1 50 S 37 39 E	Katumani Exper. Fm.
Kampi Ya Moto, rsta.	K3	Nakuru	0 07.5 S 35 56.5 E	Drummond & Hemsley, Tweedie, Vesey-FitzGerald
[Kampi ya Mpofu] see Kampi ya Pofu	T1	Musoma	prob. c.2 10 S 35 00 E	Greenway
Kampi ya Ndege, ln.	K4	Machakos	c.1 12 S 37 33 E	Bally
Kampi ya Ndizi	T7	Iringa	Not traced	von Prittwitz & Gaffron
Kampi ya Nyoka, popl., sch.	T2	Mbulu	3 18.5 S 35 35.5 E	B.D. Burtt
Kampi ya Pofu, ln.	T1	Musoma	prob. c.2 10 S 35 00 E	Greenway
Kampi ya Simba, ln.	K7	Tana River	0 07 S 38 45 E	
[Kampolongwe] see Kandelongwe	K4	Kitui	1 03 S 38 58 E	L.C. Edwards
Kampunda, popl.	T4	Ufipa	8 05 S 32 07 E	Bullock

103

Name		District	Coordinates	Collector
Kamsamba, popl.	T7	Mbeya	8 21 S 32 18 E	Richards
Kamuda, popl.	U3	Teso	1 45 N 33 31 E	Fiennes
Kamuganguzi, popl.	U2	Kigezi	1 20.5 S 30 00.5 E	
Kanuge, popl.	U3	Mbale	1 10 N 33 49 E	Snowden
[Kamukanguzi] see Kamuganguzi	U2	Kigezi	1 20.5 S 30 00.5 E	Eggeling
Kamukumu, hill	U2	Kigezi	c.1 13 S 29 38 E	Norman
[Kamukuywa] see Kamakuywa	K5	North Kavirondo	0 48 N 34 47.5 E	Bauer
Kamuli, popl.	U3	Busoga	0 57 N 33 07 E	G.H.S. Wood, Harker
Kamuma, popl., airstrip	K2	Turkana	3 42 N 34 52 E	Martin
Kamunye, hill	T1	Musoma	2 25 S 34 48 E	Greenway
[Kamuri] see Kamori	T3	Pare	3 50 S 37 42 E	Haarer
Kamwaki, fm.	K4	North Nyeri	c.0 15 N 37 12.5 E	Haarer
Kamwala, mt.	T3	Pare	3 41 S 37 38 E	J. Bally
Kamwalsi ? see Kamwaki	K4	North Nyeri	?c.0 15 N 37 12.5 E	
Kamwanga, area, sawmill	T2	Moshi	2 54.5 S 37 22 E	
Kamwanye (Sese Is.)	U4	Masaka	Not traced	Maitland
[Kamwe] see Kawi	T6	Uzaramo	6 44 S 39 14 E	Harris
Kamwenge, popl., rsta.	U2	Toro	0 12 N 30 27 E	
[Kamwete] see Kamweti	K4	South Nyeri	0 18.5 S 37 18 E–0 28.5 S 37 21 E	Bamps
Kamweti, str., valley	K4	South Nyeri	0 18.5 S 37 18 E–0 28.5 S 37 21 E	Townsend
Kamweti Forest Station	K4	South Nyeri	0 20 S 37 19 E	Robertson, Townsend
Kamweti Track	K4	South Nyeri	0 18.5 S 37 18 E–0 27 S 37 21 E	Faden
Kamwezi, popl., r.h.	U2	Kigezi	1 13 S 30 13 E	Purseglove, Eggeling
Kanaba Gap, pass	U2	Kigezi	1 15 S 29 47.5 E	Numerous
Kanam, ln.	K5	South Kavirondo	c.0 23 S 34 28.5 E	Turner
Kanam, sch.	K5	South Kavirondo	0 21 S 34 29 E	
[Kanama] see Kanamai	K7	Kilifi	3 55 S 39 47 E	
Kanamai, popl.	K7	Kilifi	3 55 S 39 47 E	Bally
Kanamuget, str.	U1	Karamoja	c.3 14 N 33 51 E	Bally
Kanamugit, area	U1	Karamoja	c.2 44 N 34 35 E	Eggeling, A.S. Thomas

Name	Region	District	Coordinates	Collector
Kanana, ln.	K7	Kwale	4 32 S 39 22 E	Goetze
Kananda, popl.	T7	Rungwe	9 20 S 33 02.5 E	Pirozynski
Kanangiye, ln.	T4	Kigoma	4 47 S 29 39 E	Allan
[Kanani] see Kanana	K7	Kwale	4 32 S 39 22 E	Braun
[Kanasi] see Kanazi	T1	Bukoba	1 28 S 31 44.5 E	Goetze
[Kanauda] see Kananda	T7	Rungwe	9 20 S 33 02.5 E	Procter
Kanazi, popl.	T1	Bukoba	1 28 S 31 44.5 E	
Kanda, area	T4/7	Ufipa/Mbeya	c.8 55 S 32 05 E	Vesey-FitzGerald
Kanda, hills	T4	Ufipa	8 27 S 31 50 E	B.D. Burtt
Kandaga, popl., area	T5	Kondoa	4 33 S 35 47 E	Peter
Kandaga, rsta.	T4	Kigoma	4 57 S 29 51 E	
[Kandajega] see Gandajega	T5	Singida	4 48 S 34 18 E–4 51 S 33 58 E	Bax, Greenway
Kandecha, dam	K7	Teita	3 22 S 38 47 E	Greenway
[Kandechwa] see Kandecha	K7	Teita	3 22 S 38 47 E	L.C. Edwards
Kandelongwe, hill	K4	Kitui	1 03 S 38 58 E	Greenway & Kanuri
Kandere Dam, pond	K7	Teita	3 22 S 38 40 E	Cribb & Grey-Wilson
Kandete, popl., str.	T7	Rungwe	9 09 S 33 48 E	Greenway
[Kandiri, Lake] see Kandere Dam	K7	Teita	3 22 S 38 40 E	Gilli
[Kandjomere] see Kanjunjumere	T7	Rungwe	9 36 S 33 55 E	
[Kandulu] see Msamala	T8	Tunduru	10 46 S 37 14 E	Vaughan
Kandwi, ridge	Z	Zanzibar	c.5 56 S 39 21 E	Bally
[Kandziko] see Kanziko	K4	Kitui	1 59 S 38 21 E	Verdcourt
Kanga, rsta.	K4	Machakos	2 46 S 38 16 E	
Kanga, f.r.	T6	Morogoro	c.6 00 S 37 43 E	Peter
Kanga, popl.	T6	Morogoro	6 01.5 S 37 44 E	Greenway
Kanga, popl.	T6	Rufiji	7 43.5 S 39 51 E	Vaughan
Kangagaani, popl.	P	Pemba	5 09.5 S 39 50 E	
Kangai, popl.	U1	Lango	1 47 N 33 05 E	Stuhlmann
Kangani, popl.	T6	Bagamoyo	6 37 S 38 59 E	Bally
[Kangata] see Langata	K4	Nairobi	c.1 20 S 36 46 E	

Name	Code	Area	Coordinates	Authority
Kantalamba, regional HQ	T4	Ufipa	8 01 S 31 40 E	Mwasumbi
Kantale, popl., f.r.	T1	Bukoba	1 11 S 31 40 E	Gillman, Ritchie
[Kantanda] see Kashanda	U2	Kigezi	0 55 S 29 59 E	Stuhlmann
[Kantare] see Kantale	T1	Bukoba	1 11 S 31 40 E	Gillman
Kantasia, fm.	T7	Mbeya	c.9 02 S 32 56 E	B.D. Burtt
Kanunga, hills	K6	Masai	1 38 S 35 55 E	
Kanungu, popl., mssn.	U2	Kigezi	0 54 S 29 47 E	Purseglove, Eggeling
Kanwe-Mayi, pool	K7	Lamu	2 27 S 40 28 E	Hooper & Townsend
[Kanyalakata] see Kasunga	T4	Ufipa	8 05 S 31 29.5 E	Vesey-FitzGerald
Kanyamkago, popl.	K5	South Kavirondo	0 57 S 34 31 E	
Kanyampara, str.	U2	Toro	0 09 N 29 51 E–0 01.5 S 29 52 E	Osmaston
[Kanyangarang] see Kanyangareng	U1/K2	Karamoja/Turkana	2 21 N 34 55 E–1 47 N 35 09 E	Champion, Wilson, Tweedie
Kanyangareng, R.	U1/K2	Karamoja/Turkana	2 21 N 34 55 E–1 47 N 35 09 E	Eggeling, Philip
Kanyansaba, str. (Ruwenzori)	U2	Toro	Not traced	Fishlock & Hancock
[Kanyanyeine] see Kanyenyeini	K4	Fort Hall	0 41 S 36 53 E	Balbo
[Kanyao] see Kunyao	K2	Turkana	1 47 N 35 03 E	A.S. Thomas
[Kanyao] see Kunyao	U1/K2	Karamoja/Turkana	c.1 45 N 35 00 E	J. Wilson
Kanyato, popl.	T4	Buha	4 27 S 30 16 E	Eggeling
Kanyekine, popl., mssn.	K4	Meru	0 07.5 S 37 40 E	W.H. Lewis
Kanyenyeini, popl., mssn.	K4	Fort Hall	0 41 S 36 53 E	Balbo
Kanyerus, str.	K2	Turkana	1 22 N 34 47 E–1 26 N 34 51 E	
[Kanyesus] see Kanyerus	K2	Turkana	1 22 N 34 47 E–1 26 N 34 51 E	Dale
[Kanyiko] see Kanziko	K4	Kitui	1 59 S 38 21 E	Bally
Kanziko, popl., r.h.	K4	Kitui	1 59 S 38 21 E	Bally
[Kao] see Kau	K7	Lamu	2 29 S 40 26 E	Sampson
Kaole, popl., area	P	Pemba	5 14 S 39 44.5 E	Vaughan
Kaongolero, ln.	T1	Bukoba	c.1 16 S 31 37 E	Procter
Kapapa, camp	T4	Mpanda	6 48 S 31 22 E	Richards, Carmichael
Kapchebelel, t.c.	K3	Elgeyo	0 17 N 35 34 E	Gilbert & Mesfin
[Kapcherwa] see Kapchorwa	U3	Mbale	1 24 N 34 27 E	

108

Name		Region	Coordinates	Authority
Kaptalamwa, popl.	K3	Elgeyo	1 06.5 N 35 26 E	Townsend
Kaptama, popl.	K5	North Kavirondo	0 53 N 34 46 E	Tweedie
Kaputir, popl.	K1	Northern Frontier	2 05 N 35 28 E	Champion, Mwangangi & Gwynne, Newbould
[Kapyen] see Kabyen	K2	Turkana	1 56.5 N 35 07 E –1 49.5 N 35 15 E	Philip
[Kara] see Kawa	T4	Ufipa	c.8 32 S 31 20 E –8 27.5 S 31 09 E	Richards
[Karachonyo] see Karachuonyo	K5	South Kavirondo	0 25.5 S 34 41.5 E	Fischer
Karachuonyo = Karachwonyo, popl.	K5	South Kavirondo	0 25.5 S 34 41.5 E	
[Karague] see Karagwe	T1	Bukoba	c.1 30 S 31 00 E	
[Karagueh] see Karagwe	T1	Bukoba	c.1 30 S 31 00 E	Grant
Karagwe, popl.	U2	Ankole	0 48 S 30 17 E	
Karagwe, distr.	T1	Bukoba	c.1 30 S 31 00 E	Numerous
		(included in Bukoba District for F.T.E.A.)		
[Karakau] see Karikau	K2	Turkana	c.1 58 N 35 12 E	Philip
[Karakshongo] see Karachuonyo	K5	South Kavirondo	0 25.5 S 34 41.5 E	Fischer
[Karalj] see Karati	K3	Naivasha	0 49.5 S 36 34.5 E –0 43 S 36 25 E	Albrechtsen
Karambi, popl.	T1	Bukoba	2 05 S 31 31 E	Procter
[Karambo] see Kirambo	T4	Ufipa	8 15.5 S 31 00 E	Richards
Karamenu, str.	K4	Kiambu	c.0 51 S 36 44 E –1 01 S 37 01.5 E	J. Stewart
Karameri, hill	K2	Turkana	1 40 N 35 03.5 E	J. Wilson
Karamoja, distr.	U1	Karamoja	c.2 30 N 34 20 E	Liebenberg, Akol
Karamoja Drift (Suam River)	U3/K3	Mbale/Trans-Nzoia	1 13 N 34 45.5 E	Tweedie
Karamule, popl.	U2	Toro	c.0 17 N 30 21 E	Bagshawe
Karanga, R.	T2	Moshi	3 06 S 37 22 E –3 30 S 37 18 E	
Karangai = Karangai Ndogo	T2	Arusha	3 27 S 36 51 E	
Karangai Ndogo, popl., vet. stn.	T2	Arusha	3 27 S 36 51 E	
Karangora, peak	U2	Toro	0 38 N 30 07 E	G. Taylor, Osmaston
Karapokot = Karasuk	K2	Turkana	c.2 00 N 35 10 E	
Karara, dam	K3	Trans-Nzoia	c.0 54 N 35 06.5 E	Symes
Karare, hill	K1	Northern Frontier	2 10 N 37 51 E	van Swinderen

Name	Code	District	Coordinates	Collectors
Karasuk, distr.	K2	Turkana (administered as part of Karamoja District by Uganda)	c.2 00 N 35 10 E	Philip
Karasuk, hills	K2	Turkana	c.2 15 N 35 10 E	
Karati, falls	K3	Naivasha	c.0 48 S 36 35.5 E	Albrechtsen
Karati, str.	K3	Naivasha	0 49.5 S 36 34.5 E–0 43 S 36 25 E	Albrechtsen, Polhill
Karati, R.	K4	South Nyeri	0 25 S 37 27 E–0 26 S 37 27 E	
Karati (Uruwira)	T4	Mpanda	c.6 27 S 31 21 E	Richards
Karatina, popl., rsta.	K4	South Nyeri	0 29 S 37 07.5 E	
Karati Plateau, area	K3	Naivasha	c.0 48 S 36 35 E	Albrechtsen
[Karato] see Karatu	T2	Mbulu	3 20 S 35 40.5 E	Bally
[Karatschongo] see Karachuonyo	K5	South Kavirondo	0 25.5 S 34 41.5 E	Fischer
Karatu, popl., p.p., t.c.	T2	Mbulu	3 20 S 35 40.5 E	Greenway, Bally, van Rensburg, Pole Evans
Karatu, t.a.	T2	Mbulu	c.3 35 S 35 37 E	
Karawa, Vet. Dept. camp	K7	Tana River	2 39 S 40 12 E	Polhill & Paulo
Karema = Urema, t.a.	T4	Mpanda	c.6 45 S 30 35 E	C.H.B. Grant
Karema, popl.	T4	Mpanda	6 49 S 30 26 E	Numerous
Karen, popl.	K4	Nairobi	1 20 S 36 43 E	Napier, C.G. Rogers, van Someren
[Karenda] see Kawende	T4	Mpanda	c.6 30 S 30 30 E	Boehm
Karenga, popl.	U1	Karamoja	3 35 N 33 40 E	J. Wilson
Karenga, peak	T7	Iringa	7 41 S 36 54 E	Harris & Pocs
Karenge, L.	U2	Ankole	c.0 54 S 30 08 E	Bagshawe, Eggeling
[Karensi] see Kirenzi	U2	Kigezi	0 51 S 29 43 E	Purseglove
Karia, area	K4	South Nyeri	0 25 S 37 05.5 E	
Kariandus, popl., rsta.	K3	Naivasha	0 26 S 36 16.5 E	
Kariandusi = Kariandus, popl., rsta.	K3	Naivasha	0 26 S 36 16.5 E	
Karibani, hill	K6	Masai	2 01.5 S 37 21.5 E	Gregory
[Karichwa Kubwa] see Kirichwa Kubwa	K4	Nairobi	c.1 17 S 36 43.5 E–1 16 S 36 48 E	
Karikau, area	K2	Turkana	c.1 58 N 35 12 E	
[Karinga] see Kiringa	K4	South Nyeri	0 14.5 S 37 17.5 E–0 33 S 37 20 E	Copley
Karisia Hills	K1	Northern Frontier	1 05 N 36 50 E	Carter & Stannard

Name		Region	Coordinates	Collector
Karita, popl.	U1	Karamoja	1 33 N 34 50 E	Leippert, A.S. Thomas, J. Wilson
[Karitha] see Karita	U1	Karamoja	1 33 N 34 50 E	A.S. Thomas
[Karokor] see Karikau	K2	Turkana	c.1 58 N 35 12 E	
Karolo, f.r.	T7	Rungwe	c.9 33 S 33 35 E	Carmichael
Karomuch, str.	U1	Karamoja	c.2 01 N 34 55 E	Dyson-Hudson
[Karongara] see Karangora	U2	Toro	0 38 N 30 07 E	Osmaston
[Karoro] see Kahororo	T1	Bukoba	1 15 S 31 50 E	Haarer
Karpeddo, t.c., str., area	K1/3	N. Frontier/Baringo	c.1 10 N 36 04 E	Pratt
[Karrakia] see Tarakia	T2	Moshi	3 04 S 37 28 E –3 00 S 37 34 E	Volkens
Karro Gudi = Karro Gudda or Gudder	K1	Northern Frontier	prob. c.3 50 N 41 45 E	Ellenbeck
Karsadera, hill	K1	Northern Frontier	2 18.5 N 37 55 E	Carter & Stannard
Karubi, mt.	T7	Rungwe	9 29 S 33 30 E	Cribb & Grey-Wilson
Karumeri, hill	K2	Turkana	1 40 N 35 03.5 E	J. Wilson
Karumo, popl. f.r.	T1	Mwanza	2 30 S 32 49 E	Numerous
Karuna, hill, fm.	K3	Uasin Gishu	0 44 N 35 28 E	Tweedie
Karura, falls	K4	South Nyeri	0 30 S 36 46 E	Townsend
Karura, for.	K4	Kiambu/Nairobi	c.1 14.5 S 36 49.5 E	Numerous
Karuru = Kalulu, R.	K4	North Nyeri	0 00 37 15 E–0 06 N 37 07 E	Copley
[Karuti] see Karute, str.	K4	South Nyeri	0 17 S 37 17 E–0 26 S 37 21 E	Fries, Townsend
Kasa, f.r.	U4	Mengo	c.0 18 N 32 02 E	Dawkins
Kasagama, old popl.	U2	Toro	0 38 N 30 16 E	
Kasagama, popl.	U4	Masaka	0 07 S 31 09 E	
[Kasagj] see Kasoge	T2	Mpanda	6 09 S 29 44 E	Harley
Kasago, L.	U3	Teso	c.1 34 N 33 45 E	
Kasaiga, str.	U2	Toro	c.0 55 N 30 15 E	Hazel
Kasakati, area	T4	Kigoma	c.5 20–5 23 S 30 10–30 20 E	Suzuki, Itani
Kasakati = Kasangati, R.	T4	Kigoma	5 20 S 29 54 E–5 26 S 29 56 E	Suzuki
Kasakela, str.	T4	Buha	c.4 37.5 S 29 36 E	Morris-Goodall, Verdcourt, Pirozynski
[Kasakela] see Kasekela	T4	Mpanda	7 27 S 31 29 E	Richards, Vesey-FitzGerald
Kasala, f.r.	U4	Mengo	c.0 16 N 32 47.5 E	Dummer

111

112

Name		Region	Coordinates	Authority
Kaseri, R.	K4	Meru	prob. c.0 05 S 37 45 E	Fries
Kasese, rsta.	U2	Toro	0 10 N 30 05 E	Maitland
Kasese, popl.	T1	Bukoba	c.1 35 S 31 07 E	Stuhlmann
Kasha, ln., str., rapids	K1	Northern Frontier	c.0 07 S 39 12 E	Richards
Kashabukashi = Kashukasi	T4	Mpanda	6 28.5 S 30 25.5 E–31 22.5 E	Snowden
[Kashambia] see Kashambya	U2	Kigezi	1 03 S 29 59 E	
[Kashambia] see Kashambya	T1	Bukoba	1 12 S 31 26 E	
Kashambya, popl., r.h.	U2	Kigezi	1 03 S 29 59 E	Haarer
Kashambya, popl.	T1	Bukoba	1 12 S 31 26 E	Norman, Snowden
Kashanda, area	U2	Kigezi	0 55 S 29 59 E	Sampson
Kashari, co.	U2	Ankole	c.0 28 S 30 33 E	H.B. Johnston
[Kashe] see Kasha	K1	Northern Frontier	c.0 07 S 39 12 E	
[Kashenji] see Kishanje	U2	Kigezi	1 18 S 29 52 E	
Kashenyi, popl., r.c.	U2	Ankole	0 35 S 30 08 E	Purseglove, Lock
Kashishi, area	T1	Mwanza	2 42 S 32 56 E	Tanner
Kashota, popl. (Rwasi Reserve)	T4	Ufipa	?8 53 S 32 15 E	Wigg
Kashoya-Kitomi, for.	U2	Ankole	c.0 15 S 30 15 E	Cree, Eggeling, Okodi
Kashukasi, str.	T4	Mpanda	6 28.5 S 30 25.5 E–31 22.5 E	Stuhlmann, Goetze
[Kasi] see Kazimzumbwi	T6	Uzaramo	6 57 S 39 01 E	
[Kasieha] see Kasiha	T4	Mpanda	6 06 S 29 46.5 E–6 07.5 S 29 43.5 E	
Kasiga, popl.	T3	Lushoto	4 50 S 38 14 E	Newbould & Harley
Kasigao = Kasigau, mt., for.	K7	Teita	c.3 50 S 38 40 E	Numerous
Kasigau, mt., for.	K7	Teita	c.3 50 S 38 40 E	Jefford et al.
Kasiha, str.	T4	Mpanda	6 06 S 29 46.5 E–6 07.5 S 29 43.5 E	Chandler
Kasilo, ln.	U3	Teso	1 32 N 33 21.5 E	Eggeling, Sanane
Kasimba, popl.	T4	Mpanda	6 22 S 31 08 E	Goetze
Kasimulo, hill	T7	Rungwe	prob. c.9 32 S 33 43 E	Osmaston
[Kasingiii] see Kasingiri	U2	Toro	0 05.5 N 29 49 E	
Kasingiri, popl.	U2	Toro	0 05.5 N 29 49 E	Maitland
[Kasingo] see Kazingo	U2	Toro	0 39 N 30 10 E	

114

Name		Region	Coordinates	Collector
Kasulu, distr.	T4	Buha	c.4 25 S 30 20 E	Procter
(considered as part of Buha District for F.T.E.A.)				
Kasulu Highlands	T4	Buha	c.4 35 S 30 03 E	Eggeling, Procter
Kasumba, popl., r.h.	U2	Ankole	0 54 S 30 58 E	Lind
Kasunga, popl.	T4	Ufipa	8 05 S 31 29.5 E	Richards
Kasvala, popl.	T4	Ufipa	8 25 S 32 07 E	Bullock
[Kasyoha, for.] see Kashoya-Kitomi	U2	Ankole	c.0 15 S 30 15 E	Symnott
Katabe, L.	T4	Mpanda	6 38 S 31 00 E	Richards
Katabi, popl.	U4	Mengo	0 06 N 32 28 E	Chandler
Katakwi, popl., mssn.	U3	Teso	1 55 N 33 58 E	Harker, Lye
Katale, popl.	T4	Kigoma	4 59 S 31 03.5 E	Lowe
Katama, str.	K3/4	Naivasha/Kiambu	c.0 59 S 36 35 E	Gardner
[Katamaju] see Gatamayu	K4	Kiambu	c.0 59 S 36 42 E	Bally
[Katamayu] see Gatamayu	K4	Kiambu	c.0 59 S 36 42 E	Numerous
Katandala, mssn.	T4	Ufipa	7 57.5 S 31 37 E	Hooper & Townsend
[Katanga] see Kangata	T3	Handeni	5 41 S 37 53 E	Bryce
Katangele = Katengele, popl.	T7	Rungwe	9 27 S 33 28 E	Cribb & Grey-Wilson
[Katare] see Katale	T4	Kigoma	4 59 S 31 03.5 E	Lowe
[Katari] see Katavi	T4	Mpanda	6 42 S 31 00.5 E	Richards
Katavi, L.	T4	Mpanda	6 42 S 31 00.5 E	Bullock, Richards, Boaler
Katavi Plains, g.r., s.w.	T4	Mpanda	c.6 55 S 31 00 E	Richards, Procter, Hooper & Townsend
[Katawi] see Katavi	T4	Mpanda		Carmichael
Kate, popl., mssn., hill	T4	Ufipa	7 51 S 31 10 E	Silungwe, Procter
Katebo, popl.	U4	Mengo	0 02 S 32 05 E	
Kateirewe, b.h.	U2	Bunyoro	1 27 N 31 50 E	B.J. Turner
Katende, for.	K4	Machakos	1 41 S 37 31 E	
Katende, hills	K4	Machakos	1 41 S 37 31 E	
[Katengato] see Kitengeto	U2	Toro	0 01.5 S 30 21 E –0 00 30 17 E	Jarrett
Katengele, popl.	T7	Rungwe	9 27 S 33 28 E	
[Katengeta] see Kitengeto	U2	Toro	0 01.5 S 30 21 E –0 00 30 17 E	

Name	Code	Region	Coordinates	Collector
Katera, popl., sawmill	U4	Masaka	0 55 S 31 39 E	Numerous
Kateruk, str./s.	K2	Turkana	c.2 10 N 35 09 E—2 30 N 35 23 E	Hemming
Katesa, P.W.D. road camp	T4	Ufipa	7 05 S 30 54.5 E	
Katesh, popl.	T2	Mbulu	4 31 S 35 23 E	Carmichael
[Kateshi] see Katesh	T2	Mbulu	4 31 35 23 E	
Kateta, hill, f.r.	U3	Teso	1 28 N 33 30.5 E	Eggeling
Katete, popl.	U2	Kigezi	0 47 S 29 45 E	Purseglove
Kathita = Kazita, R.	K4	Meru	0 04 S 37 25 E—0 14 S 38 00 E	Schelpe
Kathonsweni (near Makueni)	K4	Machakos	?c.1 50 S 37 37 E	D.B. Thomas
Katigithigiria, hills	K1	Northern Frontier	c.2 37 N 35 56 E	
Katika Mbuga	T5	Kondoa	near 4 45 S 36 03 E	B.D. Burtt
Katikekile, popl.	U1	Karamoja	2 23 N 34 50 E	J. Wilson
Katilia, for., w.h.	K1	Northern Frontier	2 07.5 N 36 07 E	Mathew
Katilo Forest (on [Karinga] Kiringa R.)	K4	South Nyeri	?c.0 23 S 37 15 E	Copley
Katimok, for.	K3	Baringo	c.0 36 N 35 47 E	Dale
Katire, popl.	Sudan *not* U1		4 47 N 32 26 E	A.S. Thomas
Katirikiki, ln.	K2	Turkana	4 14 N 35 27 E	Carter & Stannard
Katisunga, popl.	T4	Mpanda	6 55 S 31 03 E	Bullock, L. Thomas
Katoke, popl.	U2	Toro	0 45 N 30 42 E	
Katoke, mssn.	T1	Biharamulo	2 40 S 31 22 E	Gane
[Katolo] see Katoro	T1	Mwanza	3 04 S 31 55 E	
Katoma, popl.	T1	Bukoba	1 18 S 31 45.5 E	Gillman, Haarer
[Katome] see Kotome	K2	Turkana	4 42 N 35 23 E—4 19 N 35 03.5 E	Carter & Stannard
Katonga, chnl.	U4	Mengo	c.0 03 S 32 02 E	Rose
Katonga, R.	U2/4	several	0 10 N 30 36 E—0 03 S 32 01 E	Rose, Trapnell
Katoro, popl.	T1	Mwanza	3 04 S 31 55 E	Ukiriguru Res. Centre
Katovu, popl.	U4	Masaka	0 21 S 31 09 E	?Maitland
Katovu, popl.	U4	Masaka	0 24 S 31 14 E	Porter
Katse, popl., r.h.	K4	Kitui	0 30 S 38 05 E	Purseglove, Eggeling
Katulikire, popl. r.h.	U2	Bunyoro	2 00 N 32 09 E	

116

Name		Region	Coordinates	Authority
Katuma, popl.	T4	Mpanda	6 10.5 S 30 34 E	Richards
Katuma, str.	T4	Mpanda	c.6 06 S 30 36 E–6 57 S 31 14 E	Bullock
(sometimes called Kafufu, the river which flows south from L. Chada)				
Katumani, exper. fm.	K4	Machakos	1 35 S 37 15 E	D.B. Thomas, Napper
Katumba, hill	T4	Mpanda	6 20 S 31 14 E	Richards
Katumet, ln.	U1	Karamoja	3 24 N 34 45 E	Tweedie
Katundo, popl.	T4	Ufipa	prob. 8 16 S 31 26 E	Richards, Vesey-FitzGerald
Katungulu, popl., mssn.	T1	Mwanza	2 31 S 32 39.5 E	Tanner
Katunguru, popl.	U2	Ankole	0 08 S 30 03 E	Eggeling
Katupulo, popl.	T4	Ufipa	prob. c.8 10 S 31 25 E	Richards
Katwe, popl.	U2	Toro	0 08 S 29 52 E	Numerous
[Katwe] see Kibiro	U2	Bunyoro	1 41 N 31 15 E	Eggeling
Katwe, L.	U2	Toro	c.0 07.5 S 29 52 E	Scott Elliot, Eggeling, Greenway
Katwe, hill	U4	Mengo	0 47 N 32 01 E	Numerous
Kau, popl.	K7	Lamu	2 29 S 40 26 E	Dale, Gregory, Sampson
Kaujuma	T1	Bukoba	Not traced	Ford
Kauma, Kaya	K7	Kilifi	3 36.5 S 39 44 E	Musyoki & Hansen
Kautaku, hill	U1	Karamoja	2 29.5 N 34 32 E	J. Wilson
Kauwaa, peak	K4	Kitui	1 16 S 38 35.5 E	
Kauwalathe = Kagwalas, str./s.	K2	Turkana	c.3 35 N 35 04 E–3 07 N 35 38 E	Paulo
Kavali	Zaire			Stuhlmann
[Kavantanzoo] see Kavata Nzou	K4	Machakos	1 47 N 37 26 E	Mwangangi
Kavata Nzou, sch.	K4	Machakos	1 47 N 37 26 E	
Kavingo, popl.	T1	Bukoba	1 04 S 30 34 E	Stuhlmann
(There is a Kabingo marked on modern maps on the other bank of the Kagera R. in Uganda, Ankole District, but all old maps show Kavingo south of the river)				
Kavinji (near Kirere)	U2/T1	Ankole/Bukoba	Not traced	Stuhlmann
Kavinyiro = Kawinjiro, hill	T2	Masai	2 53 S 35 54 E	
Kavirondo, gulf	K5	C. Kavirondo/S. Kavirondo	c.0 15 S 34 35 E	Napier, Scott Elliot
Kavu, R.	T4	Ufipa/Mpanda	c.7 11 S 31 31 E–7 40 S 31 46 E	Richards
(upper reaches sometimes called Kafufu R.)				

117

Name		District	Coordinates	Authority
Kavura, hill	T4	Buha	4 35 S 29 42 E	Procter
Kawa, R.	T4	Ufipa	c.8 32 S 31 20 E–8 27.5 S 31 09 E	Richards, Vesey-FitzGerald
[Kawadie] see Naibor-Kawadie				
[Kawalathe] see Kauwalathe	K6	Masai	1 50.5 S 36 35 E	Glover & Cooper
[Kawamsisi] see Kwamsisi	K2	Turkana	c.3 35 N 35 04 E–3 07 N 35 38 E	
Kawanda, popl., hill	T3	Handeni	5 51 S 38 33 E	Procter
Kawanja Lodomayu	U4	Mengo	0 25 N 32 31 E	A.S. Thomas, Chandler
Kawanja ya Matheo = Kawanya ya Matheo, ln.	T2	Arusha	prob. c. 3 17S 36 56.5 E	Greenway
Kawanya ya Matteo, ln.	T2	Arusha	c.3 15.5 S 36 53 E	Pegler, Renvoize, Richards
Kawa River Falls	T4	Ufipa	c.8 28 S 31 16.5 E	Richards
Kawe = Kawi, popl.	T6	Uzaramo	6 44 S 39 14 E	Tweedie, Harris, McCusker
[Kawela] see Kwela	T4	Ufipa	8 24 S 31 58.5 E	Vesey-FitzGerald
Kawele, popl.	T4	Kigoma	4 58 S 29 43 E	Cameron
Kawempe, popl.	U4	Mengo	0 23 N 32 33 E	
Kawende, old area	T4	Mpanda	c.6 30 S 30 30 E	
Kawetire, popl.	T7	Mbeya	8 50 S 33 30 E	Procter
[Kawewe's] see Igigwa	T4	Tabora	5 26 S 32 52 E	Carnochan
Kawi, popl.	T6	Uzaramo	6 44 S 39 14 E	Tweedie
Kawinjiro = Kavinyiro, hill	T2	Masai	2 53 S 35 54 E	Peter
Kawoko, popl.	U4	Masaka	0 12 S 31 41 E	
Kawolo, rsta.	U4	Mengo	0 21 N 32 56 E	
Kaya, R.	U1/Sudan	West Nile	3 29 N 30 52 E–4 18 N 31 30 E	Newbould
Kaya Ejon, ln.	K2	Turkana	2 25 N 34 55 E	Rose
[Kayaka] see Kyaka	T1	Bukoba	1 16 S 31 25 E	
[Kayamakago] see Kanyamkago	K5	South Kavirondo	0 57 S 34 31 E	Greenway
Kayanga, popl.	T1	Bukoba	1 32.5 S 31 09 E	Procter
Kayango, hill	U3	Busoga	0 35 N 34 51 E	
Kayanja, popl.	U2	Kigezi	0 05 S 29 46 E	Lock
Kayanja, L.	U4	Masaka	0 17 S 31 52 E	Rose, Lye, Katende, A.S. Thomas

Name	Code	District	Coordinates	Collector
Kayanja, popl.	U4	Mengo	0 21.5 N 32 52.5 E	Dummer
Kayanta, popl.	U2	Ankole	0 33 S 30 17 E	
Kayata, est.	K4	Machakos	1 09 S 37 21 E	Bally, Gardner
[Kayatta] see Kayata	K4	Machakos	1 09 S 37 21 E	
Kayenzi, popl.	T1	Mwanza	2 23 S 33 06 E	Fischer, Stuhlmann
Kayenzi, popl., admin. post	T1	Mwanza	3 16 S 32 36.5 E	
[Kaynam] see Kainam	T2	Mbulu	3 55 S 35 35 E	
[Kayonje] see Kayanja	U4	Masaka	0 17 S 31 52 E	
Kayonza, popl., r.c.	U2	Kigezi	0 55 S 29 42 E	Purseglove
Kayonza, f.r.	U2	Kigezi	c.0 56 S 29 42 E	Numerous
[Kayu] see Ayo	U1/Sudan	West Nile	3 46 N 31 39 E–3 51 N 31 49 E	
Kayugi, popl.	U4	Masaka	0 18 S 31 52 E	Lye
Kayuki	T7	Rungwe	near 9 15 S 33 38.5 E	Carmichael
Kayunga, mssn.	U4	Mengo	0 01.5 N 31 58 E	
[Kazara] see Kajara	U2	Ankole	c.1 00 S 30 10 E	
[Kazeh] see Tabora	T4	Tabora	5 01 S 32 48 E	
[Kazhara] see Kajara	U2	Ankole	c.1 00 S 30 10 E	
[Kazi] see Kaazi	U4	Mengo	0 12.5 N 32 37 E	Snowden
Kazikazi, rsta.	T5	Dodoma	5 34 S 34 12 E	Rose
[Kaziko] see Kanziko	K4	Kitui	1 59 S 38 21 E	B.D. Burtt, Greenway & Polhill, Peter
Kazimzumbwi, f.r.	T6	Uzaramo	c.6 58 S 39 03 E	Procter
Kazimzumbwi, popl.	T6	Uzaramo	6 57 S 39 01 E	
Kazinga Channel	U2	Toro/Ankole	0 12 S 29 53 E–0 04 S 30 09 E	Chandler & Hancock, Eggeling
Kazinga (near Bukoba)	T1	Bukoba	near 1 20 S 31 49 E	Conrads
Kazingo, popl., mssn.	U2	Toro	0 39 N 30 10 E	Maitland, Hazel
Kazita = Kathita, R.	K4	Meru	0 04 S 37 25 E–0 14 S 38 00 E	
Kazita, popl.	T3	Lushoto	5 01 S 38 37 E	Peter
Kazita East, str.	K4	Meru	0 06.5 S 37 20 E–0 04 S 37 25 E	
Kazita West, str.	K4	N. Nyeri/Meru	0 06 S 37 19.5 E–0 04 S 37 25 E	Hanid
[Kazizumbwi] see Kazimzumbwi	T6	Uzaramo		

119

Name		Region	Coordinates	Collector
Kazo, popl.	U2	Ankole	0 03 S 30 45 E	
Kazone, Ras, pt., popl.	T3	Tanga	5 03 S 39 07.5 E	Irwin
Keben, spr.	K3	Uasin Gishu	0 08 N 35 20 E	Dale, Purseglove, Eggeling
Kebisoni, popl., r.c.	U2	Kigezi	0 51 S 30 01 E	Carter & Stannard
Kechilu Pass	K1	Northern Frontier	2 02 N 36 57 E	Tweedie
[Kedam] see Kadam	U1	Karamoja	1 45 N 34 43 E	Dale
Kedong, escarp.	K4	Kiambu	c.1 00–1 15 S 36 34 E	Numerous
Kedong Valley	K3/6	Naivasha/Masai	c.1 00 S 36 30 E–1 30 S 36 25 E	Numerous
Kedowa, rsta.	K5	Kisumu-Londiani	0 13 S 35 34 E	Verdcourt & Fraser Darling, Greenway
Keekorok, game lodge, ln.	K6	Masai	1 36 S 35 14.5 E	A.S. Thomas
[Keem] see Kem	U1	Karamoja	2 41 N 33 47 E	Peter
[Kehangani] see Kichangani	T3	Lushoto	4 51 S 38 47 E	
Kei, popl., mssn.	U1	West Nile	3 37 N 31 07 E	Dale, Eggeling
Kei, mt.	U1	West Nile	3 36 N 31 06 E	
Kei, Mt., f.r., g.r.	U1	West Nile	c.3 40 N 31 07 E	Eggeling
[Keich] see Koich	U1	West Nile	c.3 28 N 31 00 E–3 23 N 31 30 E	Peter
Keili, mt.	T3	Pare	3 47 S 37 35.5 E	Bogdan
[Keita] see Keite	K4	Machakos	c.1 44 S 37 21 E–1 45 S 37 42 E	
Keite, R.	K4	Machakos	c.1 44 S 37 21 E–1 45 S 37 42 E	
Keitherin, peak	Sudan *not* K2		4 47 N 35 16 E	Martin
[Keja] ? see Kaja	K4	Meru	prob. 0 01 S 37 44 E–0 00.5 S 37 49 E	Balbo
Kekonyokie = Olepolos, area	K6	Masai	1 29 S 36 38 E	Verdcourt
Kelele Lake, dam	K3	Laikipia	0 48 N 36 52 E	
Kelema, str.	T5	Kondoa	c.5 08 S 35 47 E	B.D. Burtt, Polhill & Paulo, Richards
[Kelemma] see Kelema	T5	Kondoa	c.5 08 S 35 47 E	Magor
Kelim, R.	U3	Mbale	c.1 22 N 34 47 E–1 34 N 34 16 E	A.S. Thomas
[Kelole] see Kelele	K3	Laikipia	0 48 N 36 52 E	Edwards
Kem, hill	U1	Karamoja	2 41 N 33 47 E	Carmichael
[Kemana] see Kimana	K6	Masai	2 48 S 37 32 E	
Kemesheka (Ruiga R. f.r.)	T1	Biharamulo	c.2 25 S 31 30 E	

Name	Code	Region	Coordinates	Collector
Kemeto, f.r.	K3	Baringo	0 37 N 35 50 E	Richards
Kemosomu = Kimosonu, gorge, str.	T2	Arusha	prob. c.3 14 S 36 50 E	J. Wilson
Kemtaku (Bokora county)	U1	Karamoja	c.2 20 N 34 20 E	Symes
Kenailmet, mine	K2	Turkana	1 59 N 35 04 E	
Kenani, rsta.	K4	Machakos	2 51 S 38 20 E	Drummond & Hemsley, Verdcourt, Scott Elliot
Kendabi, str., area	T2	Mbulu	c.4 25 S 35 18 E	Greenway
[Kendu] see Kindu	K5	South Kavirondo	c.0 21 S 34 39 E	
Kengedi, val.	T8	Lindi	Not traced	
Kengeja, popl., area	P	Pemba	5 25 S 39 44 E	Wigg
Keni, area	T2	Moshi	3 09 S 37 36 E	R.O. Williams
Kenya, mt.	K4	North Nyeri/South Nyeri/Embu/Meru	c.0 09 S 37 19 E	Haarer; Numerous
[Kenyangeti] see Kinyangesi				
Kenyatta, L. = Mukunguya	T7	Mbeya	c.7 53 S 34 45 E	Greenway & Kanuri
[Keo Hill, near Sultan Hamud] ? see Kiou	K7	Lamu	2 25 S 40 41 E	Hooper & Townsend
[Kepembawe] see Kipembawe	K4	Machakos	1 56 S 37 19 E	Bally
[Keragoya] see Kerugoya	T7	Chunya	7 39 S 33 24 E	Richards
Kere, hill	K4	South Nyeri	0 30 S 37 16 E	J. Bally
Kerebi, popl., mine	T1	Bukoba	1 06 S 31 31.5 E	Procter
Kerege, Tsetse Research Unit	K5	Central Kavirondo	0 11 S 34 10 E	Gillett
Kerekenyi, gorge	T6	Bagamoyo	6 34.5 S 39 02 E	S.A. Robertson, Harris
[Kerekese] see Kiregese	T2	Arusha	prob. c.3 14 S 36 50 E	Richards
Kerenge, popl., sisal est.	T6	Uzaramo	prob. c.7 25 S 39 20 E	Paulo
[Keria] see Karia	T3	Lushoto	5 02 S 38 32 E	Numerous
Kericho, distr.	K4	South Nyeri	0 25 S 37 05.5 E	Kerfoot
Kericho, for. stn.	K5	Kericho	c.0 35 S 35 20 E	
Kericho, popl.	K5	Kericho	0 21 S 35 22 E	
Kericho, tea est.	K5	Kericho	0 22 S 35 17 E	
	K5	Kericho	0 24 S 35 15 E	Numerous
Kerichwa Dogo = Kirichwa Ndogo, str.	K4	Nairobi	1 17 S 36 46 E–1 16 S 36 48 E	

Name	Code	Region	Coordinates	Collector
Kerichwa Kubwa, str.	K4	Nairobi	c.1 17 S 36 43.5 E−1 16 S 36 48 E	Kuchar
Kerigodwa, ln.	K6	Masai	c.1 23 S 34 47 E	Kuchar
Keringani, hill	K6	Masai	1 21 S 34 52 E	
[Keringat] see Keringet	K2	West Suk	1 13 N 35 03 E	Tweedie, W.H. Lewis
Keringet, popl., t.c., p.p.	K2	West Suk	1 13 N 35 03 E	Numerous
Kerio, R.	K1/2/3	several	c.0 10 N 35 41 E−2 59 N 36 07 E	R.M. Graham
Kerisoi (near Londiani)	K5	Kisumu-Londiani	near 0 09 S 35 35 E	Verdcourt
Kerita, for. stn.	K4	Kiambu	0 59 S 36 39 E	J. Adamson
[Kerokero] see Korokora	K7	Tana River	c.0 38 S 39 46 E	Greenway
[Kerongwe] see Kirongwe	T6	Rufiji	7 48 S 39 49.5 E	
Kerugoya, popl.	K4	South Nyeri	0 30 S 37 16 E	Glover et al.
Keshemoruo, ln.	K6	Masai	c.1 51 S 35 51 E	Mildbraed
Kesimbili, ln.	U2	Ankole	c.1 00 S 30 44 E	A.S. Thomas
[Kesm] see Kem	U1	Karamoja	2 41 N 33 47 E	Richards
[Keto] see Kito	T4	Ufipa	8 33.5 S 31 29 E	Greenway
[Ketumbaine] see Kitumbeine	T2	Masai	2 45 S 36 12.5 E	Greenway
[Ketumbane] see Kitumbeine	T2	Masai	2 45 S 36 12.5 E	Eggeling
Keyo, hill	U1	Acholi	2 50 N 32 11 E	Tanner, Eggeling
Keza, r.h.	T1	Ngara	2 47 S 30 42 E	Bacon
Khaidapalait (Dodoth county)	U1	Karamoja	Not traced	Grant
[Khali] see Mkhali	T5	Mpwapwa	6 18 S 36 46 E	Hannington
Khambe	T5	Mpwapwa/Dodoma	Not traced	G.H.S. Wood
[Khami] see Kami	U3	Mbale	c.0 29 N 34 13 E−0 35 N 33 59.5 E	
Khansee, L.	T2	Arusha	3 14 S 36 52.5 E	
Khayekhorongole (Turkwell R.)	K1/2	N. Frontier/Turkana	prob. c.1 57 N 35 22 E	Newbould
[K'hutu] see Ukutu	T6	Morogoro/Rufiji	c.7 30 S 37 45 E	Goetze
[Kiadondo] see Kyadondo	U4	Mengo	c.0 25 N 32 35 E	
[Kiaffuma qua Makolo] see Makolo				
Kiaga, popl.	T1	Mwanza	3 04 S 32 45 E	Stuhlmann
Kiagata, r.h.	K4	South Nyeri	0 36 S 37 16 E	
	T1	Musoma	1 39 S 34 07 E	Tanner

Entry	Code	Region	Coordinates	Collector
Kiagu, for.	K4	Meru	0 02 S 37 53 E	
[Kiagwe] see Kyagwe	U4	Mengo	c.0 20 N 32 50 E	
Kiamariga, hill, sw.	K4	South Nyeri	0 23 S 37 14 E	Eggeling
Kiamawa, f.r.	T1	Bukoba	1 21 S 31 43 E	
[Kiamawe] see Kiamawa	T1	Bukoba	1 21 S 31 43 E	Gillman, Wigg, Procter
[Kiambee] see Kiambere	K4	Embu	0 41.5 S 37 48.5 E	
Kiambere, hill	K4	Embu	0 41.5 S 37 48.5 E	Kirrika
Kiambere, t.a.	K4	Embu	c.0 40 S 37 50 E	
Kiambicho, for.	K4	Fort Hall	0 42 S 37 13 E	
Kiambu, distr.	K4	Kiambu	c.1 00 S 36 45 E	
Kiambu, f.r.	K4	Kiambu	c.1 11 S 36 51.5 E	
Kiambu, res.	K4	Kiambu	c.0 50 S 36 50 E	Elmer
Kiambu, popl., t.c.	K4	Kiambu	1 10.5 S 36 50 E	Numerous
[Kiamburu] see Chambura	U2	Ankole	c.0 07 S 30 04 E	Dawe
Kiamweri, peak	K3/4	Naivasha/Fort Hall	0 39 S 36 40 E	
[Kianan] see Kainam	T2	Mbulu	3 55 S 35 35 E	Eggeling
Kiandongoro, for.	K4	South Nyeri	0 27.5 S 36 50 E	Kahurananga & Kibui
[Kiango] see Kinango	K7	Kwale	4 08 S 39 19 E	Greenway
[Kiangole] see Iliangelo	T7	Iringa	prob. c.7 38 S 36 33 E	Carmichael
Kiangombe, mt.	K4	Embu	0 34 S 37 42.5 E	Dyson
Kiangwe, ln.	K7	Lamu	1 56 S 40 58 E	Battiscombe
Kianzabe, area, est.	K4	Machakos	1 10 S 37 19 E	Bally
[Kiao] see Kiau	T1	Bukoba	1 02 S 31 48 E	Gillman
Kiasa, hill	K4	Kitui	2.39.5 S 37 29 E	Hucks
Kiau, l.	T1	Bukoba	1 02 S 31 48 E	
Kibabii, coll.	K5	North Kavirondo	0 37 N 34 31 E	Greenway & Doughty
Kibafuta, popl.	T3	Tanga	5 01 S 39 02 E	Holst
Kibaha, hill	T6	Uzaramo	6 46 S 38 56 E	
Kibaha, popl.	T6	Uzaramo	6 46 S 38 55 E	Peter, Flock
[Kibaia] see Kibaya	T2	Masai	5 17.5 S 36 34 E	Fischer

123

124

Kibembawe, popl.	T6	Rufiji	7 47 S 38 03 E	
Kibembe, str.	U4	Mengo	c.0 29 N 32 41.5 E	Lye
Kibengu, popl.	T7	Iringa	8 16 S 35 42.5 E	Polhill & Paulo
Kibera, rsta.	K4	Nairobi	1 18 S 36 46 E	
Kiberashi, popl.	T3	Handeni	5 23 S 37 26 E	Hornby
Kiberege, popl., r.h.	T6	Ulanga	7 57 S 36 51.5 E	Culwick, Haerdi, Eggeling
[Kiberenge] see Kiverenge	T3	Pare	3 49 S 37 38 E	Peter
[Kibero] see Kibiro	U2	Bunyoro	1 41 N 31 15 E	A.S. Thomas, Dawe
Kibibi, popl.	U3	Busoga	0 32 N 33 07 E	G.H. Wood
Kibibi, I.	U4	Mengo	c.0 06 N 33 02 E	Maitland
Kibibi, popl.	U4	Mengo	0 14 N 32 10 E	
Kibibi North, I.	U4	Masaka	0 14 S 32 18 E	
Kibibi South, I.	U4	Masaka	0 23 S 32 12 E	
Kibigori, rsta., p.p.	K5	Kisumu-Londiani	0 04.5 S 35 03 E	Green
[Kibila] see Kiwira	T7	Rungwe	c.9 05 S 33 42 E–9 37 S 33 57.5 E	Stolz
Kibimba, sw., str.	U3	Busoga	c.0 40 N 33 51 E	G.H. Wood
Kibimjor, hills	K3	Baringo	0 16 N 35 47 E	Beentje
Kibingo, popl.	U2	Ankole	0 36 S 30 22 E	Snowden
Kibingor, popl.	K3	Baringo	0 31 N 35 52 E	Hivernel
Kibiri, str. (Kakamega Forest)	K5	North Kavirondo	c.0 12 N 34 54 E	Faden, Jonsell, Tweedie
Kibiriga, area	U2	Toro	c.0 09 N 29 58 E	A.S. Thomas
Kibiro, salt pans, hot spr.	U2	Bunyoro	1 41 N 31 15 E	A.S. Thomas
Kibiti, hill	T6	Rufiji	7 44 S 38 54 E	
Kibiti, popl.	T6	Rufiji	7 44 S 38 57 E	Shabani
Kibo, peak	T2	Moshi	3 04 S 37 21 E	Numerous
Kiboga, popl., r.h.	U4	Mengo	0 55.5 N 31 46 E	Langdale-Brown
[Kibognoto] see Kibongoto	T2	Moshi	3 12 S 37 06.5 E	Endlich
Kibogwa, peak	T6	Morogoro	7 01 S 37 43 E	Haarer
Kiboi = Elephant's Cave	K3	Trans-Nzoia	1 07 N 34 45 E	
[Kiboko] see Kelim	U3	Mbale	c.1 22 N 34 47 E–1 34 N 34 16 E	Numerous

125

Kiboko, rsta., popl.	K4	Machakos	2 11 S 37 43 E	Bogdan, Tweedie
Kiboko, R.	K4/6	Machakos/Masai	c.2 30 S 37 15 E–2 09 S 37 54 E	Numerous
Kibokwa, Ziwa, sw.	Z	Zanzibar	5 58 S 39 20 E	Vaughan
Kiboma, hill	T7	Iringa	near 8 00 S 36 00 E	Carmichael
Kibonde, ln.	Z	Zanzibar	prob. 6 01 S 39 12 E	Greenway
Kibondo, distr.	T4	Buha	c.4 00 S 31 00 E	
		(considered as part of Buha District for F.T.E.A.)		
Kibondo, popl.	T4	Buha	3 35 S 30 42 E	Numerous
Kibongoto, popl., area	T2	Moshi	3 12 S 37 06.5 E	Endlich
Kiboni, popl.	T7	Iringa	7 55 S 35 50 E	Polhill & Paulo
[Kibonoto] see Kibongoto	T2	Moshi	3 12 S 37 06.5 E	
[Kiborani] see Kibarani	K7	Kilifi	3 37 S 39 51 E	
Kiboriani, mts.	T5	Mpwapwa	6 17 S 36 29.5 E	Numerous
Kibos, rsta., Agr. Dept., t.c.	K5	Kisumu-Londiani	0 04 S 34 48.5 E	Linton
[Kiboscho] see Kibosho	T2	Moshi	3 14 S 37 19 E	Volkens, Uhlig
Kibosek, popl.	K5	Kericho	1 00 S 35 15 E	Trapnell, Glover et al.
Kibosho, hill	T2	Moshi	3 14 S 37 19.5 E	Doughty, Sanders
Kibosho = Kiwoso, popl., mssn.	T2	Moshi	3 14 S 37 19 E	
[Kiboss] see Kibos	K5	Kisumu-Londiani	0 04 S 34 48.5 E	Powell
Kibubu (near Mkuzi Katani)	T3	Pangani	near 5 20 S 38 56 E	Tanner
Kibuguni, area	T3	Tanga	c.4 55 S 39 04 E	Greenway
[Kibuje] see Ibuje	U1	Lango	1 54 N 32 24 E	
[Kibuji] see Ibuje	U1	Lango	1 54 N 32 24 E	
Kibuko, old popl.	T6	Morogoro	6 58 S 37 53 E	
Kibuko, popl.	T6	Morogoro	7 07 S 37 33.5 E	Mgaza, Moreau, Semsei
Kibuku, hill	U2	Toro	0 55 N 30 13 E	Liebenberg
Kibuku, popl.	U3	Mbale	1 03 N 33 48 E	
Kibungu, popl., area	T6	Morogoro	7 12 S 37 47 E	Haarer, Stuhlmann
[Kibungu] see Iwungu	T7	Chunya	8 26 S 32 54 E	Goetze
Kibwangule, hill	T7	Iringa	8 29.5 S 35 02 E	

Kibwele, area	T7	Iringa	8 37 S 35 21 E	Carmichael, Cribb et al.
Kibwendela, popl.	T6	Uzaramo/Morogoro	7 02 S 38 30 E	
Kibwendera = Kibwendela	T6	Uzaramo/Morogoro	7 02 S 38 30 E	Busse
Kibweni, Palace, popl.	Z	Zanzibar	6 07 S 39 13 E	R.M. Davies
Kibwesa, pt.	T4	Mpanda	6 30 S 29 57 E	Juniper et al., Mahinda
[Kibwesi] see Kibwezi	K4	Machakos	2 24.5 S 37 57.5 E	Scheffler, Dummer
Kibwezi, popl., rsta.	K4	Machakos	2 24.5 S 37 57.5 E	Numerous
Kibwezi, R.	K4	Machakos	2 19.5 S 38 07 E–2 23.5 S 38 00 E	Kassner
Kibwezi Forest	K4	Machakos	2 27 S 37 55 E	Timberlake
Kibwo, popl. (near Ikuu)	T4	Mpanda	near 6 58 S 31 10 E	Richards
[Kibwona] ? see Kwebona	Z	Zanzibar	?6 16.5 S 39 22.5 E	Vaughan
Kichangani, popl.	T3	Lushoto	4 51 S 38 47 E	Peter
Kichange, popl.	P	Pemba	5 15 S 39 44 E	Vaughan
[Kichawamba] see Kichwamba	U2	Toro	0 03 N 30 24 E	Maitland
Kichi, hills	T6	Rufiji	c.8 12 S 38 40 E	Vollesen
Kichi, t.a.	T6	Rufiji	c.8 14 S 38 46 E	Brulz & Martin
Kichich (Mathews Range)	K1	Northern Frontier	c.1 01 N 37 15 E	Newbould, Kerfoot
Kichingani, popl.	T3	Lushoto	4 51 S 38 47 E	Peter
[Kichuamba] see Kichwamba	U2	Toro	0 03 N 30 24 E	Greenway & Eggeling, Hazel
Kichuchu, camp	U2	Toro	0 21 N 29 56.5 E	Duke of Abruzzi, Eggeling
[Kichueru] see Kichuiru, area	K4	Kiambu	1 15 S 36 35 E	Deakin
Kichungwani, popl.	P	Pemba	4 57 S 39 43 E	Greenway
Kichwamba, mt.	U2	Toro	0 03 N 30 24 E	Greenway & Eggeling
Kichwamba, hotel	U2	Ankole	0 14 S 30 05.5 E	Osmaston, Eggeling
[Kiciucui] see Kichuchu	U2	Toro	0 21 N 29 56.5 E	
Kidabaga, popl.	T7	Iringa	8 07 S 35 55 E	Bridson
Kidai, ln.	T6	Morogoro	6 57 S 37 48 E	Stuhlmann
Kidamari, ln.	T7	Iringa	7 53 S 35 25 E	Greenway
Kidarege, hill	T7	Iringa	near 7 50 S 36 20 E	Carmichael
Kidareko, popl.	T3	Handeni	5 15 S 38 21 E	

127

Kidatu, popl., rsta.	T6	Ulanga	7 42 S 36 57 E	Thulin, Mhoro, Carmichael
Kidatu, area	T7	Iringa	7 40 S 36 57 E	Mhoro, Batty
Kidatu, dam	T6/7	Kilosa/Iringa	c.7 38 S 36 54 E	
Kidawe, popl., str.	T4	Kigoma	4 54.5 S 29 51 E	Procter
Kideleko, popl., mssn., mt.	T3	Handeni	5 29 S 38 01 E	Numerous
[Kideliko] see Kideleko	T3	Handeni	5 29 S 38 01 E	Faulkner
Kidenge, popl.	T6	Uzaramo	6 44 S 38 56 E	Stuhlmann
Kidepo, Nat. Pk.	U1	Karamoja	c.3 55 N 33 50 E	Harrington
Kidepo, R.	U1/Sudan	Karamoja	c.3 56 N 33 42 E	J. Wilson
Kidera, popl.	U3	Busoga	1 21 N 32 59 E	
Kidero, hills	T2/5	Mbulu/Singida	3 52 S 35 05 E	Richards
Kiderungwa, popl.	T6	Morogoro	7 28 S 37 47 E	Goetze
Kidete, popl., rsta., mssn.	T6	Kilosa	6 38 S 36 42 E	Peter, Stuhlmann, Busse
Kidete, popl.	T7	Iringa	8 12 S 35 43 E	Carmichael
Kidibaga	T7	Iringa	prob. c.7 45 S 35 00 E	Greenway
Kidichi, popl.	Z	Zanzibar	6 05.5 S 39 14 E	Faulkner, Vaughan
Kidisa (Ganze)	K7	Kilifi	c.3 32 S 39 43 E	L.B. Evans
[Kidizi] see Kizi	U2	Bunyoro	c.1 38 N 31 49 E–1 20 N 31 44 E	
Kidodi, popl., mssn., r.h., t.c.	T6	Kilosa	7 36.5 S 36 59.5 E	Eggeling, Goetze, Semsei
[Kidoko, Lager] see Kitogo	T7	Njombe	9 25 S 34 29 E	von Prittwitz & Gaffron
Kidologwai, popl.	T3	Lushoto	4 35 S 38 15 E	Drummond & Hemsley
Kidomole, popl.	T6	Bagamoyo	6 26 S 38 42 E	Semsei
Kidomoni, Ras = English Point	K7	Mombasa	4 03 S 39 41 E	
[Kidondoni] see Kinondoni	T6	Uzaramo	6 47 S 39 16 E	
Kidongo, popl.	U2	Toro	0 46 N 30 03 E	Eggeling
Kidongo, popl.	K7	Kwale	4 19 S 39 22 E	Mwangangi
Kidongole, popl.	U3	Teso	1 16 N 33 58 E	Snowden
Kidopi = Kidope, est.	T7	Iringa	8 37 S 35 16 E	Carmichael
[Kidoti] see Kidodi	T6	Kilosa	7 36.5 S 36 59.5 E	Goetze
Kidoti, popl.	Z	Zanzibar	5 48 S 39 18 E	Hildebrandt, Stuhlmann

Kidudwa, area, hill	T6	Morogoro	c.6 12 S 37 43 E	Hannington
Kidugala, area	T7	Njombe	c.9 35 S 34 41 E	
Kidugala, mt.	T7	Njombe	9 34 S 34 30 E	
Kidugala, mssn.	T7	Njombe	9 07 S 34 32 E	Goetze
Kidugala, mt.	T7	Njombe	9 34 S 34 39.5 E	
Kidugala, mts.	T7	Njombe	9 37 S 34 27 E	
[Kidung] see Kedong	K3/6	Naivasha/Masai	c.1 00 S 36 30 E—1 30 S 36 25 E	
Kidwera, popl.	U2	Bunyoro	1 48 N 32 41 E	Scott Elliot
Kidzi River, popl.	U4	Masaka	0 20 S 31 49 E	Eggeling
[Kiebbe] see Kyebe	U4	Masaka	0 56 S 31 40 E	Norman
Kiejo, mt.	T7	Rungwe	9 13 S 33 47 E	Drummond & Hemsley, A.S. Thomas
Kiemba (NW. Ulugurus)	T6	Morogoro	prob. c.6 54 S 37 39 E	Goetze
Kifaru, est.	T2/3	Moshi/Pare	c.3 31 S 37 32 E	Schlieben
Kieni, area, for.	K4	Kiambu	0 51 S 36 41 E	Bigger
[Kienzema] see Chenzema	T6	Morogoro	7 07 S 37 35 E	Dyson
[Kieyo] see Kiejo	T7	Rungwe	9 13 S 33 47 E	Greenway & Eggeling
[Kifania] see Kifanya	T7	Njombe	9 32 S 35 08 E	Goetze
Kifanya, mssn.	T7	Njombe	9 32 S 35 08 E	Richards
Kifaru = Kilima Kifaru, peak	T3	Pare	3 31 S 37 35 E	Richards, Eggeling, Procter
Kifingi, popl.	T6	Rufiji	7 51 S 39 50 E	Bally
Kifinika, volcano	T2	Moshi	3 10 S 37 29 E	Greenway
Kifu, for.	U4	Mengo	c.0 26 N 32 44 E	Volkens
Kifunfu, hill	U2	Ankole	0 36 S 30 33 E	Harris, Leuchars, Templer
[Kifunga] see Kifanya	T7	Njombe	9 32 S 35 08 E	Pollock
Kifungilo, mssn.	T3	Lushoto	4 47 S 38 22 E	Drummond & Hemsley, Faulkner, Greenway
Kifura, f.r.	T2	Moshi	3 13 S 37 20 E	Semkiwa
Kifura, popl.	T4	Buha	3 48 S 30 42 E	Procter
Kifuru, area	T6	Morogoro	7 03 S 37 41 E	Stuhlmann
Kifyulila	T7	Iringa	near 8 00 S 36 00 E	Carmichael

129

Name	Region	District	Coordinates	Collector
Kigona, for.	U4	Masaka	c.0 44 S 31 44 E	Carmichael
[Kigongo] ? see Kitongo	T1	Mwanza	?2 31 S 33 18 E	Peter
Kigongoi, popl., mt.	T3	Lushoto	4 48 S 38 45 E	Peter
[Kigonoi] see Kigongoi	T3	Lushoto	4 48 S 38 45 E	
Kigonsera, mssn.	T8	Songea	10 48 S 35 04 E	Milne-Redhead & Taylor
Kigonsera, pass	T8	Songea	10 48 S 35 02.5 E	Markham
Kigonsive (near Iringa)	T7	Iringa	near 7 46 S 35 42 E	Goetze
Kigowere = Kigobere, popl.	T6	Kilosa	6 38 S 36 45 E	B.D. Burtt
Kigula (N. Uzungwa Mts.)	T7	Iringa	Not traced	Goetze
Kigulu, co.	U3	Busoga	c.0 45 N 33 30 E	
Kigumba, popl.	U2	Bunyoro	1 49 N 32 00 E	
Kigumu (Kericho Tea Estate), str.	K5	Kericho	c.0 24 S 35 15 E	Pegler
Kigungu, popl.	U4	Mengo	0 02 N 32 25 E	
Kigwa, hill, f.r.	T4	Tabora	5 11 S 33 08 E	Boaler, Ngoundai
Kigwa, popl.	T4	Tabora	5 10 S 33 08 E	Howard, Carmichael
Kigwar, str./s.	K1	Northern Frontier	1 25 N 37 16 E–1 24 N 37 28 E	
Kigwase, hill	T3	Lushoto	4 32 S 38 46 E	Peter
[Kigwasi] see Kigwase	T3	Lushoto	4 32 S 38 46 E	
Kihabule (Busongora co.)	U2	Toro	near 0 06 N 30 03 E	Eggeling
Kihanga, f.r.	T7	Iringa	8 31 S 35 14 E	Procter
Kihanio	?T7	?Ufipa	Not traced	Muenzner
Kihanzi = Kihansi, R.	T6	Ulanga	8 29 S 35 49 E–8 24 S 36 21 E	Carmichael, Nicholson
Kihata, ln.	T7	Iringa	c.8 38 S 35 18 E	Procter
Kihesa, popl.	T7	Iringa	7 43 S 35 44 E	Procter
Kihitu, popl.	T3	Lushoto	4 35 S 38 22 E	Gillman
Kihogosi, popl.	T6	Ulanga	8 07 S 36 35 E	Haerdi
Kihohu (prob. near Lindi)	T8	?Lindi	prob. near 10 00 S 39 43.5 E	Koerner
Kihoka-Kwa-Ibrahim = Kiwugu-Kwa-Brahim = Kivugo, popl.	T6	Bagamoyo	6 25 S 38 26.5 E	Holtz
Kihonda, sisal est.	T6	Morogoro	6 44 S 37 40 E	Semsei

131

[Kihuhi] see Kihuhwi				
[Kihuhui] see Kihuhwi	T3	Tanga	5 13 S 38 41 E	Greenway
Kihuhwi, f.r.	T3	Tanga	c.5 12 S 38 39 E	Moreau, von Prinz
Kihuhwi, rsta.	T3	Tanga	5 13 S 38 41 E	Semsei
Kihuhwi, str.	T3	Tanga	c.5 13 S 38 37 E–5 07 S 38 42 E	Greenway, Grote
Kihuhwi Lands, sisal est.	T3	Tanga	c.5 10 S 38 41 E	Greenway
[Kihuiro] see Kihurio	T3	Pare	4 28 S 38 04 E	
Kihumbati = Chumbati	T8	Lindi	10 26 S 38 50 E	Schlieben
Kihunda, mt.	T3	Pare	c.4 31 S 37 55 E	Greenway
Kihunza, area, popl.	T6	Morogoro	7 08 S 37 42 E	Thulin & Mhoro
[Kihurea] see Kihurio	T3	Pare	4 28 S 38 04 E	Greenway
Kihurio, L.	T3	Pare	4 28 S 38 06 E	Numerous
Kihurio, popl., mkt.	T3	Pare	4 28 S 38 04 E	Numerous
Kihuru, hill	T8	Songea	c.10 55 S 34 56 E	Milne-Redhead & Taylor
[Kii] see Kiye	K4	Embu	0 19 S 37 24 E–0 30 S 37 26 E	Fries, D. Davis
Kijabe, popl.	K4	Kiambu	0 56 S 36 34.5 E	Numerous
Kijabe, rsta.	K4	Kiambu	0 55 S 36 35 E	
Kijaguza, popl.	U4	Mubende	0 56 N 31 10 E	
Kijanebalola, L.	U4	Masaka	c.0 43 S 31 19 E	
Kijango, old area	T3	Lushoto	c.4 56 S 38 37 E	Peter, Greenway
Kijangwani, popl.	K7	Kilifi	3 47 S 39 50 E	Frazier
[Kijauk] see Kijaur	K5	S. Kavirondo/Kericho	0 46 S 34 59 E	
Kijaur, hill	K5	S. Kavirondo/Kericho	0 46 S 34 59 E	
Kijaur, popl., sch.	K5	S. Kavirondo/Kericho	0 46 S 34 58.5 E	Bogdan
Kijawe, str.	T6	Rufiji	c.8 11 S 38 30 E	Ludanga
Kijegge, hill, for.	K4	Meru	c.0 16 S 37 57 E	Smart
Kijibweni, ln.	T3	Tanga	5 07 S 39 07 E	Faulkner
[Kijinge] see Kifingi	T6	Rufiji	7 51 S 39 50 E	Greenway
[Kijojo] see Kigogo	T7	Iringa	c.8 35 S 35 14 E–8 43 S 35 18 E	R.M. Davies
Kijomba, ridge	U2	Toro	0 07 N 29 48 E	Eggeling

Name		District	Coordinates	Authority
Kijomoro, popl.	U1	West Nile	3 10 N 30 54 E	A.S. Thomas
Kijude, est.	U4	Mengo	0 14 N 32 53 E	Dummer
Kijunga	Z	Zanzibar	Not traced	Greenway
Kijungu, popl.	T2	Masai	5 22 S 37 11 E	Vesey-FitzGerald, Procter, Leippert
Kijunjubwa, popl.	U2	Bunyoro	1 29 N 31 46 E	Eggeling
Kijura, popl.	U2	Toro	0 49 N 30 25 E	
Kikafu, bdg.	T2	Moshi	3 19 S 37 13 E	Huxley, Greenway
Kikagati, popl., mssn.	U2	Ankole	1 02.5 S 30 40 E	A.S. Thomas, Eggeling, Dale, Purseglove
Kikamba, str.	T7	Chunya	7 38 S 33 08 E–8 10 S 32 37 E	
Kikambala, popl.	K7	Kilifi	3 54 S 39 47 E	Birch, Irwin, Bally
[Kikambo] see Kikombo	T5	Mpwapwa		B.D. Burtt
[Kikambo] see Kikamba	T7	Chunya	7 38 S 33 08 E–8 10 S 32 37 E	Sanane
Kikandwa, popl.	U4	Mengo	0 28 N 32 47 E	
[Kikeri] see Kikore	T5	Kondoa	4 20 S 35 50.5 E	B.D. Burtt
[Kikeri] see Kikore	T2/5	Mbulu/Kondoa	c.4 20 S 35 50 E	
Kikingo, for.	K4	Meru	0 05 S 38 06 E	
Kikoboga, area, lodge	T6	Kilosa	7 22.5 S 37 18 E	ole Sayale, Procter
Kikoboga, R.	T6	Kilosa	7 23 S 37 12 E–7 17 S 37 06 E	Wingfield et al.
[Kikobola] see Kikoboga	T6	Kilosa	7 23 S 37 12 E–7 17 S 37 06 E	
Kikoga, popl.	T7	Iringa	7 25 S 34 59 E	
Kikoka, f.r.	T6	Bagamoyo	6 28 S 38 44 E	Mgaza, Semsei, Procter
Kikoko, popl.	K4	Machakos	1 48.5 S 37 24 E	Mwangangi
[Kikola] see Kikolo	T6	Kilosa	6 54 S 36 52 E–6 54 S 36 58.5 E	Swynnerton
[Kikole] see Kikolo	T6	Kilosa	6 54 S 36 52 E–6 54 S 36 58.5 E	Swynnerton
Kikolo, str.	T6	Kilosa	6 54 S 36 52 E–6 54 S 36 58.5 E	
Kikoma, popl.	U4	Mengo	0 40 N 32 54 E	Dummer
Kikoma, hill	T1	Biharamulo	c.2 11 S 31 38 E	Procter
Kikombo, popl.	T5	Mpwapwa	6 19 S 36 30 E	Busse
Kikombo, str.	T5	Mpwapwa	6 18 S 36 30 E–6 29 S 36 27 E	B.D. Burtt
Kikondo, hill	T7	Mbeya	c.8 59 S 33 37.5 E	Hooper & Townsend

133

Name	Code	Place	Coordinates	Collector
Kikondo, popl.	T7	Njombe	c.9 02 S 33 49 E	Richards, Cribb & Grey-Wilson
Kikoneni, popl.	K7	Kwale	4 27 S 39 19 E	Moomaw, Makin
Kikonge, popl.	U4	Mengo	0 24 N 32 12 E	
Kikongo, fy.	T1	Mwanza	2 43 S 32 53.5 E	Verdcourt
Kikongoro, f.r.	T1	Bukoba	1 11 S 31 43 E	Procter
[Kikonko] see Kikongo	T1	Mwanza	2 43 S 32 53.5 E	
Kikore, escarp.	T2/5	Mbulu/Kondoa	c.4 20 S 35 50 E	B.D. Burtt, J.D. Scott
Kikore, mt.	T5	Kondoa	4 48 S 35 40 E	
Kikore, popl., area	T5	Kondoa	4 20 S 35 50.5 E	B.D. Burtt
[Kikori] see Kikore	T2/5	Mbulu/Kondoa	c.4 20 S 35 50 E	B.D. Burtt
Kikorongo, L.	U2	Toro	0 01 S 30 01 E	Eggeling
Kikorongo, popl., r.h.	U2	Toro	0 00 30 00 E	Maitland
Kikubamiti (near Kampala)	U4	Mengo	c.0 17 N 32 39 E	Chandler
Kikube, popl.	U2	Bunyoro	1 21 N 31 14 E	
[Kikuju] see Kikuyu	K4	Kiambu	c.0 55 S 36 40 E	Fischer
Kikuko Hill	T7	Iringa	prob. c.7 40 S 35 00 E	Greenway, Richards
Kikuletwa, R.	T2	Masai/Arusha	c.3 18 S 36 47 E–3 31 S 37 18 E	Leippert
Kikuli = Kikulu	T6	Uzaramo	Not traced	Stuhlmann
Kikumbi (N. of Mashewa)	T3	Lushoto	prob. c.4 42 S 38 40 E	Peter
Kikumbuliyu	K4	Machakos	c.2 37 S 38 02 E	Scott Elliot, Gregory
[Kikundi] ? see Kitunda	T4	Tabora	?6 48 S 33 13 E	von Prittwitz & Gaffron
		(Kitunda is shown on the von Prittwitz & Gaffron route)		
Kikundi, f.r.	T6	Morogoro	c.6 52 S 37 55 E	E.M. Bruce, Busse
Kikundi, popl.	T6	Morogoro	6 52 S 37 53 E	von Prittwitz
Kikundi	T			
Kikuni, popl.	T6	Rufiji	7 47 S 39 52 E	Greenway
Kikuru, f.r.	T1	Bukoba	1 04 S 31 38 E	Gillman, Procter
Kikurungu, hill	T6	Morogoro	7 09 S 37 46 E	Stuhlmann
Kikutani, popl.	T6	Rufiji	7 53.5 S 39 51 E	Greenway
Kikuyu, t.c., rsta.	K4	Kiambu	1 15 S 36 39.5 E	Numerous

Kikuyu, popl.	T5	Dodoma	6 52 S 35 05 E	Polhill & Paulo
Kikuyu Escarpment, for.	K4	Kiambu	c.0 55 S 36 40 E	Numerous
[Kikuyuni] see Kakuyuni	K7	Kilifi	3 13 S 39 59 E	Moomaw, Verdcourt
[Kikwe] see Tikwe	T7	Chunya	c.8 02 S 33 15 E	
Kilaguni, lodge	K6	Masai	2 54 S 38 04 E	
Kilak, co.	U1	Acholi	c.3 00 N 32 00 E	Eggeling, A.S. Thomas
Kilak, f.r.	U1	Acholi	c.3 05 N 31 58 E	Eggeling, A.S. Thomas
Kilak, mt.	U1	Acholi	3 02 N 31 58 E	
Kilala, for.	K4	Machakos	1 46 S 37 32 E	
Kilambo = Kalambo, hot springs	T7	Rungwe	9 23 S 33 50 E	Stolz
Kilambo = Kalambo, hot springs	T7	Rungwe	9 33.5 S 33 49 E	
Kilambo = Kirambo, str.	T7	Rungwe	c.9 32 S 33 57 E	Stolz
Kilambo = Kirambo, sw., for.	T7	Rungwe	c.9 32 S 33 57 E	
Kilando, ln.	T4	Kigoma	c.5 37 S 29 52 E	Carmichael
Kilanga = Kwakilanga, l.	T3	Pangani	5 18.5 S 38 38 E	
Kilangali, popl.	T6	Kilosa	6 58 S 37 07 E	Fuggles Couchman
Kilasi, str.	T7	Rungwe	c.9 10 S 33 36 E	Stolz, Goetze
Kileleshwa, est.	K4	Nairobi	1 17 S 36 47.5 E	Greenway
Kilema, popl., area	T2	Moshi	3 19 S 37 30.5 E	Bally, Geilinger, Volkens
Kilembe, popl.	U2	Toro	0 12 N 30 01 E	G. Taylor, A.S. Thomas
Kilemele, L. (Lwengera valley)	T3	Lushoto	prob. 4 53 S 38 35 E	Peter
Kilengwe, popl.	T6	Morogoro	7 31 S 37 33 E	Goetze
Kilgoris, mssn., mkt., t.c.	K6	Masai	1 00 S 34 53 E	Glover et al.
Kilifi, distr.	K7	Kilifi	c.3 10 S 39 40 E	
Kilifi, popl.	K7	Kilifi	3 38 S 39 51 E	Numerous
Kilima, est.	T7	Iringa	8 36 S 35 21 E	Cribb & Grey-Wilson
Kilimafedha = Kilimafeza	T1	Musoma	2 16 S 34 55 E	
[Kilima fetha] see Kilimafedha	T1	Musoma	2 16 S 34 55 E	Greenway
Kilimafeza, mine, popl.	T1	Musoma	2 16 S 34 55 E	Greenway & Turner
[Kilimagai] see Kimigai	T5	Mpwapwa	6 29 S 36 29 E	

135

Name	Code	Region	Coordinates	Collectors
Kilinga, for.	T2	Arusha	c.3 16 S 36 50 E	Richards
Kilingeni, popl.	P	Pemba	4 56.5 S 39 43 E	Greenway
Kilkoris = Kilgoris, mssn., mkt., t.c.	K6	Masai	1 00 S 34 53 E	
Kiloli, popl.	T4	Tabora	6 50 S 33 23.5 E	Carmichael
Kilolo, popl.	T7	Iringa	8 00 S 35 51 E	Polhill & Paulo
Kilombero, R.	T6	Ulanga	c.8 24 S 36 21 E–8 31 S 37 21 E	Anderson, Ede, Schlieben, Nicholson
Kilombero, fy.	T6	Ulanga	8 11 S 36 42 E	Cribb & Grey-Wilson
Kilombero Scarp = West Kilombero Scarp, f.r.	T6/7	Ulanga/Iringa	c.7 55 S 36 30 E	Harris & Pócs
Kilomeni, popl., mssn.	T3	Pare	3 46 S 37 39 E	Peter, Haarer
[Kilomoni] see Kiomoni				
Kilosa, distr.	T3	Tanga	5 04 S 39 02.5 E	Peter
Kilosa, popl., rsta., mssn., r.h.	T6	Kilosa	c.6 50 S 37 00 E	
Kilulu, sisal est., hill	T6	Kilosa	6 50 S 36 59 E	Numerous
Kilumbi, popl.	T3	Tanga	4 47 S 39 08 E	Mohammed, Greenway
Kilunda, area, road	T4	Tabora	6 17 S 33 51 E	Eggeling
Kilunda, w.h.	T8	Kilwa	c.8 35 S 38 29 E	Nicholson
Kilungu, for.	T8	Kilwa	8 31 S 38 30 E	
Kilungu, p.p.	K4	Machakos	c.1 47 S 37 20 E	Mwangangi
Kilupa (near Mwera)	K4	Machakos	1 47 S 37 21 E	Mwangangi
[Kilwa] see Kirwa	T3	Pangani	near 5 30 S 38 57 E	Tanner
Kilwa, distr.	U2	Kigezi	1 13.5 S 29 37.5 E	Eggeling
	T8	Kilwa	c.9 20 S 38 15 E	Kirk
(this includes much of Nachingwea Distr. for F.T.E.A.)				
Kilwa Kisiwani, I.	T8	Kilwa	8 59 S 39 32 E	Braun
Kilwa Kisiwani, popl.	T8	Kilwa	8 57 S 39 31 E	Kraenzlin
Kilwa Kivinje, popl., port	T8	Kilwa	8 45 S 39 24 E	Milne-Redhead & Taylor
[Kilwa Kiwindje] see Kilwa Kivinje	T8	Kilwa	8 45 S 39 24 E	Busse
Kilwani Pond	T6	Uzaramo	6 51 S 39 14 E	Wingfield
Kima, rsta.	K4	Machakos	1 57.5 S 37 15 E	Napier, Tweedie
[Kimagai] see Kimigai	T5	Mpwapwa	6 29 S 36 29 E	B.D. Burtt, Greenway, Hornby

137

Name		District	Coordinates	Authority
[Kimagi] see Kimigai	T5	Mpwapwa	6 29 S 36 29 E	Greenway
[Kimagoi] see Kimigai	T5	Mpwapwa	6 29 S 36 29 E	Hornby
Kimaka, hill, f.r.	U3	Busoga	0 27.5 N 33 11 E	G.H.S. Wood
Kimakia, f.r., for. stn.	K4	Fort Hall	c.0 46 S 36 44 E	Numerous
Kimakia, str.	K4	Fort Hall	0 45 S 36 43 E–0 54 S 36 53 E	
[Kima-Kombo] see Kinakomba	K7	Tana River	c.1 42 S 40 07 E	Leroy
[Kimakow] see Kimakowa	T2	Masai	c.2 41 S 36 43 E–2 40 S 36 40 E	Carmichael
Kimakowa, str.	T2	Masai	c.2 41 S 36 43 E–2 40 S 36 40 E	Carmichael
Kimamba, popl., rsta.	T6	Kilosa	6 47 S 37 08 E	Numerous
Kimamba, str.	T6	Kilosa	6 40 S 37 15 E–6 50 S 37 00 E	
[Kimambira] see Kimamba	T6	Kilosa	6 47 S 37 08 E	Stuhlmann
[Kimambiro] see Kimamba	T6	Kilosa	6 47 S 37 08 E	
Kimana, game sanctuary	K6	Masai	c.2 40 S 37 32 E	
Kimana, t.c.	K6	Masai	2 48 S 37 32 E	
[Kimangalia] see Kimengelia	K6/T2	Masai/Moshi	c.2 56 S 37 28 E	C.G. Rogers
Kimangaw, popl., for.	K4	Kitui	0 31 S 38 08 E	R.M. Graham
[Kimangelia] see Kimengelia	K6/T2	Masai/Moshi	c.2 56 S 37 28 E	
[Kimango] ? = Kimangau	?K4	?Kitui	?c.0 31 S 38 08 E	Wingfield
Kimani, str.	T6	Uzaramo	6 55 S 39 03 E–6 53 S 39 16 E	Eggeling, Cribb & Grey-Wilson, Procter
Kimani, R.	T7	Njombe/Mbeya	c.9 06 S 34 16 E–8 42 S 34 12 E	Procter
Kimani Bridge	T7	Mbeya/Njombe	8 46 S 34 10 E	Vaughan
Kimara, popl., area	Z	Zanzibar	6 10 S 39 15 E	Milne-Redhead & Taylor, Richards
Kimarampaka, str., marsh	T8	Songea	c.10 41 S 35 33 E	Greenway
Kimba (SW. Ngorongoro Crater Rim)	T2	Masai	c.3 15 S 35 30 E	Semsei
Kimbinyuko (Bunduki)	T6	Morogoro	c.7 02 S 37 38 E	Cribb & Grey-Wilson, E.M. Bruce
[Kimbosa] see Kimboza	T6	Morogoro	c.7 00 S 37 48 E	
Kimboza, f.r.	T6	Morogoro	c.7 00 S 37 48 E	Numerous
Kimbwi = Kimbwe, R.	T6/7	Ulanga/Iringa	8 24.5 S 35 46 E–8 51.5 S 35 40 E	Carmichael
[Kimea] see Kimeya	T1	Bukoba	2 04 S 31 34 E	
Kimengelia, str.	K6/T2	Masai/Moshi	c.2 56 S 37 28 E	C.G. Rogers

Name		Region	Coordinates	Collector
Kimengo, popl.	U2	Bunyoro	1 34 N 31 58 E	B.J. Turner
Kimereshi Corridor	T1	Musoma	c.2 20 S 34 42 E	Greenway
Kimeya, ln.	T1	Bukoba	2 04 S 31 34 E	Ukiriguru Res. Stn.
Kimigai, sw.	T5	Mpwapwa	6 29 S 36 29 E	Hornby
Kimilili, ln.	K5	North Kavirondo	0 47 N 34 43 E	Tweedie
Kiminini, area	K3	Trans-Nzoia	0 53.5 N 34 55 E	Tweedie
Kiminini, popl.	K3	Trans-Nzoia	0 51 N 34 55.5 E	Tweedie
[Kimiramatonga] see Kimiramatonge	T7	Iringa	7 36 S 34 54 E	
Kimiramatonge, hill	T7	Iringa	7 36 S 34 54 E	Richards
[Kimirimatonge] see Kimiramatonge	T7	Iringa	7 36 S 34 54 E	Richards, Greenway
Kimiwo, str.	K3	Elgeyo	1 10 N 35 27 E–1 07 N 35 27.5 E	Townsend
[Kimoani] see Kimwani	T1	Biharamulo	c.2 11 S 31 40 E	
[Kimogai] see Kimigai	T5	Mpwapwa	6 29 S 36 29 E	
Kimondi, R.	K3	Nandi	c.0 17 N 35 05 E–0 08.5 N 35 00 E	Stuhlmann
Kimosomu = Kemosomu, gorge, str.	T2	Arusha	prob. c.3 14 S 36 50 E	Hornby
Kimothon, str.	K3	Trans-Nzoia	1 07.5 N 34 36.5 E–1 05 N 34 50.5 E	Tweedie
Kimugu, est.	K5	Kericho	0 21 S 35 19 E	Richards
Kimugung, str.	K5	Kericho	c.0 25 S 35 14 E	Tweedie
[Kimunya] see Kimunye	K4	South Nyeri	0 26 S 37 18 E	
Kimunye, tea factory	K4	South Nyeri	0 26 S 37 18 E	Perdue & Kibuwa
Kimwani, popl., old area	T1	Biharamulo	c.2 11 S 31 40 E	Bamps
Kimwarer, str., popl.	K3	Elgeyo	0 19 N 35 38 E	Gilbert & Thulin
[Kinaba] see Kanaba	U2	Kigezi	1 15 S 29 47.5 E	Procter, Stuhlmann
Kinakomba, area	K7	Tana River	c.1 42 S 40 07 E	Bally
[Kinale] see Kinari	K4	Kiambu	c.0 54 S 36 39 E	Chandler & Hancock, Purseglove
Kinambo, popl.	T4	Ufipa	8 08 S 31 58 E	Gardner, Greenway, Verdcourt
Kinandia, L., sw.	T2	Arusha	3 14 S 36 54 E	Michelmore
[Kinange] see Kinango	K7	Kwale	4 08 S 39 19 E	
Kinango, popl.	K7	Kwale	4 08 S 39 19 E	Richards, Greenway
Kinango (Lwengera valley)	T3	Lushoto	prob. c.4 55 S 38 36 E	Numerous

139

Name		Region	Location	Coordinates	Authority
Kinangop, peak	K3/4	Naivasha/Fort Hall/S.Nyeri	0 37.5 S 36 42.5 E	Numerous
Kinangop, plateau	K3	Naivasha	0 30—0 53 S 36 30—36 37 E	Numerous
[Kinani] see Kenani	K4	Machakos	2 51 S 38 20 E	
[Kinani R.] see Kimani	T7	Njombe/Mbeya	c.9 06 S 34 16 E—8 42 S 34 12 E	Brenan & Greenway
Kinanira, popl., mssn.	U2	Kigezi	1 11 S 29 36 E	
Kinaniri = Kinanili, popl.	T7	Iringa	8 06 S 35 59 E	Carmichael
[Kinankop] see Kinangop	K3	Naivasha	0 30—0 53 S 36 30—36 37 E	Greenway
[Kinanura] see Kinanira	U2	Kigezi	1 11 S 29 36 E	Linder
Kinaoni = Kinuni, popl., area	Z	Zanzibar	6 10 S 39 15 E	Vaughan, Greenway
Kinari, t.c., for.	K4	Kiambu	0 54 S 36 39 E	Greenway
Kinazini, popl.	P	Pemba	5 01 S 39 47.5 E	Greenway, R.O. Williams
Kinda, ln.	T6	Morogoro	6 09 S 37 28 E	Schlieben
Kindani, R., camp	K4	Meru	0 11.5 N 38 03 E	Ament & Magogo
Kindaruma, dam	K4	Embu/Kitui	c.0 48 S 37 48 E	Gillett & Faden
Kindaruma, hill	K4	Embu	0 48 S 37 48 E	
Kindimba Juu, hill	T8	Songea	10 42 S 34 48 E	Hay
Kindoroko, peak, f.r.	T3	Pare	3 44.5 S 37 39 E	Herring, Carmichael
Kindu, bay	K5	South Kavirondo	c.0 21 S 34 39 E	
Kindu, popl.	T6	Ulanga	7 54 S 37 41 E	
Kindu Bay, popl., t.c.	K5	South Kavirondo	0 21.5 S 34 39 E	
Kinesi Point	T1	North Mara	1 27.5 S 33 51 E	van Rensburg
Kineti, R.	Sudan *not* U1		c.4 02 N 32 47 E	A.S. Thomas
[Kinga] see Ukinga	T7	Njombe	c.9 10 S 34 15 E	Goetze
[Kingani] see Ruvu (lower reaches)	T6	Bagamoyo		Kirk, Stuhlmann
Kingarane, R.	T2	Masai	c.2 06 S 35 37 E	Carmichael
Kingerengere, str.	T6	Morogoro	7 03 S 37 50 E—7 01 S 37 54 E	
Kingika = Kinjika Diude, mt.	T7	Njombe	9 13 S 34 02 E	Goetze
Kingo, popl.	T3	Lushoto	4 49.5 S 38 22.5 E	Braun
Kingobo, mt.	T7	Njombe	9 28 S 34 04 E	Goetze
Kingolwira, rsta.	T6	Morogoro	6 47 S 37 45.5 E	B.D. Burtt, Welch

[Kingoni] see Ruvu (lower reaches)	T6	Bagamoyo	0 18 N 32 33.5 E	Hildebrandt
[King's Lake] see Kabaka's Lake	U4	Mengo	c.4 45 S 38 20 E	Hancock & Chandler
Kinguelo (W. Usambaras)	T3	Lushoto	8 30 S 38 34 E	Greenway
Kingupira, for.	T8	Kilwa	8 29 S 38 32.5 E	Rodgers
Kingupira, res.stn., Selous Game Reserve	T8	Kilwa	8 28 S 38 33 E	Nicholson, Rodgers, Ludanga, Vollesen
Kingupira, w.h.	T8	Kilwa	3 14 S 36 54 E	Vollesen
[Kiniandia] see Kinandia	T2	Arusha	4 45 S 36 03 E	Greenway & Kanuri
[Kiniasse] see Kinyassi	T5	Kondoa	0 48 S 36 16 E	Townsend
Kinja Nurseries, est.	K3	Naivasha	9 13 S 34 02 E	Goetze
Kinjika Diude, mt.	T7	Njombe	9 28 S 34 04 E	
Kinjobo = Kingobo, mt.	T7	Njombe	c.0 45 S 29 45 E	Norman
[Kinkisi] see Kinkizi	U2	Kigezi	c.0 45 S 29 45 E	Purseglove
Kinkizi, co.	U2	Kigezi	0 20 N 38 08 E–0 13 N 38 18 E	J. Adamson, Hamilton
Kinna, str./s.	K4	Meru	0 20 N 38 10.5 E	Verdcourt
Kinna, hill	K4	Meru	0 19 N 38 12.5 E	Adamson
Kinna, mkt.	K4	Meru	6 53 S 37 45 E	Numerous
Kinole, ridge	T6	Morogoro	6 47 S 39 16 E	Vaughan, Procter
Kinondoni, popl.	T6	Uzaramo	0 39 S 30 28 E	
Kinoni, popl., r.h., t.c.	U2	Ankole	0 06 S 31 04 E	
Kinoni, hill, popl., mssn.	U4	Masaka	0 31 S 31 15 E	Sangster
Kinoni, hill	U4	Masaka	Not traced	Geilinger
Kinumwe (Umba steppe)	T3	Tanga	1 30 S 31 06.5 E	Stuhlmann
Kinuni, popl.	T1	Bukoba	6 10 S 39 15 E	
Kinuni, popl., area	Z	Zanzibar	1 38 N 31 35 E	Liebenberg
Kinyala, popl.	U2	Bunyoro	c.0 57 N 36 01 E	
Kinyang, area, str.	K3	Baringo	c.7 53 S 34 45 E	Richards
Kinyangesi, area	T7	Mbeya	c.7 53 S 34 45 E	Greenway
[Kinyangeti] see Kinyangesi	T7	Mbeya	7 36 S 34 50 E–7 39 S 34 53 E	Greenway, Richards
Kinyantupa, str., area	T7	Iringa	c.7 33 S 34 53 E	Richards
Kinyantupa, escarp.	T7	Iringa		

141

142

Kipangani, popl.	P	Pemba	4 58 S 39 43 E	Bullock, Richards, Nicholson
Kipangani, popl.	P	Pemba	5 07 S 39 42 E	Nicholson
[Kipangati] see Chipangati	T4	Mpanda	7 29 S 31 53 E	
Kipangati (Selous Game Reserve)	T6	Ulanga	c.9 00 S 37 30 E	
Kipange, popl.	P	Pemba	4 57 S 39 44 E	Greenway
Kiparbara = Kibarbara, ln.	T2	Masai	5 15 S 37 20 E	Jaeger
Kipassa, popl.	T7	Rungwe	9 15.5 S 33 41.5 E	
Kipatimu, mssn.	T8	Kilwa	8 30 S 38 56 E	Meyer
Kipayo, est.	U4	Mengo	0 15 N 32 46 E	Dummer
[Kipegere] see Kiberege	T6	Ulanga	7 57 S 36 51.5 E	Cribb & Grey-Wilson
[Kipela] see Kipera	T6	Kilosa	6 57 S 36 56 E	Swynnerton
Kipembawe, popl., game camp	T7	Chunya	7 39 S 33 24 E	Numerous
Kipembe = Kipemba, str.	T7	Iringa	8 36 S 35 40 E–8 40 S 35 36 E	Carmichael
Kipendale	T8	Tunduru	Not traced	Tanner
[Kipengate] see Chipangati	T4	Mpanda	7 29 S 31 53 E	
Kipengere, mt. range	T7	Njombe	c.9 10 S 34 15 E	Goetze, Richards, Stolz
Kipera, ln.	T6	Kilosa	6 57 S 36 56 E	B.D. Burtt, Swynnerton
Kiperu, tea est., area	T7	Iringa	8 36.5 S 35 18 E	Polhill & Paulo
Kipili, popl., r.h.	T4	Ufipa	7 26.5 S 30 36 E	Bullock, B.D. Burtt, Procter
Kipini, popl., p.p.	K7	Lamu	2 31.5 S 40 31.5 E	Greenway & Rawlins, Sampson
Kipipiri, mt., for.	K3	Naivasha	c.0 26 S 36 32 E	Numerous
Kipipiri, popl.	K3	Naivasha	0 27 S 36 29.5 E	
Kipiri, w.h. (Mikumi Nat. Park)	T6	Kilosa	Not traced	Greenway & Kanuri
Kipkabus, rsta.	K3	Uasin Gishu	0 18 N 35 30 E	Numerous
Kipkarren, R.	K3	Uasin Gishu/Nandi	0 25 N 35 20 E–0 39 N 34 50.5 E	Dale, W.H. Lewis, Brodhurst Hill
Kipkarren River, rsta.	K3	Uasin Gishu	0 37 N 34 58.5 E	Brodhurst Hill, Dale, Napier
Kipkirima Farm	K3	Trans-Nzoia	0 51 N 34 49 E	Tweedie
Kipkulkul, est.	K3	Trans-Nzoia	0 58 N 34 46.5 E	Tweedie
Kipkunurr, for.	K3	Elgeyo	c.1 07 N 35 33 E	Townsend
Kipkunurr, peak	K3	Elgeyo	1 01 N 35 30.5 E	

143

Name		Region	Coordinates	Collector
Kipleleo, mt.	K6	Masai	1 06 S 35 16 E	Glover et al.
[Kipogoro] see Chipogolo	T5	Mpwapwa	6 52 S 36 02 E	Polhill & Paulo
Kipogoro (NW. Ulugurus)	T6	Morogoro	prob. c.6 54 S 37 39 E	Schlieben
Kipompo, sw. (near Kinambo)	T4	Ufipa	near 8 08 S 31 58 E	Michelmore
Kiponzero = Kiponzelo, popl.	T7	Iringa	7 58 S 35 22 E	Childs
[Kips] see Kipsing	K1/4	N. Frontier/N. Nyeri	0 29 N 37 07 E–0 38 N 37 20.5 E	
Kipsain, R.	K3	Trans-Nzoia	c.1 08 N 35 06 E	Tweedie
Kipsait, peak	K3	Elgeyo	1 08 N 35 21.5 E	Townsend
Kipsigis Land Unit, area	K5	Kericho	c.0 32 S 35 10 E	
Kipsing, R.	K1/4	N. Frontier/N. Nyeri	0 29 N 37 07 E–0 38 N 37 20.5 E	
Kipsonoi, mkt.	K5	Kericho	0 42 S 35 13 E	Gray, Honoré
Kipsonoi, R.	K5/6	Kericho/Masai	0 30.5 S 35 34 E–0 42.5 S 35 06 E	Royston
Kiptaberr, peak	K3	Elgeyo	1 07 N 35 19 E	
Kiptaber-Kapkanyar, f.r.	K3	Elgeyo	c.1 10 N 35 13 E	Tweedie
Kipterit = Kimugung, str.	K5	Kericho	c.0 25 S 35 14 E	Pegler, Stewart, Copley
Kiptiget, str.	K5	Kericho	c.0 33 S 35 14 E	Tweedie
Kiptogot Bridge	K3	Trans-Nzoia	1 09 N 34 46 E	Tweedie
Kiptogot Waterfall	K3	Trans-Nzoia	1 09 N 34 44 E	Tweedie
Kiptogot Weir	K3	Trans-Nzoia	1 09 N 34 50 E	Tweedie
Kiptoi Farm	K3	Trans-Nzoia	1 00 N 35 14.5 E	
Kiptuiya, ln.	K3	Nandi	0 12 N 35 00 E	Tallantire
[Kipumbwe] see Kipumbwi	T3	Pangani	5 38 S 38 53.5 E	Numerous
Kipumbwi, pop., r.h.	T3	Pangani	5 38 S 38 53.5 E	
[Kipumpvi] see Kipumbwi	T3	Pangani	5 38 S 38 53.5 E	
[Kipumpwi] see Kipumbwi	T3	Pangani	5 38 S 38 53.5 E	
Kipundi, mt.	T7	Iringa	8 04 S 35 22 E	Goetze
[Kipunga] see Kipungu	T8	Kilwa	8 43 S 39 21 E	Busse
Kipungu, hill, old popl.	T8	Kilwa	8 43 S 39 21 E	
[Kiputaland] see Kiputa's	T7	Rungwe	9 19 S 33 35 E	Stolz
Kiputa's, popl.	T7	Rungwe	9 19 S 33 35 E	

144

[Kiquarr] see Kigwar			
[Kirahura] see Kiruhura			
Kirala, f.r.	K1	Northern Frontier	c.1 25 N 37 16 E−1 24 N 37 28 E	J. Bally
Kirambo, popl.	U2	Ankole	0 13 S 30 51 E	Harker
Kirambo = Kilambo, str.	U4	Masaka	c.0 37 S 31 46.5 E	Bunzinya
Kirambo = Kilambo, sw., for.	T4	Ufipa	8 15.5 S 31 00 E	Richards
Kirando, popl.	T7	Rungwe	c.9 32 S 33 57 E	
Kirangi, hill	T7	Rungwe	c.9 32 S 33 57 E	
Kirangwona	T4	Ufipa	7 25.5 S 30 36 E	B.D. Burtt
Kirao, popl.	T3	Pare	3 59 S 37 43 E	
Kirau = Kirao, popl.	T7	Iringa	Not traced	von Prittwitz & Gaffron
Kirawira, guard post	K7	Kwale	4 35.5 S 39 22 E	R.M. Graham
Kiregese, f.r.	K7	Kwale	4 35.5 S 39 22 E	R.M. Graham
Kireka, popl., mssn., hill	T1	Musoma	2 10 S 34 09 E	Greenway
[Kirema] see Kirerema	T6	Uzaramo	prob. c.7 25 S 39 20 E	
[Kiremia] see Kirimia	U4	Mengo	0 20 N 32 39 E	Eggeling
[Kirenguë] see Kilengwe	U4	Mengo	0 18 N 32 46 E	
[Kirengwe] see Kilengwe	U2	Toro	c.0 46 N 30 06 E	A.S. Thomas
[Kirengwe] see Kirongwe	T6	Morogoro	7 31 S 37 33 E	
Kirenzi, hill	T6	Morogoro	7 31 S 37 33 E	
Kirere, popl.	T6	Rufiji	7 48 S 39 49.5 E	
Kirerema, popl.	U2	Kigezi	0 51 S 29 43 E	
Kirgum = Kurkum, w.h.	U2	Ankole	c.0 50 S 30 32 E	Stuhlmann
Kiriamet, ln.	U4	Mengo	0 18 N 32 46 E	Dummer
[Kiriandongo] see Kiryandongo	K1	Northern Frontier	2 36 N 37 30 E	Herlocker
Kiriangu	K2	Turkana/West Suk	1 29.5 N 35 03 E	Tweedie
Kirichwa Kubwa = Kerichwa Kubwa, str.	U2	Bunyoro	1 53 N 32 03 E	Hill
Kirichwa Ndogo, str.	T7	Iringa	prob. c.8 05 S 36 00 E	Carmichael
Kirigye, f.r.	K4	Nairobi	c.1 17 S 36 43.5 E−1 16 S 36 48 E	Bally
Kirima	K4	Nairobi	c.1 17 S 36 46 E−1 16 S 36 48 E	Bally
	U4	Mengo	0 11 N 33 16.5 E	
	Zaire		0 10 S 29 30 E	Stuhlmann

145

Kirima, popl.	U2	Kigezi	0 52 S 29 45 E	Purseglove
Kirima, hill	K4	Meru	0 07.5 N 37 57 E	Verdcourt & Polhill
[Kirimatonge] see Kimiramatonge	T7	Iringa	7 36 S 34 54 E	Greenway & Kanuri
Kirimia, str.	U2	Toro	c.0 46 N 30 06 E	Osmaston
Kirimiri, hill, for.	K4	Embu	0 25 S 37 33 E	Fries
Kirimon Sitha, ln.	K1	Northern Frontier	c.0 45 N 36 55 E	Bally
Kirimumie drainage line	T2	Arusha	prob. c.3 15 S 36 54 E	Greenway
Kiringa, R.	K4	South Nyeri	0 14.5 S 37 17.5 E–0 33 S 37 20 E	S.A. Robertson
Kirinyaga, distr.	K4	South Nyeri	c.0 30 S 37 20 E	
Kiriri, popl.	U4	Mengo	0 12 N 32 03 E	
Kiritiri, popl.	K4	Embu	0 41.5 S 37 39 E	S.A. Robertson
Kirks Bridge	K3	Trans-Nzoia	0 54 N 34 56 E	Bally
Kiroka, popl.	T6	Morogoro	6 50 37 48.5 E	Busse, Hoyle, Schlieben
Kirongwe, popl.	T6	Rufiji	7 48 S 39 49.5 E	Greenway
Kirua, area	T2	Moshi	3 09.5 S 37 35 E	Haarer
Kiruhura, popl., r.h.	U2	Ankole	0 13 S 30 51 E	Harker
Kirumba, popl.	U4	Masaka	0 24 S 31 22 E	Lye
Kirumba, str.	T5	Dodoma	6 02 S 34 08 E–6 07 S 34 22 E	
Kirumuma, hill (Arusha Nat. Park)	T2	Arusha	Not traced	Vesey-FitzGerald
Kirunde, mt.	T7	Njombe	c.9 30 S 34 12 E	Goetze
Kirungule, popl.	T5	Dodoma	6 00 S 35 41 E	Busse
Kiruru, area	T3	Pare	c.3 42 S 37 34 E	Haarer
Kiruruma, str.	U2	Toro	c.0 28 N 30 04 E	A. Johnston
Kiruruma, R.	T2	Mbulu	c.3 13 S 35 43 E–3 26 S 35 49.5 E	Greenway
[Kirurume] see Kiruruma	U2	Toro	c.0 28 N 30 04 E	
Kirusha = Kirushya, popl., r.h.	T1	Ngara	2 30.5 S 30 32 E	
Kirushya, popl., r.h.	T1	Ngara	2 30.5 S 30 32 E	Tanner
Kirwa, mine	U2	Kigezi	1 13.5 S 29 37.5 E	Chandler & Hancock, Eggeling
Kiryamuli, popl.	U3	Busoga	0 33.5 N 33 11 E	G.H. Wood
Kiryana, hill	U2	Bunyoro	c.1 31 N 31 53 E	B.J. Turner

Name	Region	District/Area	Coordinates	Collector/Reference
Kiryandongo, popl.	U2	Bunyoro	1 53 N 32 03 E	Langdale-Brown, Eggeling, Purseglove
Kiryanga, popl.	U4	Mubende	1 06 N 31 04 E	Eggeling
Kisa, I.R.L.C.S. station	T4	Ufipa	8 01 S 31 54 E	Michelmore
[Kisagara] see Kisangara	T3	Pare	c.3 44 S 37 35 E	
Kisaki, popl., r.h.	T6	Morogoro	7 28 S 37 36 E	B.D. Burtt, Goetze (Nos. 125–130), Greenway
Kisala, popl.	U4	Mubende	1 01 N 31 09 E	
Kisala, mt.	T6	Kilosa	6 25.5 S 36 59 E	Mabberley
Kisalwoko, hill	U4	Masaka	0 09 S 31 09 E	Lye
Kisambya, hill (near Kipembawe)	T7	Chunya	near 7 39 S 33 24 E	Hoyle
Kisamis, area	K6	Masai	1 30 S 36 36 E	Glover et al.
[Kisanero] see Kisangiro	T3	Pare	3 38 S 37 33.5 E	Greenway
Kisanga, popl.	T4	Tabora	5 32 S 32 51.5 E	Ukiriguru Res. Stn.
Kisanga, mssn.	T6	Kilosa	7 21 S 36 44 E	Eggeling
Kisangara, est.	T3	Pare	c.3 44 S 37 35 E	Grundy, Haarer
[Kisangile] see Kisangire	T6	Uzaramo	7 26 S 38 40 E	Goetze
Kisangire, popl.	T6	Uzaramo	7 26 S 38 40 E	
Kisangiro, rsta.	T3	Pare	3 38 S 37 33.5 E	Numerous
[Kisangu] see Kisanku	T4	Mpanda	7 25 S 31 46 E–7 34 S 31 54 E	Richards
Kisanku, str.	T4	Mpanda	7 25 S 31 46 E–7 34 S 31 54 E	Tweedie
Kisano, sw.	K3	Trans-Nzoia	1 09 N 34 45 E	
[Kisapu] see Kishapu	T1	Shinyanga	3 37 S 33 52 E	Numerous
[Kisarawe, distr.] see Uzaramo	T6	Uzaramo	c.7 10 S 38 50 E	
Kisarawe, f.r.	T6	Uzaramo	c.6 50 S 39 02 E	Numerous
Kisarawe, popl., t.c.	T6	Uzaramo	6 54.5 S 39 04.5 E	Numerous
Kisaru, popl., r.h.	U2	Bunyoro	1 15 N 31 01 E	Eggeling
Kisasa, popl., w.h.	U4	Masaka	0 19 S 31 58 E	Lye
Kisauke, popl.	T6	Bagamcyo	6 12 S 38 46 E	Kraenzlin, Peter
Kisawasawa, popl.	T6	Ulanga	7 53 S 36 52 E	Haerdi, Cribb & Grey-Wilson
[Kischogo] see Kishogo	T1	Bukoba	1 23.5 S 31 37.5 E	Braun

[Kisem Kazi] see Kizimkazi			
[Kisenbani] see Kisimbani			
Kisengi, popl.	T4	Tabora	5 20 S 33 33 E	Bally, Lawton
[Kiserawa] see Kisarawe			
[Kiserawe] see Kisarawe	T6	Uzaramo	c.6 54 S 39 02 E	Eggeling
Kiserian, R.	T6	Uzaramo		Vaughan
[Kiserien] see Kiserian	K6	Masai	c.1 25 S 36 38 E–1 23 S 36 49 E	
Kiseru, R.	K6	Masai	c.1 25 S 36 38 E–1 23 S 36 49 E	Napper
[Kisesa] see Kisesi	T2/3	Masai/Handeni	5 24 S 37 18 E–5 43 S 37 25 E	
Kisesi = Kissessa, popl.	T1	Mwanza	2 33 S 33 02.5 E	
Kisesse, mt.	T1	Mwanza	2 33 S 33 02.5 E	Tanner
Kisesse, popl.	T5	Kondoa	4 26 S 35 50 E	
Kishanda, popl., t.c.	T5	Kondoa	4 27 S 35 50 E	B.D. Burtt
Kishanda Valley	T1	Bukoba	1 43 S 31 34 E	Haarer, Gillman, Procter
Kishanje = Kishanzhe, mssn.	T1	Bukoba	1 15 S 30 48 E	Procter
Kishanzhe = Kishanje, mssn.	U2	Kigezi	1 18 S 29 52 E	H.B. Johnston
Kishapu, popl.	U2	Kigezi	1 18 S 29 52 E	
[Kishimungu] see Kishumundu	T1	Shinyanga	3 37 S 33 52 E	B.D. Burtt
Kishiuwi, hill, W. Usambaras	T2	Moshi	3 17 S 37 22 E	Leedal
Kishogo, popl.	T3	Lushoto	prob. 4 46 S 38 20 E	G.R. Williams
Kishumundu, mssn., popl.	T1	Bukoba	1 23.5 S 31 37.5 E	Braun
Kisi, popl.	T2	Moshi	3 17 S 37 22 E	
Kisi, hill	U4	Mengo	0 03.5 N 32 32 E	Lye
Kisi, popl., mssn., r.h.	T4	Ufipa	7 13.5 S 31 01.5 E	
Kisian, mkt.	T4	Ufipa	7 12 S 31 02 E	Bullock
Kisieni, ln.	K5	Central Kavirondo	0 04.5 S 34 40 E	Agnew et al.
[Kisigao] see Kasigau	K5	North Kavirondo	0 14 N 34 52 E	Perdue & Kibuwa
[Kisigau] see Kasigau	K7	Teita	c.3 50 S 38 40 E	
Kisigo, R.	K7	Teita	c.3 50 S 38 40 E	
Kisii, popl.	T5	Dodoma	c.5 57 S 34 47 E–7 03 S 35 50 E	Greenway
	K5	South Kavirondo	0 41 S 34 46 E	Numerous

148

Name	Region code	Region	Coordinates	Collector/authority
Kisii, distr., t.a.	K5	South Kavirondo (considered as part of South Kavirondo for F.T.E.A.)	c.0 35 S 34 45 E	Paulo, Sensei, Hawthorne
Kisiju, popl.	T6	Uzaramo	7 24 S 39 20 E	
Kisikiti	T5	Kondoa	prob. c.5 00 S 36 45 E	B.D. Burtt
Kisima, L.	K1	Northern Frontier	0 56.5 N 36 46 E	D.C. Edwards
Kisima, hill	K1	Northern Frontier	0 59 N 36 49 E	
Kisima Farm	K3	Laikipia	0 30 N 36 45 E	Carter & Stannard
[Kisima Farm] see Ngare Ndare Ranch				
Kisima, popl.	K4	North Nyeri	0 07 N 37 25 E	Hornby, Verdcourt
[Kisima cha Munge] ? see Kisima cha Mungu	T5	Mpwapwa	7 02 S 36 00 E	
[Kisima cha Munge] ? see Munge	T2	Masai	5 18 S 36 56 E	Geilinger
Kisima cha Mungu, popl.	T2	Masai	3 05 S 35 40 E	Geilinger
Kisima cha Mungu	T2	Mbulu	c.4 51 S 35 28 E	B.D. Burtt
Kisimani Mafia, popl.	T6	Rufiji	7 57 S 39 36 E	Greenway
Kisimbani = Kizimbani	Z	Zanzibar	6 05 S 39 16 E	Faulkner
Kisimiri, str., est.	T2	Arusha	3 11 S 36 46 E – 3 07 S 36 53 E	Richards
Kisimisi, for.	T2	Masai	c.2 04.5 S 35 36 E	Carmichael
[Kisimkasi] see Kizimkazi	Z	Zanzibar	6 27 S 39 28 E	Stuhlmann
[Kisim Kazi] see Kizimkazi	Z	Zanzibar	6 27 S 39 28 E	Faulkner
Kisinga, area	U2	Ankole	c.0 39 S 30 37 E	Carmichael
Kisinga, area	T7	Iringa	c.7 50 S 36 05 E	
Kisingarugara, f.r.	T7	Iringa	c.7 50 S 35 55 E	Fraser
Kisinge, popl.	T7	Iringa	c.7 35 S 35 47 E	Polhill & Paulo
[Kisinika] see Kifinika	T2	Moshi	3 10 S 37 29 E	Volkens
Kisinsi, pt.	U4	Mengo	0 14 N 32 38.5 E	Lye
Kisiriri, popl.	T5	Singida	4 12 S 34 26 E	Polhill & Paulo
Kisiro, popl.	U3	Busoga	0 43 N 33 49.5 E	Langdale-Brown
[Kisisi] see Kisizi	U2	Kigezi	1 00 S 29 57 E	Norman
Kisisi, R.	T4	Tabora	5 25 S 33 02 E – 5 13 S 32 42 E	

Name	Grid	District	Coordinates	Collector
[Kisseru] see Kiseru	T2/3	Masai/Handeni	5 24 S 37 18 E–5 43 S 37 25 E	Busse
Kissessa = Kisesi, popl.	T1	Mwanza	2 33 S 33 02.5 E	Tanner
[Kissesse] see Kisesse	T5	Kondoa	4 26 S 35 50 E	Goetze
[Kissinga] see Kisinga	T7	Iringa	c.7 50 S 36 05 E	
[Kissumbe] see Kizumbi	T1	Shinyanga	3 43 S 33 25 E	
Kissungwe	T7	Rungwe	Not traced	Stolz
[Kistungulu] see Kitungulu	T4	Ufipa	8 29 S 31 17 E	Muenzner
[Kisuani] see Kisiwani	T3	Pare	4 08 S 37 57 E	Greenway
Kisubi, bay	U4	Mengo	c.0 06 N 32 33 E	Chandler
Kisubi, mssn., sch.	U4	Mengo	0 07 N 32 32 E	Eggeling, Lye, Chandler
[Kisula] see Kizala	T3	Lushoto	4 55 S 38 40 E	Buchwald
Kisulutini, mssn.	K7	Kilifi	3 56 S 39 34 E	W.E. Taylor
[Kisumbe] see Kizumbi	T1	Shinyanga	3 43 S 33 25 E	
[Kisumbi] see Kizumbi	T1	Shinyanga	3 43 S 33 25 E	B.D. Burtt
Kisumu, port, rsta., airfield	K5	Kisumu-Londiani	0 06 S 34 45 E	Numerous
Kisumu-Londiani, distr.	K5	Kisumu-Londiani	c.0 12 S 35 15 E	
[Kisungu] see Kizungoa	T3	Pare	c.3 57 S 37 43 E	Haarer
Kisungu, popl.	T4	Ufipa	7 45 S 31 31 E	Bullock
Kisuyasuwi	T2	Masai	c.2 05 S 35 35 E	Carmichael
Kisuzi, popl.	T4	Buha	4 12 S 30 05 E	Peter
Kiswago, popl.	T6	Ulanga	8 54 S 36 18 E	Rees
Kiswani, area	T3	Pare	c.4 05 S 37 48 E	
[Kiswani] see Kisiwani	T3	Pare	4 08 S 37 57 E	
[Kiswani] see Daluni	T3	Lushoto	4 46.5 S 38 46 E	B.D. Burtt
Kiswaya, area	T5	Kondoa	c.4 52 S 35 32 E	Chandler
[Kitabe] see Katabi	U4	Mengo	0 06 N 32 28 E	Milne-Redhead & Taylor
[Kitae] see Kitai	T8	Songea	10 45 S 35 09 E	Lye & Katende
Kitagata, L.	U2	Toro	0 03.5 S 29 58 E	Snowden
Kitagata, popl.	U2	Ankole	0 40 S 30 09 E	
[Kitagwenda] see Kitakwenda	U2	Toro	c.0 00–0 25 N 30 25–31 00 E	

Name	Code	Region	Coordinates	Collector
Kitekero = Ailsa, old farm	T7	Mbeya	8 54 S 33 59 E	Carmichael
Kiten, mt.	U1	Acholi	3 32 N 33 25 E	Eggeling
Kitenden, R.	K6/T2	Masai/Moshi	c.2 57 S 37 14 E–2 47 S 37 17 E	Carmichael
Kitenden, area	T2	Moshi	2 52 S 37 15 E	Eggeling, Procter
Kitendio, mt.	K4	Kitui	1 35 S 38 12 E	Bally
Kitengeto, str.	U2	Toro	c.0 01 S 30 17–30 20 E	Jarrett
Kitengya, for.	U2	Toro	c.0 43 N 30 03 E	Eggeling
Kitere, L., popl.	T8	Mikindani	10 21 S 39 42 E	May
Kiterera, popl.	U3	Busoga	0 22 N 33 32 E	
Kitesa = Kiteza, hill	T8	Songea	11 07 S 34 51 E	
Kiteta, mkt.	K4	Machakos	1 32 S 37 29.5 E	
Kitete, popl.	T4	Ufipa	7 33 S 31 22 E	Richards
Kiteza, hill	T8	Songea	11 07 S 34 51 E	Mgaza
Kitgum, popl., mssn., r.h.	U1	Acholi	3 18 N 32 52 E	Numerous
Kitgum, mt.	U1	Acholi	3 17.5 N 32 53 E	
Kitgum Matidi, mssn., r.h.	U1	Acholi	3 17 N 33 03 E	Purseglove
Kithembe, hill, popl.	K4	Machakos	1 45.5 S 37 23 E	Mwangangi
Kithunguri, popl.	K4	Meru	0 05 S 37 41 E	
Kitikuyu, str.	K4	Kiambu	0 50 S 36 37 E–0 52 S 36 45 E	H.M. Gardner
[Kitindiri] ? see Kitendio	K4	Kitui	?1 35 S 38 12 E	Bally
Kitingi, str.	T2	Mbulu	4 26 S 35 26 E–4 26 S 35 32 E	Hukui
[Kitini] see Kitiri	K3	Naivasha	0 32.5 S 36 32.5 E–0 31.5 S 36 28 E	D. Leakey
Kitiri, str.	K3	Naivasha	0 32.5 S 36 32.5 E–0 31.5 S 36 28 E	
[Kitisiro] see Kitisuru	K4	Nairobi	1 14 S 36 46 E	Verdcourt
Kitisuru, popl.	K4	Nairobi	1 14 S 36 46 E	
Kititimu, popl., exper. fm.	T5	Singida	4 50 S 34 47 E	van Rensburg
Kitivo, area	T3	Lushoto	c.4 36 S 38 12 E	
Kitivo, f.r.	T3	Lushoto	4 47 S 38 18 E	
Kitivo, area	T3	Lushoto	4 39.5 S 38 31 E	Mgaza, Ngoundai, Willan
Kitivo, area	T3	Lushoto	4 46 S 38 35.5 E	

153

Kitui, distr.	K4	Kitui	c.1 30 S 38 45 E	
Kitui, popl., airstrip	K4	Kitui	1 22 S 38 00.5 E	Numerous
Kitulanghalo, hill	T6	Morogoro	6 41 S 37 57 E	Welch
Kitulo = Elton Plateau	T7	Njomìe/Mbeya	c.9 00 S 33 50 E	Harris, Prins, Cribb & Grey-Wilson
Kitulo, Farm, popl.	T7	Njomìe	9 05 S 33 58 E	Cribb & Grey-Wilson
[Kitulo] see Kitulu	T8	Lindi	10 01 S 39 42 E	Busse
Kitulu, hill	T8	Lindi	10 01 S 39 42 E	Busse
Kitumbeine, mt., f.r.	T2	Masai	2 53 S 36 13 E	Greenway, Richards
Kitumbeine, popl., mssn.	T2	Masai	2 45 S 36 12.5 E	
Kitunda, L.	U4	Masaka	c.0 38 S 31 44 E	
Kitunda, popl.	T4	Tabora	6 48 S 33 13 E	Boaler
Kitunda, popl., fy.	T8	Lindi	10 00 S 39 45 E	Busse
Kitundu, hills	T6	Morogoro	6 51 S 37 45 E	E.M. Bruce, Schlieben
Kitungole, popl.	T1	Bukoba	1 17 S 31 17 E	Stuhlmann
Kitungulu, popl.	T4	Ufipa	8 29 S 31 17 E	
Kituoni, popl.	Z	Zanzibar	6 07.5 S 39 22 E	Vaughan
Kitupa (near Mwera)	T3	Pangani	near 5 29 S 38 57 E	
Kituria, R. (near L. Sundu)	T4	Ufipa	near 8 31.5 S 31 38.5 E	Richards
[Kitushi] see Titushi	T1	Musoma	c.2 37 S 34 46 E	Greenway & Turner
[Kituta] see Ikuka	T7	Iringa	7 28 S 34 58 E–7 38 S 34 50 E	Richards
Kitutu, t.a.	K5	South Kavirondo	0 35 S 34 48 E	Buxton, Spranger
Kituwaba Farm	K3	Trans-Nzoia	0 54.5 N 34 49 E	Tweedie
Kituza, est.	U4	Mengo	0 15.5 N 32 47 E	Griffiths
[Kitwa] see Kitwai	T2	Masai		Procter
Kitwai, plains, mbuga	T2	Masai	c.4 50 S 37 05 E	
Kitwai, g.r.	T2	Masai	c.5 00 S 37 20 E	
Kitwa Pembe, hill	K7	Lamu	2 28 S 40 42.5 E	
Kitwe, popl.	U2	Ankole	0 08 S 30 30 E	Welch
Kitwe, popl.	T1	Bukoba	1 18 S 31 44 E	Haarer
Kitwe, res.	T2	Masai	c.5 00 S 37 20 E	Vesey-FitzGerald

155

Name	Region	District	Coordinates	Collector(s)
[Kitwei] see Kitwe	T2	Masai	c.5 00 S 37 20 E	Vesey-FitzGerald
[Kitwei] see Kitwai	T2	Masai	c.4 50 S 37 05 E	Jaeger
Kityerera, popl.	U3	Busoga	0 19 N 33 29 E	Harris, Webb
Kiu, t.c., rsta.	K4/6	Machakos/Masai	1 54 S 37 09.5 E	Napier, Polhill & Paulo, Verdcourt
[Kiuhuhwi] see Kihuhwi	T3	Tanga	5 13 S 38 41 E	Grote
[Kiuhui] see Kihuhwi	T3	Tanga		
Kiumba, hill, popl.	K5	South Kavirondo	0 30 S 34 23 E	Law
Kiumba, popl.	T3	Lushoto	5 05.5 S 38 38.5 E	Greenway, Lommel, Verdcourt
Kiumbi, str.	K4	Machakos	c.2 11 S 37 53 E –2 18 S 37 48 E	
Kiume, hill	K7	Tana River	0 04 S 38 39 E	
Kiumu, hill	U3	Busoga	c.0 35 N 33 08 E	G.H.S. Wood
Kiunga I. = Kiungamwina	K7	Lamu	1 46 S 41 30 E	
Kiunga, popl.	K7	Lamu	1 44.5 S 41 29.5 E	Oxford Univ., Rawlins
Kiunga Mudirate, ln.	K7	Lamu	c.1 57 S 41 18 E	Numerous
Kiungamwina, I.	K7	Lamu	1 46 S 41 30 E	Rawlins
[Kiunga Mwini] see Kiungamwina	K7	Lamu	1 46 S 41 30 E	
Kiutu, hill	T2	Arusha	3 21 S 36 43 E	
Kiuu, hill	K4	Machakos	1 47 S 37 14.5 E	
Kivata, popl.	U2	Toro	0 33 N 30 15 E	Scott Elliot
Kivere, tea est., for.	T7	Iringa	8 39 S 35 14 E	Paget-Wilkes, Gilchrist
Kiverenge, mt., area	T3	Pare	3 49 S 37 38 E	Peter, Bally
[Kivesoweri] see Kyosoweri	U3	Mbale	1 20 N 34 40.5 E	
[Kiviamet] see Akiriamet	K3	Baringo	c.1 20 N 35 43 E	
Kivindani, popl.	T3	Tanga	5 11 S 39 06 E	Faulkner
Kivingo, popl., t.c.	T3	Lushoto	4 25 S 38 26 E	Greenway, Gillman
[Kivinzo] see Kivingo	T3	Lushoto	4 25 S 38 26 E	Greenway
[Kivira] see Kiwira	T7	Rungwe		
Kiviruviru	T5	Kondoa	c.5 00 S 36 00 E	B.D. Burtt
Kivugo, popl.	T6	Bagamoyo	6 25 S 38 26.5 E	
Kivukoni, popl.	T6	Ulanga	8 12 S 36 42 E	Rees

Name	Code	Region	Coordinates	Collector
Kivumba, popl.	T4	Buha	4 36 S 30 40 E	Peter
Kivumoni, for., t.a.	K7	Kwale	c.4 13 S 39 29 E	Drummond & Hemsley, Magogo & Glover
Kivumoni, hill	K7	Kwale	4 13 S 39 30 E	
[Kivungilo] see Kifungilo	T3	Lushoto	4 47 S 38 22 E	Mgaza
Kivuvu, est., popl.	U4	Mengo	0 24 N 32 47 E	Dummer
Kiwafu, popl.	U4	Mengo	0 13 N 32 48 E	Dummer
[Kiwa Kavu] see Ziwa Kavu	T2	Arusha	3 15 S 36 52.5 E	Greenway
Kiwala, popl., est.	U4	Masaka	0 19 S 31 49 E	Chandler, Eggeling
Kiwala, est.	U4	Mengo	0 28.5 N 32 54.5 E	Dummer, Irwin, Lye
[Kiwambale, camp	K7	Kwale	4 35 S 39 23 E	Allan
[Kiwambali] see Kiwambale	K7	Kwale	4 35 S 39 23 E	
Kiwanda, area	T3	Handeni	c.5 26 S 38 29 E	Fischer
Kiwanda, mssn.	T3	Tanga	5 03 S 38 43 E	Greenway
[Kiwangolo] see Kibwangule	T7	Iringa	8 29.5 S 35 02 E	Carmichael
Kiwangwa, hill, popl.	T6	Bagamoyo	6 22 S 38 35 E	Procter
Kiwani, popl.	P	Pemba	5 23.5 S 39 46 E	Greenway
[Kiwanji Nyamateo] see Kawanya ya Matteo	T2	Arusha	c.3 15.5 S 36 53 E	Greenway
[Kiwapu] see Kiwafu	U4	Mengo	0 13 N 32 48 E	Dummer
Kiwara ? = Kirawira	T1	Musoma	?2 10 S 34 09 E	Richards
[Kiwara] see Kiwira	T7	Rungwe		Greenway
Kiwatule, popl.	U4	Mengo	0 22 N 32 37.5 E	Liebenberg, Snowden
Kiwawe, L., sw.	T6	Rufiji	c.7 52 S 39 49 E	Greenway
Kiwengwa, popl., area	Z	Zanzibar	6 00 S 39 23 E	Faulkner, Greenway, Oxtoby
[Kiwera] see Kiwira	T7	Rungwe		Greenway & Hoyle, Brenan & Greenway
Kiwere, popl.	T7	Mbeya	8 34 S 34 29 E	Gilchrist, Zimmermann
Kiwira, fishing camp	T7	Rungwe	9 02 S 33 38 E	Greenway, Procter, Eggeling
Kiwira, popl.	T7	Rungwe	9 10.5 S 33 32 E	Eggeling, Davies
Kiwira, R.	T7	Rungwe	c.9 05 S 33 42 E–9 37 S 33 57.5 E	Numerous
Kiwoko, popl.	U4	Mengo	0 50.5 N 32 22 E	Langdale-Brown

158

Name		Region	Coordinates	Authority
Koboko, co.	U1	West Nile	c.3 30 N 31 00 E	Alonzie
Koboko, popl, mssn., r.h.	U1	West Nile	3 25 N 30 58 E	Numerous
[Kocemaluk] see Kochemaluk	U1	Karamoja	2 11 N 34 21 E	Dyson-Hudson
Kochemaluk, mt.	U1	Karamoja	2 11 N 34 21 E	
Kocholia, hills	K5	North Kavirondo	c.0 39 N 34 21 E	
Kocholia, popl., sch.	K5	North Kavirondo	0 37 N 34 21.5 E	Tweedie
Kodi (Kachagalou)	K2	Turkana	c.2 19 N 35 03 E	Osmaston
Kodino	T5	Kondoa	Not traced	Ruffo
Kodit, ln.	K2	Turkana	1 58 N 35 14 E	J. Wilson
Kodonia, str.	U1	Karamoja	c.2 25 N 34 40 E	
Kodunyo (near Mt. Moroto)	U1	Karamoja	2 29 N 34 42 E	Eggeling
[Kofole] see Kufole	K1	Northern Frontier	3 22 N 39 47 E	Bally & Smith
Koga, popl.	T4	Tabora	6 14 S 32 26 E	Carmichael
Kogatende, guard post	T1	Musoma	1 35 S 34 56 E	Greenway
[Kogh] see Koh	K2	West Suk	1 24 N 35 30 E	
[Kogombe] see Kigombe	T3	Tanga	c.5 19 S 39 01 E	Milne-Redhead & Taylor
Koh, ln.	K2	West Suk	1 24 N 35 30 E	Canfield
[Koia] see Melka Koja	K1	Northern Frontier	1 12 N 38 58.5 E	J. Adamson
Koich, str.	U1	West Nile	c.3 28 N 31 00 E–3 23 N 31 30 E	Eggeling
[Koitagotch] see Koitogoch	K3	Trans-Nzoia	1 08 N 34 41 E	Tweedie
Koitilial, popl.	K3	Elgeyo	0 56 N 35 37 E	Bally
[Koitobbos] see Koitobos	K3	Trans-Nzoia	1 07.5 N 34 36 E	Hedberg
Koitobos, peak	K3	Trans-Nzoia	1 07.5 N 34 36 E	Tweedie
Koitobos, str.	K3	Trans-Nzoia	1 05 N 34 48 E–0 55 N 35 06 E	Tweedie
[Koitoboss] see Koitobos	K3	Trans-Nzoia	1 07.5 N 34 36 E	
Koitogoch, hill	K3	Trans-Nzoia	1 08 N 34 41 E	
Koitogor, mt.	K1	Northern Frontier	0 36 N 37 33.5 E	Hooper & Townsend
Koja, pen.	U4	Mengo	0 07 N 32 43 E	Snowden
Koka, popl., r.h.	U1	West Nile	3 33 N 31 20 E	Greenway & Eggeling
Koki, popl. (Gani area)	U1	Acholi	c.2 55 N 32 05 E	Grant

Name		Region	Coordinates	Collector
Kome, I.	T1	Mwanza	2 22 S 32 28 E	Makwilo, Carmichael
[Komel] see Kome	U4	Mengo	c.0 06 S 32 45 E	Carmichael
[Komi] see Kome	T1	Mwanza	2 22 S 32 28 E	
Komolo, mt., sw.	U1	Karamoja	2 43 N 34 15 E	Harker, Liebenberg
[Komosiy] see Kamothing	U1	Karamoja	2 12 N 34 29 E–2 02 N 34 29 E	Dyson-Hudson
Kondeland, old area	T7	Rungwe	c.9 20–9 30 S 33 50–34 00 E	Goetze, Stolz
Kondoa, distr.	T5	Kondoa	c.5 00 S 35 50 E	Numerous
Kondoa, popl., r.h., airstrip	T5	Kondoa	4 54 S 35 47 E	Stuhlmann
Kondoa, mssn.	T6	Kilosa	6 49 S 37 03 E	Geilinger, Schellhaase
[Kondoa-Irangi] see Kondoa	T5	Kondoa	4 54 S 35 47 E	Peter
[Konduchi] see Kunduchi	T6	Uzaramo	6 40 S 39 13 E	Eggeling
[Kondwe] see Nkondwe	T4	Kigoma	5 52 S 30 52 E	Gillett
Kongao, mkt.	K5	Central Kavirondo	0 11 S 34 12 E	Peter
[Konge] see Nkonge	T3	Lushoto	5 00.5 S 38 27 E	Greenway
Kongei, popl.	T3	Lushoto	4 49 S 38 22 E	Tweedie, Lucas, Knight
Kongelai, popl., p.p.	K2	West Suk	1 29 N 35 00 E	Newman
Kongogo, hill	T5	Kondoa/Dodoma	5 39 S 35 36 E	Cribb & Grey-Wilson, Holst
[Kongoi] see Nkongoi	T3	Lushoto	4 46 S 38 31 E	Versell, Trelawny
[Kongolai] see Kongelai	K2	West Suk	1 29 N 35 00 E	Haarer
Kongolero, L.	T1	Bukoba	? c.1 34 S 31 31 E	Lucas
[Kongoli] see Kongelai	K2	West Suk	1 29 N 35 00 E	Bogdan
[Kongondi Rock] see Kangondi Rock	K4	Kitui	1 04 S 37 41 E	Greenway, Hooper & Townsend
Kongoni, fm.	K3	Naivasha	0 49.5 S 36 15.5 E	Fries
Kongoni, str.	K4	North Nyeri	c.0 06 N 37 07 E	Anderson, Martin, Leippert
Kongwa, pasture res. stn.	T5	Mpwapwa	6 04 S 36 23 E	Numerous
Kongwa, popl., mssn.	T5	Mpwapwa	6 11 S 36 25 E	von Prittwitz
Konta, mt.	T5	Dodoma	Not traced	Bogdan, Polhill & Paulo
Konza, rsta.	K4/6	Machakos/Masai	1 44.5 S 37 07.5 E	Faden
Koobi Fora, area	K1	Northern Frontier	c.3 57 N 36 13 E	Wimbush, Bally
Koora Plains	K6	Masai	c.1 46 S 36 26 E	

Name	Region	District	Coordinates	Collector
Kotido, popl., mssn.	U1	Karamoja	3 01 N 34 06 E	Numerous
Kotipe, str.	U1	Karamoja	c.2 22 N 34 10 E	
Kotome, R.	K2	Turkana	4 42 N 35 23 E–4 19 N 35 03.5 E	Carter & Stannard
Kouongo (near Namanyere)	T4	Ufipa	near 7 31 S 31 03 E	Swynnerton
Kowop, mt.	K1	Northern Frontier	c.1 56.5 N 36 48 E	Adamson
[Kozara] see Kizara	T3	Lushoto	c.4 55 S 38 40 E	Holst
Kua, popl.	T6	Rufiji	8 00 S 39 46 E	Greenway
Kubaru, ln.	K1	Northern Frontier	c.3 57 N 41 29 E	Bally & Carter
Kuchelebai = Kacheliba, popl., p.p.	K2	Turkana	1 29.5 N 35 01 E	
Kufile, popl., cave well	Z	Zanzibar	6 25 S 39 29 E	Vaughan
Kufole, hill, w.h.	K1	Northern Frontier	3 22 N 39 47 E	Bally & Carter
Kugota (Selous Game Reserve)	T6	Ulanga	Not traced	Nicholson
[Kugunguli] see Kagunguli	T1	Mwanza	2 00 S 33 04 E	Uhlig
Kui, I.	K7	Lamu	1 49 S 41 26 E	Rawlins
[Kuiwa] see Kuywa	K5	North Kavirondo	0 48.5 N 34 36 E–0 28 N 34 38 E	Tweedie
Kuja, popl., mkt.	K5	South Kavirondo	0 48.5 S 34 34 E	Napier
[Kuja] see Gucha	K5	South Kavirondo	c.0 38 S 35 00 E–0 55 S 34 08 E	Glasgow, Napier
Kuju, popl., r.h.	U3	Teso	2 02.5 N 33 37 E	
Kuka, hills	T1	Musoma	1 44 S 35 13 E	Greenway & Turner
Kuku, R.	T4	Ufipa	8 07 S 31 30 E–8 11 S 31 24 E	Richards
Kukui, ln.	K4	Kiambu	c.1 05 S 36 42 E	Kassner
Kukuu, popl.	P	Pemba	5 26 S 39 42 E	Greenway
Kulal, mt., for.	K1	Northern Frontier	c.2 43 N 36 55 E	Numerous
Kula Mawe, t.c.	K1	Northern Frontier	0 34.5 N 38 12 E	Bally & Smith
Kulank, waterpan	K1	Northern Frontier	1 40 S 40 35 E	Gillett
Kulasi, popl.	T3	Lushoto	4 50 S 38 34.5 E	Peter
[Kulemuzi] see Mkulumuzi	T3	Tanga	5 07 S 38 45 E–5 04 S 39 04 E	Peter
Kuliafiri, old popl.	U2	Toro	0 10 N 30 07 E	Scott Elliot
Kulikila, hill	K7	Teita	3 43 S 38 40 E	Bally
Kumba, popl.	U2	Kigezi	1 08 S 29 54 E	Lye, Eggeling, Greenway

163

Kumba, sw.	U2	Kigezi	c. 1 15 S 29 59 E	Greenway
Kumba, str., sw.	T3	Lushoto	c. 4 49 S 38 38 E	Drummond & Hemsley, Peter
Kumba = Mkumba, popl.	T4	Ufipa	prob. c.8 35 S 31 15 E	
Kumbamtoni, popl.	T3	Pangani	5 24 S 38 57.5 E	Tanner, Carmichael
[Kumbeni] see Kombeni	Z	Zanzibar	6 15 S 39 16 E	Stuhlmann
[Kumbira] see Lumbila	T7	Njombe	9 34.5 S 34 07.5 E	Goetze
Kumi, co.	U3	Teso	c. 1 30 N 33 50 E	
Kumi, popl., rsta.	U3	Teso	1 30 N 33 56 E	Chandler, Maitland, Snowden
Kunduchi, popl., harbour	T6	Uzaramo	6 40 S 39 13 E	Peter, Vaughan, Bjornstad, Harris
Kungului = Kungurui, popl., fm.	T3	Lushoto	4 38.5 S 38 28 E	Mgaza, Shabani
Kunguya, popl.	T5	Dodoma	5 37 S 34 52 E	B.D. Burtt
Kungwe, mt.	T4	Mpanda	6 07 S 29 48 E	Harley, Newbould et al.
Kungwe Bay, f.r.	T4	Kigoma	5 37 S 29 52 E	Carmichael
Kunyao, popl.	K2	Turkana	1 47 N 35 03 E	Tweedie
Kunyao, str.	U1/K2	Karamoja/Turkana	c. 1 45 N 35 00 E	J. Wilson, A.S. Thomas, Tweedie
Kurasini, mssn.	T6	Uzaramo	6 50 S 39 18 E	Vaughan
Kurasini, popl.	T6	Uzaramo	6 51 S 39 17 E	Polhill & Paulo, Rawlins
Kurawa = Karawa	K7	Tana River	2 39 S 40 12 E	Werner
[Kurawe] see Kurawa	K7	Tana River	2 39 S 40 12 E	Semsei
[Kurekese] see Kiregese	T6	Uzaramo	c. 7 25 S 39 20 E	Greenway
Kurekwe	Z	Zanzibar	Not traced	
Kurkum, w.h.	K1	Northern Frontier	2 36 N 37 30 E	Procter
Kuru Barata, popl., w.h.	K1	Northern Frontier	0 43 N 38 21 E	Brunt
Kuruhita, ln.	T4	Buha	c. 4 30 S 30 10 E	
Kururuma (Busia to Mumias, alt. 4200 ft.)	K5	Central Kavirondo	Not traced	Greenway
Kurutini, popl.	T6	Uzaramo	6 57 S 39 09 E	Greenway
Kusare, hill	T2	Arusha	c. 3 13.5 S 36 53 E	Carmichael
Kusare, L.	T2	Arusha	3 13.5 S 36 53 E	
[Kuuragwa, for., R.] see Rungwa	T4	Mpanda	c. 7 00 S 32 15 E	

(Carmichael Nos. 950–953 were collected at Sipa in Rungwa River F.R., *not* Kuuragwa – typing error)

[Kuwalath] see Kauwalathe			
[Kuwalathe] see Kauwalathe			
Kuywa, R.	K2	Turkana	c.3 35 N 35 04 E – 3 07 N 35 38 E	Paulo
Kwa Besa (near Mwera)	K2	Turkana	c.3 35 N 35 04 E – 3 07 N 35 38 E	
Kwa Bikare (near Bujenzi)	K5	North Kavirondo	0 48.5 N 34 36 E – 0 28 N 34 38 E	Greenway
[Kwa Brahim] see Kivugo	T3	Pangani	near 5 29 S 38 57 E	Tanner
Kwa Chiropa	T4	Buha	near 4 27 S 29 58 E	Peter
Kwa Demu, popl.	T6	Bagamoyo	6 25 S 38 26.5 E	Holtz
Kwadihombo, popl.	T6	Morogoro	6 15 S 37 32 E	Hannington
Kwafungo, hill, sisal est.	K7	Kilifi	3 42 S 39 36 E	Mapperley
[Kwafungu] see Kwafungo	T6	Morogoro	6 16 S 37 33 E	
Kwafunta, popl.	T3	Tanga	c.5 16 S 38 43 E	Faulkner
[Kwafuto] see Kwafunta	T3	Tanga	c.5 16 S 38 43 E	
Kwaga, popl.	T3	Tanga	5 13 S 38 34.5 E	Semsei
[Kwagogo] see Kwa Ngoga	T3	Tanga	5 13 S 38 34.5 E	Procter
Kwahata = Gwahata, ln.	T4	Buha	4 50 S 29 58 E	Engler
Kwai, popl., vet. stn.	T2/3	Moshi/Pare	3 39 S 37 34 E	Carmichael
Kwaisagat, area	T2	Mbulu	3 56 S 35 39 E	Numerous
Kwaisagat, camp	T3	Lushoto	4 44 S 38 21 E	Dale
[Kwa Kalila] see Mwakaleli	K2	Turkana	c.2 09 N 35 17 E	Mortimer
Kwa Kihingi, popl.	K3	Elgeyo	1 15 N 35 10 E	
Kwakikumba (prob. Matumbi area)	T7	Rungwe	9 09 S 33 49 E	Busse
Kwakilanga, I.	T8	Songea	10 30 S 36 03 E	Busse
Kwakilanga, str.	T8	Kilwa	Not traced	
[Kwa Kuchinja] see Kwa Kuchinja	T3	Pangani	5 18.5 S 38 38 E	
Kwa Kuchinja, popl.	T3	Pangani	c.5 17 S 38 36.5 E	
Kwale, distr.	T2	Mbulu	3 40 S 35 57 E	
Kwale, I.	T2	Mbulu	3 40 S 35 57 E	Hornby, Milne-Redhead & Taylor
Kwale, popl., For. Dept.	K7	Kwale	c.4 10 S 39 10 E	
Kwale, popl.	T3	Tanga	4 58 S 39 09.5 E	
	K7	Kwale	4 10.5 S 39 27 E	Numerous
	T3	Tanga	4 57.5 S 39 08.5 E	Greenway

165

Name	Code	District	Coordinates	Collector
Kwamkunde, popl., hill	T3	Tanga	5 12.5 S 38 50 E	Peter, Verdcourt, Braun
Kwamkuyo, str., falls	T3	Lushoto	c.5 04 S 38 38 E	Numerous
[Kwamkuyu] see Kwamkuyo	T3	Lushoto	c.5 04 S 38 38 E	Peter
Kwamkwe (near Hale)	T3	Pangani	near 5 19 S 38 37 E	Semsei
Kwamndolwa, popl.	T3	Lushoto	5 06 S 38 31 E	Busse
Kwa Mpanda	T8	Kilwa	prob. c.9 35 S 37 35 E	Milne-Redhead & Taylor
Kwamponjore, val.	T8	Songea	c.10 45 S 35 35 E	
[Kwa Mschusa] see Kwa Mshusa	T3	Lushoto	4 46 S 38 29 E	Drummond & Hemsley, Peter, Engler
[Kwamshemshi] see Kwashemshi	T3	Lushoto	5 02.5 S 38 29 E	Cribb & Grey-Wilson, L. Tanner
Kwamshundi, ln.	T3	Lushoto	c.4 48 S 38 30 E	Holst
Kwa Mshusa, popl.	T3	Lushoto	4 46 S 38 29 E	Holst
[Kwa Mshuza] see Kwa Mshusa	T3	Lushoto	4 46 S 38 29 E	Procter
Kwamsisi, popl	T3	Handeni	5 51 S 38 33 E	
Kwamsusa, str.	T3	Lushoto	c.4 58 S 38 22 E	Peter, Bogner
Kwamtili, popl.	T3	Lushoto	4 56 S 38 44 E	Busse
Kwa Mtira, popl.	T8	Tunduru	11 33 S 36 55 E	B.D. Burtt
Kwa Mtoro, popl., t.c.	T5	Kondoa	5 14 S 35 25 E	
Kwamtoro = Kwa Mtoro, popl., t.c.	T5	Kondoa	5 14 S 35 25 E	
[Kwa Muala] see Kamwala	T3	Pare	3 41 S 37 38 E	Procter
Kwamushai, prob. = Hemashai	T3	Lushoto	prob. 4 46.5 S 38 17 E	
Kwamweleti, popl.	T3	Lushoto	4 47.5 S 38 16 E	Stuhlmann
[Kwangaviassi] see Ngono	T1	Bukoba	c.1 58 S 31 37 E–1 08 S 31 35 E	
Kwa Ngoga, old rsta.	T2/3	Moshi/Pare	3 39 S 37 34 E	
Kwa Ngomba, old popl.	T8	Songea	10 57 S 34 45 E	Busse
Kwa Ngomba, popl.	T5	Mpwapwa	c.6 16 S 36 43 E	Busse
Kwania, co.	U1	Lango	c.1 55 N 32 45 E	
Kwania, L.	U1	Lango	1 45 N 32 45 E	
Kwaraha, volcano	T2	Mbulu	4 14 S 35 48 E	
Kware, str.	T2	Moshi	3 16 S 37 10 E–3 27 S 37 12.5 E	Volkens
Kware, popl.	T2	Moshi	3 17 S 37 09 E	

167

Name		Region	Coordinates	Collector
[Kwa Schamba] see Chamba				
Kwa Sengiwa, popl.	T8	Tunduru	11 34 S 36 58 E	Busse
Kwashemshi, popl., sisal est.	T3	Pare	3 44 S 37 45 E	Uhlig
Kwa Sikumbi (Rondo—Noto)	T3	Lushoto	5 02.5 S 38 29 E	Busse
Kwasimba	T8	Lindi	Not traced	Semsei
Kwa Sonda (near Msalala)	T4	Mpanda	Not traced	Hannington
Kwasossa = Kwamsusa, str.	T4	? Nzega/Kahama	Not traced	Buchwald
[Kwa Ssenlanga] see Kwa Sulanga	T3	Lushoto	c.4 58 S 38 22 E	Busse
Kwa Sulanga, popl.	T3	Handeni	5 24 S 38 08 E	Busse
Kwata, I.	T3	Handeni	5 24 S 38 08 E	
Kwata, popl., t.c., sawmill	P	Pemba	5 22 S 39 35 E	Greenway
[Kwatango] see Kwetango	T3	Lushoto	4 54 S 38 34 E	Drummond & Hemsley
[Kwatangu] see Kwetangu	T3	Tanga	5 01 S 38 46 E	
[Kwa Tschiropa] see Kwa Chiropa	T3	Lushoto	4 45 S 38 41 E	Peter
Kwatumbili, hill	T6	Morogoro	6 15 S 37 32 E	
Kwa Wasiri	T3	Lushoto	4 53 S 38 47 E	Peter
Kwazinga, popl., area	T6	Kilosa	6 45 S 37 07 E	Busse
[Kwa Zuranga] see Kwa Sulanga	T3	Lushoto	4 46 S 38 20 E	Drummond & Hemsley, Greenway
Kwebao, for.	T3	Handeni	5 24 S 38 08 E	
Kwebona, popl.	T3	Lushoto	4 45 S 38 19 E	G.R. Williams
Kwediboma, f.r.	Z	Zanzibar	6 16.5 S 39 22.5 E	
Kwediboma, popl., mssn.	T3	Handeni	c.5 27 S 37 33 E	
[Kwedihombo] see Kwadihombo	T3	Handeni	5 26 S 37 35 E	
Kwegangaga	T6	Morogoro	6 16 S 37 33 E	Peter
Kwegoba, peak	T3	Lushoto	Not traced	Ross
Kwegoka, peak	T6	Morogoro	6 03 S 37 30 E	Thulin & Mhoro
Kwehangala, old pl.	T3	Lushoto	4 33 S 38 16.5 E	Drummond & Hemsley
Kwejoka, ln.	T3	Lushoto	4 51 S 38 26 E	Drummond & Hemsley
Kwekangara (Mlinga mt.)	T3	Lushoto	c.4 50 S 37 55 E	Procter
Kwela, L.	T3	Lushoto	c.5 05 S 38 45 E	Greenway
	T4	Ufipa	8 24 S 31 58.5 E	Numerous

Name		Region	Coordinates	Collector
[Kwela] see Mwela	T7	Mbeya	8 55 S 33 49 E	Cribb & Grey-Wilson, Wingfield
Kwemandala, area	T3	Lushoto	4 35 S 38 13 E	Drummond & Hemsley
Kwendoghoi	T3	Lushoto	Not traced	Shabani
Kwengoma, popl.	T3	Handeni	5 30 S 37 59 E	
Kwesimu	T3	Lushoto	Not traced	Semsei
[Kwesoweri] see Kyosoweri	U3	Mbale	1 20 N 34 40.5 E	
Kwetango, popl., old area	T3	Tanga	5 01 S 38 46 E	
Kwetango = Kwetangu, old area	T3	Lushoto	4 45 S 38 41 E	Peter
Kwetangu, old area	T3	Lushoto	4 45 S 38 41 E	
[Kwewaga] see Kyiwaga	U4	Mengo	c.0 05 N 32 30 E	
Kwial, hill	K1	Northern Frontier	c.3 32 N 38 14 E	Dawkins
Kwibasa ? see Kipassa	T7	Rungwe	?9 15.5 S 33 41.5 E	Gillett
[Kwihala] see Kwihara	T4	Tabora	5 03 S 32 44 E	Stolz
Kwihara, popl.	T4	Tabora	5 03 S 32 44 E	Peter
Kwikonge Buki	K5	Kavirondo	Not traced	
Kwimba, distr.	T1	Kwimba	c.3 00 S 33 15 E	Templer
Kwimba, t.a.	T1	Kwimba	c.3 00 S 33 10 E	Marshall
Kwiro, f.r.	T6	Ulanga	8 41 S 36 41 E	Rounce, Staples
Kwiro, popl., mssn.	T6	Ulanga	8 40 S 36 41 E	Cribb et al.
[Kwko.] see Kwamkoro	T3	Lushoto	5 08 S 38 36 E	Haerdi, Eggeling, Cribb et al.
[Kwole] see Bululu	T3	Lushoto	c.5 03 S 38 23 E–5 04.5 S 38 33 E	Zimmermann
[Kwologongi] see Kwalukonge	T3	Lushoto	c.4 55 S 38 08 E–4 57 S 38 16 E	Peter
Kyabakara, camp	U2	Ankole	c.0 11 S 30 13 E	Synnott
Kyabana, ln.	U4	Mengo	0 08 N 32 56 E	Dummer
Kyabandara, ln.	U2	Toro	0 16 N 30 19 E	Osmaston
Kyadondo, co.	U4	Mengo	c.0 25 N 32 35 E	Dummer, Liebenberg
Kyagwe, co.	U4	Mengo	c.0 20 N 32 50 E	Numerous
Kyahara, f.r. (Mwenge Co. North)	U2	Toro	c.0 50 N 30 40 E	St. Clair-Thompson, A.M.S. Smith
Kyai, for.	K4	Machakos	1 43 S 37 32 E	
Kyaka, co.	U2	Toro	c.0 30 N 31 00 E	Osmaston, Snowden

Name		District	Coordinates	References
Kyaka, popl., mssn., t.c.	T1	Bukoba	1 16 S 31 25 E	Lind, Gillman, Procter
[Kyamahunga] see Kyamahungu	U2	Ankole	0 27 S 30 07 E	Eggeling
Kyamahungu, popl.	U2	Ankole	0 27 S 30 07 E	Purseglove
Kyamatu, popl.	K4	Kitui	1 34 S 38 22 E	Kuchar
[Kyambu] see Kiambu	K4	Kiambu	1 10.5 S 36 50 E	Eggeling, Purseglove
[Kyamuhunga] see Kyamahungu	U2	Ankole	0 27 S 30 07 E	Reakes Williams
Kyamutwara, area	T1	Bukoba	c.1 20 S 31 48 E	Harker
Kyanamugera, popl.	U4	Mengo	0 26 N 31 50 E	Gibson
Kyanamukaka, popl.	U4	Masaka	0 30 S 31 42 E	W.H. Lewis
Kyanga Camp, Kalinzu For.	U2	Ankole	c.0 25 S 30 00 E	Sangster
Kyanite Mine, Tsavo Park	K7	Teita	3 16 S 37 53.5 E	Lock
[Kyanjojo] see Kyenjojo	U2	Toro	0 37 N 30 38 E	
Kyarumba, popl., mssn., r.h.	U2	Toro	0 08 N 29 57 E	
Kyasanduka, L.	U2	Ankole	0 17.5 S 30 03 E	
[Kyasoweri] see Kyosoweri	U3	Mbale	1 20 N 34 40.5 E	
Kyebe, popl.	U4	Masaka	0 56 S 31 40 E	Drummond & Hemsley, Purseglove, A.S. Thomas
Kyegegwa, popl., r.h., mssn., t.c.	U2	Toro	0 29 N 31 04 E	St. Clair-Thompson
[Kyegudi] see Kijude	U4	Mengo	0 14 N 32 53 E	?Dummer
Kyeke	U4	Masaka	Not traced	Purseglove
Kyela, popl., p.p.	T7	Rungwe	9 35 S 33 51 E	Cribb & Grey-Wilson, Wingfield
Kyembogo, Farm Institute	U2	Toro	0 42 N 30 20 E	Sturdy
Kyemeire, popl., sch.	U3	Busoga	0 28 N 33 41 E	G.H.S. Wood
Kyenjojo, popl.	U2	Toro	0 37 N 30 38 E	Purseglove
Kyere, popl.	U3	Teso	1 29 N 33 36 E	Chandler
Kyerwa, popl.	T1	Bukoba	1 22 S 30 47 E	Ford, Procter
[Kyes Oewri] see Kyosoweri	U3	Mbale	1 20 N 34 40.5 E	
Kyesowe, ln.	U1	Karamoja	1 47 N 34 43 E	G.H.S. Wood
[Kyesoweri] see Kyosoweri	U3	Mbale	1 20 N 34 40.5 E	Eggeling, Tweedie
[Kyewaga] see Kyiwaga	U4	Mengo	c.0 05 N 32 30 E	Maitland, Dawkins

Name		Region	Coordinates	Authority
[Kyikukama] see Jekukumia	T2	Arusha	3 14 S 36 45.5 E–3 15 S 36 49.5 E	Richards
Kyimbila, mssn.	T7	Rungwe	9 17 S 33 39 E	Stolz, Cribb & Grey-Wilson
[Kyimoani] see Kimwani	T1	Biharamulo	c.2 11 S 31 40 E	Dawkins
Kyiwaga, f.r.	U4	Mengo	c.0 05 N 32 30 E	For. Dept.
[Kymbila] see Kyimbila	T7	Rungwe	9 17 S 33 39 E	Fiennes, Michelmore, Verdcourt
Kyoga, co.	U1	Lango	c.1 40 N 32 45 E	Eggeling, Norman
Kyoga, L.	U1/3/4	several	c.1 30 N 33 00 E	Maitland, Purseglove
Kyosoweri, popl., r.c.	U3	Mbale	1 20 N 34 40.5 E	Makerere College
Kyotera, popl., t.c.	U4	Masaka	0 38 S 31 32.5 E	Dummer
Kyrinye (near Jinja)	U3	Busoga	near 0 26 S 33 13 E	Verdcourt
[Kyude] see Kijude	U4	Mengo	0 14 N 32 53 E	Peter
Kyulu, rsta.	K4	Machakos	2 56 S 38 24 E	
[Kyumba] see Kiumba	T3	Lushoto	5 05.5 S 38 38.5 E	
Laangada Damari (Yaida Valley)	T2	Mbulu	c.3 55 S 35 10 E	Richards
Labai, hills	T1	Maswa	2 50 S 34 40 E	Greenway
Labilo, hill	T4	Buha	4 00 S 31 15 E	Bullock
Laboli = Labori, r.h.	U3	Teso	1 27 N 33 15 E	
Laboli = Labori, popl., fy.	U3	Teso	1 28 N 33 14 E	
Laboot, popl., str.	K5	North Kavirondo	1 01 N 34 37.5 E	Hooper & Townsend
Laborary	T2	Arusha	c.3 15 S 36 54 E	Greenway
Labori = Laboli, r.h.	U3	Teso	1 27 N 33 15 E	Maitland
Labori = Laboli, popl., fy.	U3	Teso	1 28 N 33 14 E	
Labot, popl.	K3	Elgeyo	1 04 N 35 25 E	
Labur, mt.	K2	Turkana	4 25 N 35 48.5 E	Martin
Labwor, co.	U1	Karamoja	c.2 40 N 33 45 E	A.S. Thomas, Tweedie, J. Wilson
Labwor, area	U1	Karamoja	c.2 30 N 33 45 E	J. Wilson, Eggeling
Labwor Hills, mts., f.r.	U1	Karamoja	c.2 43 N 33 47 E	St. Clair-Thompson
Labworomo = Labworomor, hill	U1	Acholi	3 07 N 32 24 E	
Lac Tula, R.	K7	Tana River	c.0 42 S 38 45 E–0 50 S 39 51 E	

Name	Code	Region	Coordinates	Collectors
Ladonga, popl., mssn., r.h.	U1	West Nile	3 24 N 31 05 E	Eggeling, Langdale-Brown
[La Doriak] see Loodo Ariak				
Ladwong, mt.	K6	Masai	c.1 40 S 36 30 E	
Lafiti, popl.	U1	Acholi	3 06 N 32 19 E	
Lagadema, hill	K7	Kilifi	2 56 S 39 57 E	Dale
[Lagaja] see Lgarya	K4	Meru	0 34 N 38 06 E	Bally & Smith
Lagarja = Lgarya, L.	T2	Masai	3 00 S 35 02 E	A. Moore
[Lager Kidoko] see Lager Kitogo	T2	Masai	3 00 S 35 02 E	Greenway, A. Moore, Richards
[Lager Kikundi] ? see Kitunda	T7	Njombe	9 25 S 34 29 E	von Prittwitz & Gaffron
	T4	Tabora	?6 48 S 33 13 E	von Prittwitz & Gaffron
(Kitunda is shown on the von Prittwitz & Gaffron route)				
Lager Kitogo	T7	Njombe	9 25 S 34 29 E	
Lagh Dera, str./s.	K1/Somalia	Northern Frontier	c.0 15 N 40 15 E–0 15 N 42 17 E	Gillett & Gachathi
Laghi, valle dei (valley of the Lakes*)	U2	Toro	c.0 21 N 29 53 E	Duke of Abruzzi
(*probably Lakes Kitandara)				
Lagh Telangor, str./s.	K1	Northern Frontier	c.0 35 N 39 25 E	Bally & Smith
Lag Ola, str./s.	K1	Northern Frontier	c.3 56 N 40 50 E	Gillett, Bally & Smith, Bally & Carter
Lagumishero, peak, area, fm.	T2	Moshi	c.2 51 S 37 11 E	Carmichael
Laikipia, distr.	K3	Laikipia	c.0 30 N 36 30 E	Numerous
Laikipia, escarp.	K3	Baringo/Laikipia	c.0 25 N 36 08 E	Routledge, J. Thompson, Wimbush
Lairobi = Olairobi, area	T2	Masai	c.3 13 S 35 27 E	Bally
Laisamis, t.c., airstrip	K1	Northern Frontier	1 35.5 S 37 48 E	Bally, T. Adamson, Delamere
[Laisamus] see Laisamis	K1	Northern Frontier	1 35.5 S 37 48 E	J. Adamson
Laiteruk, hill	K1	Northern Frontier	1 44 N 35 43 E	
Laitokitok, p.p., r.h., airstrip	K6	Masai	2 56 S 37 30.5 E	Numerous
[Laitong] see Leitong	T2	Arusha	3 17 S 36 55 E	Greenway
Lak Bogal, str./s.	K1	Northern Frontier	c.2 00 N 39 00 E–0 45 N 41 00 E	
Lak Bor, str./s.	K1	Northern Frontier	c.2 55 N 39 29 E–1 18 N 40 40 E	
Lake Jilori, sw.	K7	Kilifi	3 12 S 39 54 E	
Lake Manyara National Park, g.r.	T2	Mbulu	c.3 22–3 38 S 35 45 E	
Lalachat, str.	U1	Karamoja	1 54 N 34 43.5 E–1 52 N 34 29 E	Tweedie

Name	Grid	District	Coordinates	Collector
Lalachat, popl.	U1	Karamoja	1 59 N 34 37 E	Eggeling
Lalaikitalak, ln.	T2	Masai	4 35 S 36 26 E	Peterson
Lalalei ? see Lowa Lalale	K1	Northern Frontier	?2 04.5 N 36 48 E	J. Adamson
Lalarok, area	K1	Northern Frontier	c.2 12 N 36 46 E	Verdcourt
Lale, popl., fy.	U3	Teso	1 40 N 33 27 E	
Lali, hills	K4/7	Kitui/Kilifi	3 00 S 39 17 E	T. Adamson, Parker, Bally
Lali Hills, popl.	K7	Kilifi	3 04 S 39 16.5 E	D. Wood
Lamadel	K1	Northern Frontier	c.1 09 S 41 12 E	J. Adamson
[Lamagai] see Mbuga ya Larmakau	T2	Mbulu	c.4 10 S 36 10 E	Richards
[Lamani] see Lamunyane	T2	Masai	c.2 28 S 35 30 E	Greenway
[Lambari] see Lombori	T2	Masai	2 37 S 36 08 E	Carmichael
Lambu, I.	U4	Masaka	0 19 S 32 02 E	Philip
Lambwe = Olambwe, str./s.	K5	South Kavirondo	0 28.5 S 34 18 E–0 43 S 34 12 E	Lewis, Napier
Lambwe Valley, for.	K5	South Kavirondo	c.0 40 S 34 17 E	Makin, Glover, Lewis
Lambwe Valley, g.r.	K5	South Kavirondo	c.0 37 S 34 15 E	
Lamia, R.	U2/Zaire	Toro	0 24 N 29 57 E–0 51 N 29 59 E	Eggeling
Lamogi, popl.	U1	Acholi	2 50 N 32 10 E	G.R. Williams
Lamok, str.	K3	Uasin Gishu	c.0 29 N 35 12 E	
Lamu, distr.	K7	Lamu	c.2 05 S 41 00 E	Numerous
Lamu, I.	K7	Lamu	c.2 17 S 40 52 E	Greenway
Lamu, popl.	K7	Lamu	2 16 S 40 54 E	
[Lamungani] see Lamunyane	T2	Masai	c.2 28 S 35 30 E	Vesey-FitzGerald
Lamunyane = Lamuniane, Hills	T2	Masai	c.2 28 S 35 30 E	Scheffler
Lamura, gorge	T2	Arusha	prob. c.3 14 S 36 50 E	Vesey-FitzGerald
[Lamuru] see Limuru	K4	Kiambu	1 07 S 36 38.5 E	
Lamuru = Lamura	T2	Arusha	prob. c.3 14 S 36 50 E	
Lamwo, co.	U1	Acholi	c.3 30 N 32 40 E	
Lananiken, Mathews Range	K1	Northern Frontier	Not traced	J. Bally
Lanconi, Mhinduro	T3	Lushoto	c.4 57 S 38 46 E	Bogner
[Landorossi] see Londorossi	T2	Moshi	2 58.5 S 37 13 E–2 59 S 37 06.5 E	Carmichael

174

Name	Code	Region	Coordinates	Collector
Laridabach, str./s., w.h.	K1	Northern Frontier	c.2 48 N 36 46 E	J. Adamson
Lari Swamp, sw.	K4	Kiambu	1 02 S 36 38.5 E	Bogdan, C.F. Elliot, Greenway
Larmakau, Mbuga ya	T2	Mbulu	c.4 10 S 36 10 E	
Laroda, popl.	T2	Masai	3 13.5 S 35 30.5 E	B.D. Burtt, Peter
Laropi, popl., r.h., t.c.	U1	West Nile	3 34 N 31 49 E	Brasnett, Eggeling, Purseglove
Lasa, mt.	T3	Lushoto	4 35 S 38 05 E	Greenway
Lasiti, mt.	T3	Pare	4 28 S 37 41.5 E	Peter
[Lassa] see Lasa	T3	Lushoto	4 35 S 38 05 E	B.D. Burtt
[Latagipi Swamp] see Lotikipi Plain	K2	Turkana	c.4 18 N 34 42 E	
Latakwen (Ndoto Mts.)	K1	Northern Frontier	Not traced	Newbould
[Lateruk] see Laiteruk	K1	Northern Frontier	1 44 N 35 43 E	
[Latome] see Lotome	U1	Karamoja	2 22 N 34 32 E	
[Lauka] see Luuka	U3	Busoga	c.0 45 N 33 20 E	Karani
[Lawiri] see Yawiri	T7	Njombe	c.9 26 S 34 04 E	
Lebetero, hills	K6	Masai	1 47 S 35 57 E	van Someren
Le Breton Road	K3	Trans-Nzoia	c.1 07 S 34 49 E	Tweedie
Lebugombisso = Legumbisso, ln.	K1	Northern Frontier	c.1 09 S 41 12 E	J. Adamson
Ledingombe, area, popl.	T6	Kilosa	7 04.5 S 36 37.5 E	Meyer
[Legaruti] see Lekuruki	T2	Arusha	3 15.5 S 36 55 E	Greenway
Legas, str., ridge	K1	Northern Frontier	c.1 15 N 36 41 E–1 01 N 36 43 E	Kerfoot
[Legeri] see Leserin	K6	Masai	1 47 S 35 42 E	Kuchar
[Leggas] see Legas	K1	Northern Frontier	c.1 15 N 36 41 E–1 01 N 36 43 E	
Legoman, popl., area	K6	Masai	c.0 35 S 35 55 E	
Legumbisso, ln.	K1	Northern Frontier	c.1 09 S 41 12 E	J. Adamson
[Leikipia] see Laikipia	K3	Laikipia	c.0 30 N 36 30 E	
Leitokitok, spr., ln.	T2	Masai	c.3 12 S 35 37 E	
Leitong, ln.	T2	Arusha	3 17 S 36 55 E	Greenway
Lekandiro, L.	T2	Arusha	3 12.5 S 36 53.5 E	Richards
Lekuruki, hill	T2	Arusha	3 15.5 S 36 55 E	Richards
Lekuruki, popl.	T2	Moshi	3 16 S 36 58 E	Richards

Lelan, f.r.	K3	Elgeyo	1 17 N 35 27 E	Bridson
Lelwa, popl.	T3	Lushoto	4 30 S 38 30 E	Raadts, Hanid & Nuernbergk
Lemagrut, cone	T2	Masai	3 10 S 35 21 E	Jaeger, Greenway, Newbould
[Lemalok] see Lomolok	K1	Northern Frontier	1 10 N 37 24 E	J. Adamson, Bally
Lembeni, popl., rsta.	T3	Pare	3 47 S 37 37 E	Numerous
Lembus, for.	K3	Ravine	c.0 08 N 35 40 E	R.M. Graham
Lemek, peak	K6	Masai	1 04.5 S 35 23 E	
Lemek, str. s.	K6	Masai	1 08 S 35 30 E—1 08 S 35 20.5 E	Hornby
Lemek, ln.	K6	Masai	1 05.5 S 35 23 E	
Lemesikio, popl.	T2	Masai	1 56.5 S 35 40 E	Verdcourt
[Lemesule] see Lumesule	T8	Tunduru/Masasi	10 16 S 37 30 E—11 14 S 38 06 E	Milne-Redhead & Taylor
Lemisikio, str.	K6	Masai	c.1 34 S 35 34 E	
[Lemoko] see Lemek	K6	Masai		Glover
Lemosho, area, str., glades	T2	Moshi	c.3 02 S 37 07 E	Robertson, Vesey-FitzGerald
[Lemunge] see Munge	T2	Masai		Peter
Lemuta, hill, b.h.	T2	Masai	2 42 S 35 16 E	Oteke
Lendanai, popl.	T2	Masai	4 05 S 37 09 E	
[Lendenai] see Lendanai	T2	Masai	4 05 S 37 09 E	Carmichael
[Lendiya] see Lendoiya	T2	Arusha	3 12 S 36 54 E	Greenway
Lendoiya, sw., Is.	T2	Arusha	3 12 S 36 54 E	Richards
Lendu, area	U1	West Nile	c.2 27 N 30 47 E	Chancellor, Eggeling, Stuhlmann

(N.B. earlier collections may be from Zaire as Lendu area then extended much further down, west of Lake Albert)

[Lengai] see Rongai, for.	T2	Moshi	3 06 S 37 02 E	Richards
Lengai, Oldonyo, mt.	T2	Masai	2 45 S 35 54 E	Richards
Lengalilla, hill	K4	North Nyeri	0 03.5 N 37 24 E	Schelpe
Lengebai, area	K6	Masai	1 09 S 35 10 E	Glover et al.
[Lengejabu] see Lengijawe	T2	Arusha	3 12.5 S 36 36 E	Carmichael
[Lengidjawe] see Lengijawe	T2	Arusha	3 12.5 S 36 36 E	Peter
[Lengijabi] see Ol Donyo Lengiyabe	T2	Masai	4 20 S 36 56 E	Procter
Lengijawe, hill	T2	Arusha	3 12.5 S 36 36 E	

Lengiyabe, Ol Donyo, hill	T2	Masai	4 20 S 36 56 E	
[Lengoman] see Legoman	K6	Masai	0 35 S 35 55 E	
Lensayu, w.h.	K1	Northern Frontier	2 52 N 39 16 E	Bally
Lenyoro = Lenyora	K6	Masai	c.1 53 S 35 49 E	Bally
Leo, Koich R.	U1	West Nile	Not traced	Eggeling
Leopard Hill, ln.	T2	Arusha	3 16.5 S 36 56 E	Richards
Leopard Point = Leopard Hill	T2	Arusha	3 16.5 S 36 56 E	Greenway
Leopard Rock, g.r. H.Q.	K4	Meru	0 14 N 38 12 E	Numerous
Leopards Cove (near Dar es Salaam)	T6	Uzaramo	near 6 48 S 39 15 E	Batty
Lerai, conservation centre, for.	T2	Masai	3 13 S 35 30 E	
Lereko, N., peak	K3/4	Naivasha/S. Nyeri	0 33.5 S 36 40 E	Bally, Napier
Lereko, S., peak	K3/4	Naivasha/S. Nyeri	0 35 S 36 40.5 E	
[Leri] see Lerai	T2	Masai	3 13 S 35 30 E	Greenway
[Leroghi Forest] see Lorogi Forest	K1	Northern Frontier	c.1 00 N 36 50 E	Kerfoot, Rammell
[Leroghi Plateau] see Lorogi Plateau	K1	Northern Frontier	c.1 00 N 36 35 E	
[Leroki Forest] see Lorogi Forest	K1	Northern Frontier	c.1 00 N 36 50 E	
[Leroki Plateau] see Lorogi Plateau	K1	Northern Frontier	c.1 00 N 36 35 E	Leakey
Lerong (W. side Lemagrut)	T2	Masai	c.3 10 S 35 17 E	Newbould
[Lesagoi] see Losergoi	K1	Northern Frontier	c.2 08 N 36 38 E	J. Adamson
Leserin, hills	K6	Masai	1 47 S 35 42 E	
Leseru, rsta.	K3	Uasin Gishu	0 35 S 35 10.5 E	Brodhurst Hill
Lesirikan, w.h.	K1	Northern Frontier	1 48 N 36 58.5 E	Carter & Stannard
[Lesokonoi] see Losokonoi	T2	Arusha	c.3 15 S 36 52 E	Richards
[Lessago] see Losergoi	K1	Northern Frontier	c.2 08 N 36 38 E	J. Adamson
Lesser Kiboko = Kiboko (part), R.	K4/6	Machakos/Masai	c.2 30 S 37 15 E–2 09 S 37 54 E	
[Letema Pool] see Retima	T1	Musoma	2 17 S 34 46 E	Greenway
[Levessabi] see Lwessabi	T1	Mwanza	2 00 S 33 04 E	Conrads
Lewa, popl., mssn.	T3	Lushoto	5 06 S 38 24 E	Archbold
Leya, R.	U1	West Nile	c.3 35 N 31 35 E	Greenway & Eggeling
Ley's Farm Turning	K3	Trans-Nzoia	1 03 N 34 56.5 E	Tweedie

178

Name	Grid	Location	Coordinates	Collector
Likoni, fy.	K7	Mombasa	4 05 S 39 40 E	Milne-Redhead & Taylor, Migeod
Likuyu, str.	T8	Songea	c.10 45 S 35 13 E	Richards
Limestone Gorge	T7	Mbeya	prob. c.8 57 S 33 13 E	Dummer, Napier
[Limoru] see Limuru	K4	Kiambu	1 07 S 36 38.5 E	
[Limur] see Nyimur R.	U1	Acholi	c.3 42 N 32 36 E	
Limuru, rsta., popl., t.c.	K4	Kiambu	1 07 S 36 38.5 E	Numerous
Limutet, hill	K5	Kisumu-Londiani	0 07 S 35 31.5 E	
[Limutit] see Limutet	K5	Kisumu-Londiani	0 07 S 35 31.5 E	
Lindi, creek, harbour	T8	Lindi	c.10 00 S 39 44 E	Drummond & Hemsley
Lindi, distr.	T8	Lindi	c.10 00 S 39 00 E	Gillman
(includes part of Nachingwea Distr. for F.T.E.A.)				
Lindi, popl.	T8	Lindi	10 00 S 39 43.5 E	Numerous
[Lindu] see Lendu	U1	West Nile	c.2 27 N 30 47 E	Greenway & Eggeling
[Liondo] see Liando	T6	Ulanga	8 42 S 36 47 E	Schlieben
Lipanga, mts.	T7	Njombe	c.9 06 S 34 03 E	Goetze
[Lipange] see Lipanga, mts.	T7	Njombe	c.9 06 S 34 03 E	Goetze
[Lipanje] see Lipanga, mts.	T7	Njombe	c.9 06 S 34 03 E	
[Lipanye] see Lipanga, mts.	T7	Njombe	c.9 06 S 34 03 E	Goetze
Lipumba, popl.	T8	Songea	10 51 S 35 01.5 E	Stenhouse
Lipumburu, popl.	T8	Masasi	10 59 S 38 51 E	Gillman
Lira, popl., r.h., f.r.	U1	Lango	2 15 N 32 55 E	Numerous
[Lirima] see Ririma	U3	Mbale		
Liroro, R.	T7	Njombe	c.9 15 S 34 04 E	Gillman
Lisekese, popl.	T8	Masasi	10 42 S 38 47 E	Richards
Lisingita, area	T2	Masai	prob. c.2 45 S 36 15 E	Kerfoot
[Lisirika] see Lesirikan	K1	Northern Frontier	1 48 N 36 58.5 E	Thulin & Mhoro
Lisitu, popl.	T7	Njombe	9 39 S 34 39 E	Prizov
[Litalikj] see Sitalike	T4	Mpanda		
Liteho, popl., D.O. post	T8	Newala	10 31 S 39 31 E	Eggeling, Hay
Litembo, mssn.	T8	Songea	10 58 S 34 50 E	Semsei

Name		Region	Coordinates	Collectors
Litenga, hill	T8	Songea	10 45 S 35 16 E	Milne-Redhead & Taylor
[Litengo] see Litenga	T8	Songea	10 45 S 35 16 E	Eggeling, Lynes
[Lithoni] see Litoni	T7	Njombe	9 26 S 34 50 E	Eggeling
Litobu	U4	?	Not traced	Godman
		(probably Masaka–Mbarara)		
Litoni, f.r.	T7	Njombe	9 26 S 34 50 E	Procter
[Little Jambe] see Little Yambe	T3	Tanga	5 05 S 39 10 E	Faulkner
Little Kasenene, hill	U2	Bunyoro	near 1 44 N 31 28 E	Dawkins
Little Meru, peak	T2	Arusha	3 12.5 S 36 46.5 E	Vesey-FitzGerald
Little Momela, L.	T2	Arusha	3 13.5 S 36 53.5 E	Greenway
Little Ruaha, R.	T7	Iringa	8 19 S 35 12 E–7 17 S 35 28 E	
Little Yambe, I.	T3	Tanga	5 05 S 39 10 E	Faulkner
Litungura, popl.	T8	Tunduru	c.10 41 S 36 34 E	Milne-Redhead & Taylor
Lituno's = Kwa Lituno, popl.	T8	Songea	10 41 S 35 22 E	
Liuli, mssn.	T8	Songea	11 05 S 34 37 E	Hay, Nobbs
[Livalonde] see Livalonge	T7	Iringa	8 41 S 35 10.5 E	Congdon
Livalonge, tea est.	T7	Iringa	8 41 S 35 10.5 E	Perdue & Kibuwa
Livingstone, f.r.	T7	Mbeya/Rungwe/Njombe	c.9 05 S 33 48 E	Procter, Cribb & Grey-Wilson
Livingstone, mts.	T7	Njombe	c.9 25 S 34 15 E–10 15 S 34 45 E	Numerous
		(Goetze used in a wider sense)		
[Livule] see Liwale	T6	Morogoro	6 04 S 37 32.5 E–6 17 S 37 41 E	
Liwale, str.	T6	Morogoro	6 04 S 37 32.5 E–6 17 S 37 41 E	
Liwale, popl.	T8	Kilwa	9 46 S 37 56 E	Numerous
[Liwale] see Liwale Mkubwa	T8	Kilwa	c.9 51 S 37 42 E–9 08 S 38 17 E	Magogo & Rose Innes
[Liwale Makubwa] see Liwale Mkubwa	T8	Kilwa	c.9 51 S 37 42 E–9 08 S 38 17 E	Busse, Lommel, Rose Innes
Liwale Mkubwa, R.	T8	Kilwa	c.9 51 S 37 42 E–9 08 S 38 17 E	
Liwengula, f.r.	T8	Lindi	9 59 S 39 32 E	
Liwiri-Kiteza, f.r.	T8	Songea	c.11 10 S 34 50 E	Milne-Redhead & Taylor, Semsei, Eggeling
Liwiti = Libiti, hills	T8	Kilwa	9 05 S 39 10 E	Gillman
[Liwuale] see Liwale	T6	Morogoro	6 04 S 37 32.5 E–6 17 S 37 41 E	Holtz

[Liwule] see Liwale				
Liyoma, area	Morogoro	T6	6 04 S 37 32.5 E–6 17 S 37 41 E	
Llewellyn's Farm	Mwanza	T1	2 48 S 32 57 E	Tanner
Loaltian = Supuko Looltian	Trans-Nzoia	K3	1 07 N 34 50 E	Tweedie
[Loapetiethe] ? see Lokapeliethe	Naivasha/Nakuru	K3	0 02–0 18 S 36 15–36 21 E	Piers
Lobelia Corner	Karamoja	U1	2 52 N 34 29 E	Dale
Lobo, hill	Trans-Nzoia	K3	1 08.5 N 34 42.5 E	Tweedie
Loboi, popl.	Musoma	T1	1 58 S 35 13 E	Greenway
[Loboret] see Loburet	Baringo	K3	0 21 N 36 04 E	Tweedie
[Loboret] see Loburet	Turkana	K2	c.4 10 N 35 21 E–4 14 N 35 08 E	
Lobot Crossroads	Turkana	K2	c.4 12 N 35 12 E	
Loburet, area	Elgeyo	K3	1 05 N 35 24 E	Tweedie
Loburet, str./s.	Turkana	K2	c.4 12 N 35 12 E	
Loburiangiromo	Turkana	K2	c.4 10 N 35 21 E–4 14 N 35 08 E	
Loburre, gorge	Turkana	K2	Not traced	J. Wilson
Lochinimwoi, str.	Karamoja	U1	3 44 N 33 57 E	Brasnett
[Lochoi] ? see Lochowai	Turkana	K2	prob. c. 1 50 N 36 20 E	Mwangangi
Lochomon, R.	Karamoja	U1	2 11 N 34 31 E	A.S. Thomas
Lochowai, hill	Karamoja	U1	2 25 N 34 08 E–2 20 N 34 02.5 E	J. Wilson
Lodare, for.	Karamoja	U1	2 11 N 34 31 E	
[Lodokeminet] see Lodoketemit	Masai	T2	c.2 04.5 S 35 36 E	Carmichael
[Lodoketeminit] see Lodoketemit	Karamoja	U1	c.2 17 N 34 39 E	Kerfoot
Lodoketemit, str.	Karamoja	U1	c.2 17 N 34 39 E	
Lodumaiyu = Kawanja Lodomayu	Karamoja	U1	c.2 17 N 34 39 E	Kerfoot
Lodwar, popl., airstrip	Arusha	T2	prob. c.3 17 S 36 56.5 E	Greenway
Lodwar Cone, hill	Turkana	K2	3 07 N 35 36 E	Numerous
Lofia, popl.	Turkana	K2	3 10 N 35 35 E	Numerous
Lofia, str.	Iringa	T7	7 43 S 36 44 E	
[Lofio] see Lofia	Iringa	T7	7 45 S 36 34 E–7 38 S 36 47 E	Goetze
[Lofya] see Lofia	Iringa	T7	7 45 S 36 34 E–7 38 S 36 47 E	Goetze
	Iringa	T7	7 43 S 36 44 E	Carmichael

181

Name	Region	Area	Coordinates	Collector(s)
[Lokadema] see Lagadema	K4	Meru	0 34 N 38 06 E	Bally & Smith
Lokapeliethe, hill	U1	Karamoja	2 52 N 34 29 E	A.S. Thomas
Lokathir, hill	U1	Karamoja	2 08 N 34 38 E	
[Lokbor] see Lak Bor	K1	Northern Frontier	c.2 55 N 39 29 E–1 18 N 40 40 E	Dale
Lokibenet, area	K2	Turkana	c.1 48 N 35 07 E	Tweedie
Lokichar, str.	K1	Northern Frontier	c.3 02 N 36 07 E	Hemming
Lokichar, popl.	K1	Northern Frontier	2 22.5 N 35 40 E	Hemming, Mathew, Carter & Stannard
[Lokichoggio] see Lokichokio	K2	Turkana	4 12 N 34 21 E	Poultney
Lokichokio, p.p., airstrip	K2	Turkana	4 12 N 34 21 E	
Lokie, sw.	T2	Arusha	3 15.5 S 36 53 E	Richards, Greenway
[Lokiperengitome] see Lokiporangatome	U1	Karamoja	2 20 N 34 21 E	J. Wilson
Lokiporangatome, popl.	U1	Karamoja	2 20 N 34 21 E	
Lokitanyala, r.c., sand R.	U1/K2	Karamoja/Turkana	2 22.5 N 34 56 E	
Lokitanyala, popl.	K2	Turkana	2 22 N 34 55 E	
Lokitau ? = Lokathir	U1	Karamoja	?2 08 N 34 38 E	A.S. Thomas
Lokitaung, p.p., well, gorge	K2	Turkana	4 16 N 35 46 E	Numerous
Lokitela Farm	K3	Trans-Nzoia	1 00 N 34 52 E	Tweedie
[Lokitonyala] see Lokitanyala	U1,K2	Karamoja/Turkana	2 22.5 N 34 56 E	J. Wilson
[Lokitoyale] see Lokitanyala	U1,K2	Karamoja/Turkana	2 22.5 N 34 56 E	J. Wilson
[Lokitunyalla] see Lokitanyala	U1,K2	Karamoja/Turkana		Champion
Lokolio = Lokolyo, wells		Sučan *not* Kenya		Dale
Lokori, popl., mssn., airstrip	K1	Northern Frontier	1 56 N 36 01 E	Mathew, Hemming, Mwangangi
Lokuloko, w.h.	K1	Northern Frontier	1 59 N 37 55.5 E	Sato
Lokung, popl., r.h.	U1	Acholi	3 36 N 32 42 E	Purseglove
Lokung, f.r.	U1	Acholi	c.3 34 N 32 43 E	Greenway & Hummel
Lokwien, campsite	K2	West Suk	1 55.5 N 35 22 E	Lye
[Lolaikaimaishi] see Lolaikumaishi	T2	Masai	4 51 S 37 12 E	Page-Jones
Lolaikumaishi = Loloigumaishi, hills	T2	Masai	4 51 S 37 12 E	
Lolbene, mt.	T2	Masai	3 57 S 37 07 E	Page-Jones, Moreau
Loldaika, mts.	K4	North Nyeri	c.0 12 N 37 07 E	J. Bally, Lawton

Lomolok, mt.	K1	Northern Frontier	1 10 N 37 24 E	J. Adamson
Lomoru Itae, w.h.	K2	Turkana	4 22 N 35 33 E	Carter & Stannard
Lomuleng, w.h., r.h.	Sudan *not* U1		3 57 33 00 E	A.S. Thomas
Lomunga, popl.	U1	West Nile	3 24.5 N 31 17 E	Brooks
Lomunyenakwan, str./s.	K2	Turkana	3 31 N 35 17 E–3 31 N 35 26 E	Carter & Stannard
Lomusio, popl. (Kaabong—Kamion)	U1	Karamoja	3 31 N 34 08 E–3 43 N 34 12 E	Eggeling
Lomut, popl., sch.	K2	West Suk	1 24 N 35 34 E	
Lomwaga, mt.	U1	Acholi	3 49 N 32 54 E	Eggeling, Greenway & Hummel
Lomwe = Romwe, popl.	T3	Handeni	5 40 S 37 41.5 E	
Londergess, hills	T2	Masai	4 40 S 37 02 E	Vesey-FitzGerald
Londiani, mt.	K5	Kisumu-Londiani	0 07 S 35 44 E	
Londiani, popl., rsta.	K5	Kisumu-Londiani	0 10 S 35 35.5 E	Numerous
[Londorosi] see Londorossi	T2	Moshi	2 58.5 S 37 13 E–2 59 S 37 06.5 E	Carmichael
Londorossi, str.	T2	Moshi	2 58.5 S 37 13 E–2 59 S 37 06.5 E	
Londorossi, Glades, for.	T2	Moshi	2 57 S 37 09 E	Richards
Londuka Gorge (E. of Meru)	T2	Arusha	prob. c.3 15 S 36 49 E	Richards
[Longai] see Rongai, for.	T2	Moshi	3 06 S 37 02 E	Richards
Longalat ? see Logelat	U1	Karamoja	?3 33 N 33 39 E–3 20 N 33 44 E	A.S. Thomas
[Longidjawe] see Lengijawe	T2	Arusha	3 12.5 S 36 36 E	Magogo & Estes, Peter
Longido, mt., f.r.	T2	Masai	2 41 S 36 42 E	Numerous
Longido, popl., customs post, mssn.	T2	Masai	2 44 S 36 42 E	Numerous
Longil, L., sw.	T2	Arusha	3 15.5 S 36 53 E	Richards, Greenway & Kanuri
[Longili] see Lonyili	U1/Sudan	Acholi/Karamoja	c.3 45 N 33 38 E	J. Wilson
[Longomwagandi] see Lango ya Mwagandi	K7	Kwale	4 14 S 39 25 E	
[Longongare] see Ngongongare	T2	Arusha	3 17 S 36 54 E	Greenway
Longonot, mt., crater	K3	Naivasha	0 55 S 36 27 E	Numerous
Longonot, rsta.	K3	Naivasha	0 53 S 36 29.5 E	Numerous
Longosa, hill	T1	Musoma	2 03 S 35 13 E	Greenway
[Longusa] see Longuza	T3	Lushoto/Tanga	5 03 S 38 42 E	Numerous
Longuza, f.r.	T3	Lushoto/Tanga	5 03 S 38 42 E	Numerous

186

Name	Code	Region	Coordinates	Source
Longuza, old pl.	T3	Lushoto/Tanga	5 06 S 38 42 E	Numerous
Longuza, peak	T3	Lushoto	5 06 S 38 41 E	
Lonyili, mts.	U1	Acholi/Karamoja	c.3 45 N 33 38 E	J. Wilson
Lonyo Lengiyo, mt.	K1	Northern Frontier	c.1 25 N 37 14 E	
Loodo Ariak, area	K6	Masai	c.1 40 S 36 30 E	
Loolmalassin, mts.	T2	Mbulu/Masai	c.3 03 S 35 50 E	Jaeger, Bally
Loonguru-m-yu, for.	T2	Arusha	prob. c.3 12 S 36 42 E	Richards
Loosuate, hills	K6	Masai	1 57 S 36 02 E	
Looya, str.	K2	Turkana	c.2 25 N 35 00 E	
Lopet Plateau	K1	Northern Frontier	1 30 N 36 45 E	Carter & Stannard
Loporokocho, popl.	U1	Karamoja	1 32 N 34 44 E	Eggeling
Lopua, R.	U1	Karamoja	c.3 33 N 34 19 E –3 21 N 34 17 E	Brasnett
Lordariak, hill	K6	Masai	1 40 S 36 31.5 E	
[Loreko] see Lereko	K3/4	Naivasha/S. Nyeri		
Lorengedwat, popl.	U1	Karamoja	2 22 N 34 35 E	Bally
[Lorengeppi] see Lorengipe	K2	Turkana	c.2 33 N 34 58 E	Eggeling
[Lorengidwat] see Lorengedwat	U1	Karamoja	2 22 N 34 35 E	Bally
[Lorengikipe] see Lorengkipi	K2	Turkana	c.2 33 N 34 58 E	Eggeling
Lorengikipi, dam	U1	Karamoja	2 24 N 34 00 E	J. Wilson
Lorengipe, str.	K2	Turkana	c.2 33 N 34 58 E	Bogdan, Hemming, Newbould
[Lorgasalik] see Ol Lorgosailie	K6	Masai	1 42 S 36 26 E	
[Lorgosailic] see Ol Lorgosailie	K6	Masai	1 42 S 36 26 E	Numerous
[Lorgosailie] see Ol Lorgosailie	K6	Masai	1 42 S 36 26 E	Numerous
Lorian Plain, area	K1	Northern Frontier	c.0 50 N 39 15 E	Bally
Lorian Swamp, sw.	K1	Northern Frontier	c.0 45 N 39 30 E	Carmichael
Lorien	T2	Masai	c.2 07 S 35 38 E	Dale
Lorienaton = Lorienetom		Sudan *not* Kenya (although shown on certain maps as such)		
Loriu Plateau	K1	Northern Frontier	c.2 20 N 36 23 E	Mathew
Loro, popl.	U1	Karamoja	2 12 N 34 53 E	J. Wilson, Tweedie
Lorogi Forest, for.	K1	Northern Frontier	c.1 00 N 36 50 E	Carter & Stannard

Name	Code	Region	Coordinates	Collector
Lorogi Plateau, plat.	K1	Northern Frontier	c.1 00 N 36 35 E	Kerfoot
[Lorogum] see Lorugumu	K2	Turkana	2 53 N 35 15 E	Popov
[Lorogumo] see Lorugumu	K2	Turkana	2 53 N 35 15 E	Martin
[Lorogumu] see Lorugumu	K2	Turkana	2 53 N 35 15 E	Bally
[Loroki Forest] see Lorogi Forest	K1	Northern Frontier	c.1 00 N 36 50 E	
[Loroki Plateau] see Lorogi Pleteau	K1	Northern Frontier	c.1 00 N 36 35 E	
Lorosuk, mt.	K2	Turkana	2 16 N 35 07 E	J. Wilson
[Lorsuate] see Loosuate	K6	Masai	1 57 S 36 02 E	Kuchar
Lorugumu, popl., w.h.	K2	Turkana	2 53 N 35 15 E	Leippert
Loruk, t.c.	K3	Baringo	0 43 N 36 02 E	Carter & Stannard
Lorukumu = Lorugumu	K2	Turkana	2 53 N 35 15 E	
Losergoi, area	K1	Northern Frontier	c.2 08 N 36 38 E	
[Losetete] see Lossitete	T2	Mbulu	3 07 S 35 50 E–3 14 S 35 56 E	Carmichael
Losho (Loliondo f.r.)	T2	Masai	2 06 S 35 37 E	Carmichael
Losidok = Lothidok, hill	K2	Turkana	3 21 N 35 47 E	Mortimer
[Losirua] see Olosirwa	T2	Masai	3 03.5 N 35 46.5 E	
Loskitok, mt.	T2	Masai	5 17 S 37 23 E	
[Loskitu] see Loskitok	T2	Masai	5 17 S 37 23 E	B.D. Burtt
[Losogoni] see Losogonoi	T2	Masai	2 26 S 35 32.5 E	Greenway
Losogonoi, area, popl.	T2	Masai	2 26 S 35 32.5 E	Greenway
Losokonoi, ln.	T2	Arusha	c.3 15 S 36 52 E	Richards
Lossitete, str., area	T2	Mbulu	3 07 S 35 50 E–3 14 S 35 56 E	
Lossogonoi, area, plateau	T2	Masai	c.4 00 S 37 15 E	Elliot
Lotaes ? see Lotarr	K1	Northern Frontier	?2 24 N 36 38 E	J. Adamson
[Lotagipi Swamp] see Lotikipi Plain	K2	Turkana	c.4 18 N 34 42 E	
Lotarr, w.h.	K1	Northern Frontier	2 24 N 36 38 E	Champion
Lothagam, hill	K1	Northern Frontier	2 56 N 36 04 E	
Lothidok = Losidok, hill	K2	Turkana	3 21 N 35 47 E	
Lotikipi Plain, area	K2	Turkana	c.4 18 N 34 42 E	
[Lotisan] see Lotizan	U1	Karamoja	c.3 00 N 34 27 E	A.S. Thomas

Name		Region	Coordinates	Collector
Lotizan, str.	U1	Karamoja	c.3 00 N 34 27 E	Eggeling
Lotofwa, L., sw. (near Old Ukuti)	U1	Acholi	near 3 41 N 33 32 E	Verdcourt, Lye, Tweedie
Lotome, popl., mssn.	U1	Karamoja	2 22 N 34 32 E	
[Lototuru] see Lotuturu		Acholi	3 44 N 32 55 E	
Lotuke, Mt.		Sudan *not* Uganda	4 07 N 33 47 E	J. Wilson
Lotuturu, popl., r.h.	U1	Acholi	3 44 N 32 55 E	
Lowaweregoi, mt.	K1	Northern Frontier	1 13.5 N 37 09 E	Newbould
Lower Kabete, popl., t.c.	K4	Kiambu	1 14.5 S 36 43 E	
Lower Sagana Falls	K4	N. Nyeri/S. Nyeri	c.0 15 S 37 12.5 E	Schelpe
Lowire = Luwiri, old popl., hill	T1	Mwanza	3 08 S 32 47 E	Stuhlmann
[Loya] see Loiya	K2	Turkana	c.3 15 N 34 42 E	Tweedie
[Loya] see Looya	K2	Turkana	c.2 25 N 35 00 E	Eggeling
Loya, popl.	T4	Tabora	4 58 S 33 55 E	Richards
Loyaro, mt.	U1	Karamoja	3 55 N 33 55 E	A.S. Thomas
Loyeni, I.	K1	Northern Frontier	2 47 N 36 41.5 E	
Loyoro, popl., r.c.	U1	Karamoja	3 21 N 34 16 E	Numerous
Loyoro, str.	U1	Karamoja	c.3 20 N 34 14 E	
Loyoroit, str.	U1	Karamoja	2 47 N 33 45 E–2 52 N 33 38 E	
[Loyoru] see Loyoro	U1	Karamoja	3 21 N 34 16 E	Tweedie
Lozut, ln.	U1	Karamoja	c.3 34 N 34 16 E	A.S. Thomas
Luago, str.	T4	Mpanda	6 29 S 31 29 E–6 23.5 S 31 23 E	A.S. Thomas
[Luala] see Luwala	U4	Mengo	0 27 N 33 07 E	Richards
[Luale Liwale] see Liwale	T6	Morogoro	6 04 S 37 32.5 E–6 17 S 37 41 E	Dummer
[Luanbabya] see Wambabya	U2	Bunyoro	c.1 26 N 31 08 E	
Luangwa, str.	T7	Rungwe	c.9 03 S 33 36 E	Dawe
Luato, old area	T5	Dodoma	c.6 22 S 35 35 E	Goetze
Luato = Luatu, popl.	T5	Dodoma	6 27 S 35 40 E	Busse
Luatu, popl.	T5	Dodoma	6 27 S 35 40 E	
Lubafu = Rubafu, popl.	T1	Bukoba	1 02 S 31 50 E	Gillman, Haarer
Lubaga, f.r.	T1	Shinyanga	3 36 S 33 25 E	Carmichael, Soncta

Name	Region	Location	Coordinates	Collector
[Lubale] see Rubaare	U2	Ankole	1 01 S 30 11 E	H.B. Johnston

(There are several Lubale's on modern maps but this one seems most probable for this collector)

Name	Region	Location	Coordinates	Collector
Lubale, popl., mssn.	U2	Ankole	0 38 S 30 07 E	
[Lubando] see Rubondo	T1	Mwanza	c.2 20 S 31 52 E	Carmichael
[Lubanj] see Lubanyi	U3	Busoga	0 38 N 33 05 E	G.H.S. Wood
Lubanyi, hill, f.r.	U3	Busoga	0 38 N 33 05 E	
[Lubare] see Rubaare	U2	Ankole	1 01 S 30 11 E	Eggeling
Lubili = Lubiri, popl.	T1	Mwanza	3 03 S 32 47 E	Tanner
Lubiri, popl.	T1	Mwanza	3 03 S 32 47 E	
Lubiri Mbuga, marsh	T5	Mpwapwa	6 07 S 36 35 E	Anderson
Lubolo, hill	U3	Busoga	1 08 N 33 33.5 E	G.H. Wood
[Lubondo] see Rubondo	T1	Mwanza	c.2 20 S 31 52 E	Carmichael
Lubongolo, for.	T7	Iringa	c.8 03 S 36 04 E	Mwasumbi
Lubowa, est.	U4	Mengo	0 14 N 32 34 E	
Lubugwe, str.	T4	Mpanda	6 14 S 29 51 E–6 15 S 29 44 E	Mahinda, Mgaza
Lubulungu, str.	T4	Mpanda	6 10 S 29 50 E–6 09 S 29 44 E	Jefford et al.
Lubwa's old popl.	U3	Busoga	0 25 N 33 21 E	Scott Elliot, Whyte
Lucando = Lukandu, popl.	T6	Kilosa	6 32 S 36 49 E	B.D. Burtt
Luchokoa (near Mongoloma)	T5	Kondoa	near 4 54 S 35 37 E	B.D. Burtt
Lueka = Lueki R. (Matengo Hills)	T8	Songea	Not traced	Zerny

(probably Luhekea R. or Luwika R.)

Name	Region	Location	Coordinates	Collector
Luemba, popl.	T7	Mbeya	8 57 S 32 50 E	Goetze
Luembai, mt.	T6	Kilosa	6 38 S 36 52 E	B.D. Burtt
Luembo = Luemba, popl.	T7	Mbeya	8 57 S 32 50 E	
[Luempunu] see Lwempunu	U2	Bunyoro	1 08.5 N 30 55 E	Eggeling
[Luengera] see Lwengera	T3	Lushoto/Tanga	4 53 S 38 27 E–5 11 S 38 31 E	Peter, Geilinger, Semsei, Holst
[Lufijo] see Lufirio	T7	Rungwe	c.9 01 S 33 48 E–9 32 S 33 58 E	Stolz
[Lufira] see Lufirio	T7	Rungwe	c.9 01 S 33 48 E–9 32 S 33 58 E	Richards
Lufirio, R.	T7	Rungwe	c.9 01 S 33 48 E–9 32 S 33 58 E	Goetze, Cribb & Grey-Wilson

(headwaters may be in Njombe or Mbeya District)

Name	Code	District	Coordinates	Collector
Lugusida, str.	K5	North Kavirondo	0 15 N 34 58 E–0 17 N 34 48 E	
Lugusida Bridge, ln.	K5	North Kavirondo	0 14 N 34 54 E	
Luhangulo = Luhangale, popl.	T6	Morogoro	7 01 S 37 46 E	Stuhlmann
Luhanyando, popl.	T6	Ularga	9 27 S 36 48 E	Rees
Luhanyando, R.	T6	Ulanga	9 50 S 36 40 E–9 25 S 36 48 E	
Luhekea, R.	T8	Songea	11 06 S 34 53 E–11 16 S 34 49 E	Milne-Redhead & Taylor
Luhembe, R.	T6	Kilosa	7 15 S 36 50 E–7 43 S 37 04 E	
Luhigi, str.	T7	Njombe	c.9 31 S 34 50 E	Goetze
[Luhila] see Luhira	T8	Songea	c.10 41 S 35 43 E	Semsei
Luhimba, str.	T8	Songea	c.10 26 S 35 43 E–10 30 S 35 37 E	Milne-Redhead & Taylor
Luhira, R.	T8	Songea	c.10 41 S 35 43 E	Milne-Redhead & Taylor, Semsei
[Luhiri] see Luhira	T8	Songea	c.10 41 S 35 43 E	
[Luhiza] see Luhizha	U2	Kigezi	1 02 S 29 47 E	Purseglove
Luhizha, popl.	U2	Kigezi	1 02 S 29 47 E	Purseglove, Dawkins
Luhoho	Zaire *not* Tanzania			Babault (Apr. 1937)
Luhombero, R.	T6	Ulanga	9 18.5 S 36 26 E–8 24 S 37 11 E	Rees, Haerdi
[Luhota] see Luhoto	T7	Iringa	7 48 S 35 43 E	Cribb & Grey-Wilson
Luhoto, hill	T7	Iringa	7 48 S 35 43 E	Lynes, Polhill & Paulo
Luhuga, ln.	T5	Morogoro	c.7 12 S 37 45 E	Thulin & Mhoro
Luhungu, mt.	T6	Morogoro	6 55 S 37 40 E	Winkler
Luiche, rsta.	T4	Kigoma	4 55 S 29 44 E	
Luiche, R.	T4	Buha/Kigoma	c.4 38 S 29 46 E–4 58 S 29 44 E	Peter
Luiche, R., escarp.	T4	Ufipa	8 09 S 31 38 E–7 39 S 31 44 E	Michelmore
[Luichi] see Luiche	T4	Buha/Kigoma		von Trotha
[Luichi] see Luiche	T4	Ufipa	8 09 S 31 38 E–7 39 S 31 44 E	Richards
Luiga, tea est.	T7	Iringa	8 35 S 35 15 E	Pegler, Renvoize, Procter
Luilo, popl.	T7	Njombe	10 23 S 34 44 E	Thulin & Mhoro
[Luiri Kitesa] see Liwiri-Kiteza	T8	Songea	c.11 10 S 34 50 E	Milne-Redhead & Taylor
Luisenga, L.	T7	Iringa	8 37 S 35 21 E	Polhill & Paulo
Luisenga, str.	T7	Iringa	c.8 36 S 35 18 E	Polhill & Paulo, Cribb et al.

191

Name	Region	District	Coordinates	Collector
[Lujara] ? see Kidugala, mt.	T7	Njombe	c.9 34 S 34 30 E	Goetze
Lukandamila, popl.	T4	Mpanda	6 12 S 29 54 E	Kakeya
Lukange, area	T6	Morogoro	7 16 S 37 44 E	Haarer
Lukaya, popl.	U4	Masaka	0 08 S 31 53 E	Lye
Lukaya (Rubondo I.)	T1	Mwanza	c.2 20 S 31 52 E	Carmichael
[Lukemba] see Lukimwa	T8	Songea	10 33 S 36 23 E–11 24 S 35 55 E	
[Lukenia] see Lukenya	K4	Machakos	1 28.5 S 37 04 E	
Lukenya, club	K4	Machakos	1 29 S 37 03 E	Busse
Lukenya, hill	K4	Machakos	1 28.5 S 37 04 E	Lucas, J.G. Williams
Lukenya, road	K4	Machakos	c.1 27 S 37 08 E	Numerous
Lukimwa, R.	T8	Songea	10 33 S 36 23 E–11 24 S 35 55 E	Busse
Lukiri, popl.	U2	Ankole	0 14 S 30 26 E	?Snowden
[Lukiri] see Rukiri	U2	Ankole	0 17 S 30 21 E	
Lukoga, f.r.	T6	Ulanga	near 8 08 S 36 41 E	Semsei
Lukohe, hill	U2	Bunyoro	1 53.5 N 31 38 E	Eggeling
[Lukoki] see Lukose	T7	Iringa	8 12 S 36 01 E–7 28 S 36 31 E	
Lukolansala, str.	T4	Mpanda	c.6 00 S 31 00 E	Goetze
Lukoma, popl.	U4	Mengo	0 30 N 32 17 E	Moreau
Lukoma, bay	T8	Songea	11 25 S 34 53 E	?Bellingham
(probably Malawi, Likoma, I.)				
Lukondo	*Rwanda not Tanzania*		c.2 25 S 29 40 E	
Lukonge, popl.	U3	Mbale	0 56 N 34 13 E	Mildbraed
Lukose, R.	T7	Iringa	8 12 S 36 01 E–7 28 S 36 31 E	Maitland
Lukosi, popl.	T3	Lushoto	4 40 S 38 17.5 E	B.D. Burtt, Procter
[Lukosi] see Lukose	T7	Iringa	8 12 S 36 01 E–7 28 S 36 31 E	Procter
[Lukosse] see Lukose	T7	Iringa	8 12 S 36 01 E–7 28 S 36 31 E	Goetze
[Lukota] see Luhoto, hill	T7	Iringa	7 48 S 35 43 E	Lynes
[Lukozi] see Lukosi	T3	Lushoto	4 40 S 38 17.5 E	Drummond & Hemsley
[Lukozi] see Lukose	T7	Iringa	8 12 S 36 01 E–7 28 S 36 31 E	
Lukula, old popl, airstrip	T6	Ulanga	9 24 S 36 50 E	Vollesen

Name		Location	Coordinates	Collector
Lukula, str.	T6	Ulanga	9 19 S 36 34 E–9 16 S 36 59.5 E	Schlieben
Lukuledi, popl., mssn.	T8	Masasi	10 34 S 38 47.5 E	Numerous
Lukuledi, R.	T8	Masasi/Lindi	c.10 30 S 38 33 E–10 05 S 39 42 E	Richards, Milne-Redhead & Taylor
Lukumburu, popl.	T7	Njombe	9 44.5 S 35 08.5 E	Richards
Lukungu, R. (near L. Sundu)	T2	Ufipa	near 8 31.5 S 31 38.5 E	
Lukuresi, L.	T2	Arusha	3 13.5 S 36 52 E	
[Lukwamigo] see Lukwemigo	U2	Toro	0 33 N 31 10 E	
Lukwangule, plateau	T6	Morogoro	7 08 S 37 38 E	Numerous
Lukwemigo, popl.	U2	Toro	0 33 N 31 10 E	Lankester
Lula, old popl.	T7	Iringa	c.7 38 S 35 57 E	Goetze
Lulanda = Lulando, popl., f.r.	T7	Iringa	8 36 S 35 37 E	Cribb et al.
Lulando, popl., f.r.	T7	Iringa	8 36 S 35 37 E	
Lulanguru, rsta.	T4	Tabora	5 06 S 32 38.5 E	Braun
Lulanji Farm	T7	Iringa	8 40 S 35 13 E	Gilchrist
Lulindi, popl. (near Mchinjiri)	T8	Lindi	near 10 08 S 39 11 E	Eggeling
Luis, w.h.	K1	Northern Frontier	4 01.5 N 40 20 E	Gillett
[Lululua] see Lugulua	T3	Lushoto	4 37 S 38 12 E	Holst
Luluzi, hill	T7	Iringa	prob. c.8 05 S 36 00 E	Carmichael
[Lumakali] see Rumakali	T7	Njombe/Rungwe	c.9 06 S 33 52 E–9 20 S 33 55 E	
Lumakarya, f.r.	T7	Njombe	c.9 12 S 34 03 E	Stolz
Lumakarya, = Rumakali, R.	T7	Njombe/Rungwe	c.9 06 S 33 52 E–9 20 S 33 55 E	Semsei, Eggeling
[Lumakira] see Lumakarya	T7	Njombe/Rungwe	c.9 06 S 33 52 E–9 20 S 33 55 E	
Lumba = Luemba, popl.	T7	Mbeya	8 57 S 32 50 E	Goetze, Siame
Lumbila, popl.	T7	Njombe	9 34.5 S 34 07.5 E	Gilli
[Lumbira] see Lumbila	T7	Njombe	9 34.5 S 34 07.5 E	
Lumbuka, str., sw.	U3	Busoga/Mbale	0 21 N 33 53 E–0 30 N 33 53 E	Michelmore
Lumbuye, str., sw.	U3	Busoga	c.1 00–1 06 N 33 27 E	G.H. Wood
Lumbwa, t.a.	K5	Kericho	c.0 30 S 35 10 E	Numerous
Lumbwa, popl., rsta.	K5	Kisumu-Londiani	0 12 S 35 28 E	Numerous
Lumbye, popl.	T4	Mpanda	6 16 S 29 54 E	Jefford & Newbould

194

Name		District	Coordinates	Collector
Lupa, R.	T7	Chunya	c.7 34 S 33 16 E–8 39 S 33 12 E	Goetze
Lupa = North Lupa, f.r.	T7	Chunya	c.8 02 S 33 15 E	Boaler
Lupa Market, popl.	T7	Chunya	8 38 S 33 14 E	
Lupanga, peak	T6	Morogoro	6 52 S 37 42.5 E	Numerous
Lupa Ntande = Ntande	T7	Chunya	8 10 S 33 16 E	Vesey-FitzGerald
[Lupasa] see Lupaso	T8	Masasi	10 56 S 38 54.5 E	Gerstner
Lupaso, popl.	T8	Masasi	10 56 S 38 54.5 E	Gerstner
Lupata, popl.	T7	Rungwe	9 18 S 33 51 E	R.M. Davies
Lupa Tingatinga, popl.	T7	Chunya	8 00 S 33 15 E	Procter
Lupembe, f.r.	T7	Njombe	c.9 14 S 35 08 E	Procter, Eggeling
Lupembe, mssn., str.	T7	Njombe	9 15 S 35 15 E	Schlieben, van Rensburg
Lupembe, hill	T8	Songea	11 00 S 34 56 E	Eggeling, Milne-Redhead & Taylor, Zerny
Lupeme, tea est.	T7	Iringa	c.8 35 S 35 20 E	Pegler, Renvoize
Lupingu, mssn.	T7	Njombe	10 04 S 34 32 E	Gilli
Lupiro, popl.	T6	Ulanga	8 23 S 36 40 E	Cribb & Grey-Wilson, Haerdi, Anderson
[Lurachi] see Larachi	K1	Northern Frontier	c.2 46 N 36 51 E	
Lusahanga, popl.	T1	Biharamulo	2 54 S 31 11 E	Amshoff
Lusahungu = Lusahanga, popl.	T1	Biharamulo	2 54 S 31 11 E	Tanner, Ford, Procter
[Lusanga] see Lusungo	T7	Rungwe	9 30 S 33 58 E	Richards
Lusangalala, area	T6	Morogoro	7 06 S 37 44 E	
[Lusangu] see Lusungo	T7	Rungwe	9 30 S 33 58 E	Richards
Lusanje, popl.	T7	Rungwe	9 11 S 33 47 E	Hepper et al.
[Lusaunga] see Lusahanga	T1	Biharamulo	2 54 S 31 11 E	Carmichael
Lushamba (Buhindi, f.r.)	T1	Mwanza	c.2 25 S 32 15 E	Carmichael
Lushambia	T1	Bukoba	Not traced	Braun
Lushoi Farm (Naro Moru)	K4	North Nyeri	c.0 10 S 37 01 E	Tweedie
Lushoto, distr.	T3	Lushoto	c.4 40 S 38 10 E	
Lushoto, f.r.	T3	Lushoto	c.4 47 S 38 17 E	Richards
Lushoto, popl., mssn.	T3	Lushoto	4 47.5 S 38 17.5 E	Numerous
Lusiro, est.	K4	Kiambu	c.1 07 S 36 45 E	Bally & Smith

Name	Grid	District	Coordinates	Collector
[Lussangeli] see Luhangala				
Lussegwa, for.	T6	Morogoro	7 01 S 37 46 E	Stuhlmann
	T6	Morogoro	6 58.5 S 37 48 E	Stuhlmann
[Lussimbi] see Mkalinzi				
Lusungo, area	T4	Buha	4 37 S 29 44 E	Peter
Lusungo, popl.	T7	Rungwe	c.9 28 S 33 52 E	Richards
	T7	Rungwe	9 30 S 33 58 E	Richards
[Lusungu] see Lusungo				
[Lusunguri] see Lusunguru				
Lusunguru, f.r.	T6	Morogoro	6 05 S 37 41 E	Drummond & Hemsley, Mgaza,
	T6	Morogoro	6 05 S 37 41 E	van Someren
Luswiswe, R.	T7	Rungwe	c.9 17 S 33 28 E–9 22 S 33 33 E	Stolz
Lutamba, L.	T8	Lindi	10 02.5 S 39 28.5 E	Numerous
Lutembe, bay	U4	Mengo	c.0 10 N 32 35 E	Lind
Lutindi, popl.	T3	Lushoto	5 04.5 S 38 22 E	Holst
Lutindi, f.r.	T3	Lushoto	c.4 53 S 38 38 E	Greenway, Liebusch, Peter
Lutindi, mt.	T3	Lushoto	4 54 S 38 37 E	Purseglove, Eggeling
Lutobo, hill	U2	Ankole	1 09 S 30 08 E	Eggeling
Lutoboka, f.r.	U4	Masaka	c.0 18 S 32 17.5 E	Osmaston, Eggeling, Sangster
Lutoto, L.	U2	Ankole	c.0 20 S 30 06 E	Numerous
Lutoto, popl., r.c.	U2	Ankole	0 20 S 30 06 E	Kertland
[Lututuro] see Lotuturu				
[Lututuru] see Lotuturu				
Lutyen Pass (? Loitanit)	U1	Acholi	3 44 N 32 55 E	Eggeling, Forbes
	U1	Acholi	3 44 N 32 55 E	
	U1	Karamoja	?2 43 N 34 42 E	Michelmore
Luuka, co.	U3	Busoga	c.0 45 N 33 20 E	Karani
Luunga, popl.	U4	Masaka	0 17 S 31 58 E	Lye
Luvia, I.	U3	Busoga	0 03 N 33 28 E	G.H. Wood
Luvunya, f.r.	U3	Busoga	c.0 27 N 33 51 E	G.H. Wood
Luwala, popl.	U4	Mengo	0 27 N 33 07 E	Dummer
Luwalisi = Ruwarisi, R.	T7	Rungwe	c.9 16 S 33 38 E–9 30 S 33 50 E	R.M. Davies
Luwanda = Ruwanda, est.	T7	Rungwe	prob. c.9 20 S 33 45 E	
Luwegu, R.	T6/8	Ulanga/Songea	c.10 00 S 36 02 E–8 31 S 37 23 E	Schlieben

[Luwelo] see Luwero	U4	Mengo	0 51 N 32 29 E	Maitland
Luwengi, popl.	T7	Njombe	c.9 36 S 35 07 E	Milne-Redhead & Taylor
[Luwera] ? see Lwera	U4	Masaka	c.0 06 S 31 56 E	Hansford
Luwero, popl., mssn., t.c.	U4	Mengo	0 51 N 32 29 E	Maitland, Eggeling, Langdale-Brown
Luwii	T7	Njombe	near 9 00 S 35 15 E (fide V-F)	Vesey-FitzGerald
Luwika, R.	T8	Songea	11 06 S 34 48 E–11 13 S 34 43 E	
Luwila, R.	T5	Dodoma	5 30 S 34 58 E–5 50 S 35 05 E	
[Luwira Kiteza] see Liwiri-Kiteza	T8	Songea	c.11 10 S 34 50 E	Semsei
Luwiri, old popl., hill	T1	Mwanza	3 08 S 32 47 E	
[Luwiri Kitesa] see Liwiri-Kiteza	T8	Songea	c.11 10 S 34 50 E	Milne-Redhead & Taylor
[Luwondo] see Rubondo	T1	Mwanza	2 20 S 31 52 E	Herring
Luzinga, popl.	U3	Busoga	0 42 N 33 12 E	
Luzira, hill	U4	Mengo	0 37.5 N 31 56 E	Snowden
Luziwi, R.	T7	Mbeya	c.8 58 S 33 24 E	Wingfield
Lwabiyata, popl., mssn., r.h.	U4	Mengo	1 31 N 32 22 E	H.B. Johnston
[Lwakaka] see Lwakhakha	U3	Mbale	0 48 N 34 22.5 E	Snowden
[Lwakaka] see Lwakhakha	U3/K5	Mbale/N. Kavirondo	c.1 06 N 35 01 E–0 41 N 34 17 E	Snowden
Lwakhakha, popl.	U3	Mbale	0 48 N 34 22.5 E	Snowden
Lwakhakha, R.	U3/K5	Mbale/N. Kavirondo	c.1 06 N 35 01 E–0 41 N 34 17 E	
Lwala, for.	U1	Karamoja	c.3 43 N 34 00 E	
Lwala, popl., mssn., t.c.	U3	Teso	1 52 N 33 16 E	
Lwamatuka (N. Mengo)	U4	Mengo	prob. c.1 19 N 32 28 E	Langdale-Brown
Lwampanga, popl.	U4	Mengo	1 30 N 32 30.5 E	Langdale-Brown, Eggeling
Lwamunda, for.	U4	Mengo	0 19 N 32 22 E	Dawkins
Lwanda, r.h.	U4	Masaka	0 40 S 31 29 E	H.B. Johnston
Lwandani, str.	K7	Kilifi	3 46 S 39 39 E–3 55 S 39 42 E	Faden
Lwankima, for.	U4	Mengo	c.0 24 N 32 59 E	Dawkins
[Lwanyando Juu] see Luhanyando (upper reaches)	T6	Ulanga	9 50 S 36 40 E–9 25 S 36 48 E	Rees
Lwanyonyi, popl., mssn.	U4	Mengo	0 22 N 32 47.5 E	Harker

197

Lwasamaire = Rwashamaire, popl., r.h.	U2	Ankole	0 50 S 30 08 E	Snowden, A.S. Thomas, Eggeling
[Lwasamarie] see Lwasamaire	U2	Ankole	0 50 S 30 08 E	
Lwashamaire = Rwashamaire	U2	Ankole	0 50 S 30 08 E	Eggeling
Lwemiyaga, popl., r.h.	U4	Masaka	0 06 N 31 06 E	
Lwempunu, ln.	U2	Bunyoro	1 08.5 N 30 55 E	Gillman
[Lwengala] see Liwengula	T8	Lindi	9 59 S 39 32 E	
Lwengera, R.	T3	Lushoto/Tanga	4 53 S 38 27 E–5 11 S 38 31 E	Numerous
Lwentaama, popl.	U4	Mubende	0 32 N 31 17 E	Katende
[Lwenyando Juu] see Luhanyando (upper reaches)	T6	Ulanga	9 50 S 36 40 E–9 25 S 36 48 E	Rees
Lwera, area	U4	Masaka	c.0 06 S 31 56 E	Conrads
Lwessabi, ln.	T1	Mwanza	2 00 S 33 04 E	Richards
Lwimo, str.	T7	Njombe	9 53 S 34 40 E	Mwasumbi
[Lwize] see Luiche	T4	Ufipa	8 09 S 31 38 E–7 39 S 31 44 E	Vesey-FitzGerald
Lwomukoma, drainage line	T4	Ufipa	c. 8 24 S 31 58 E	Styles
[Lyamunda] see Lwamunda	U4	Mengo	0 19 N 32 22 E	Greenway
[Lyamungo] see Lyamungu	T2	Moshi	3 11 S 37 16 E	Greenway, Ivens, Wallace
Lyamungu, coffee res. stn.	T2	Moshi	3 14 S 37 15 E	
Lyamungu, foothill	T2	Moshi	3 11 S 37 16 E	
Lyantonde, popl., r.h.	U4	Masaka	0 24 S 31 09 E	Harker, Lye, Trapnell
Lyela, sawmill	T4	Tabora	6 18 S 33 37 E	Groome
[Lykipia] see Laikipia	K3	Laikipia	c.0 30 N 36 30 E	Thomson
[Lyontonde] see Lyantonde	U4	Masaka	0 24 S 31 09 E	
Maandaa (Juani I.)	T6	Rufiji	c.8 00 S 39 47 E	Greenway
Maanyi, popl.	U4	Mengo	0 16 N 31 58 E	
Mabale, mssn.	U4	Mubende	1 03 N 30 55 E	A.S. Thomas
Mabale, popl.	T1	Mwanza	2 51 S 32 08 E	Tanner, Procter
Mabaloni Rocks, hill	K4	Fort Hall	1 05 S 37 25.5 E	Bally, Verdcourt
Mabamba, popl.	T4	Buha	3 36 S 30 29 E	Procter

Name		District	Position	Collector(s)
Mabanduka (E. Usambara Mts.)	T3	Lushoto	Not traced	Orgenes John
[Mabangala] see Bangala, popl.	T7	Chunya	prob. c.8 22 S 32 55 E	Richards
[Mabaonde] see Maboonde	K3	Trans-Nzoia	c.1 00 N 34 57 E	
[Mabatini] see Songa	T3	Tanga/Pangani	5 16 S 38 39 E	
Mabawe, popl., r.h., t.c.	T1	Ngara	2 35.5 S 30 30 E	Paulo
Mabega, mt.	T6	Morogoro	6 05 S 37 27 E	Tanner
[Mabenera] see Makenera	T7	Iringa	7 50 S 35 38 E	
Mabiembe (? near L. Nyasa)	T7	?Rungwe	Not traced	Thulin & Mhoro, Schlieben
Mabira, popl.	T1	Bukoba	1 14 S 30 56 E	R.M. Davies
Mabira Forest, f.r.	U4	Mengo	c.0 30 N 32 57 E	Haarer
Mabogwe, popl.	T4	Buha	4 05 S 31 10 E	Numerous
Maboko I. ("15 miles W. of Kisumu")	K5	Central Kavirondo	?c.0 11 S 34 33 E	Hornby
Mabokweni, popl., t.c., area	T3	Tanga	5 01.5 S 39 03.5 E	Peter
Maboonde Farm	K3	Trans-Nzoia	c.1 00 N 34 57 E	Tweedie
[Mabrei] see Mwalesi	T7	Rungwe	9 12 S 33 39 E –9 23 S 33 48 E	
Mabuimerafuru, mssn.	T1	Musoma	1 52 S 33 33 E	R.M. Davies
Mabuki, popl.	T1	Kwimba	2 59 S 33 11 E	Kihongo
Mabungo, popl., r.c.	U2	Kigezi	1 18 S 29 41 E	Numerous
Mabungo, hill	T2	Moshi	3 24 S 37 30.5 E	Gilbert
[Mabungu] see Mabungo		Moshi	3 24 S 37 30.5 E	
[Macha] see Masha	U2	Ankole	0 40 S 30 49 E	
[Machako] see Machakos	K4	Machakos	1 31 S 37 16 E	
Machakos, distr.	K4	Machakos	c.2 00 S 37 40 E	Eggeling
Machakos, popl.	K4	Machakos	1 31 S 37 16 E	Kassner
Machame = Mashame, area	T2	Moshi	c.3 11 S 37 13 E	Numerous
Machame = Mashame, popl., mssn.	T2	Moshi	3 12 S 37 14 E	Huxley, Moreau, Carmichael
[Machami] see Machame	T2	Moshi		
[Machaso] see Machazo	T4	Kigoma	4 54 S 29 45 E	Haarer
[Machata] see Mchata	T4	Ufipa	8 13 S 31 34 E	Peter, Procter
Machazo, popl.	T4	Kigoma	4 54 S 29 45 E	Peter

Name	Code	District	Coordinates	Collector
Machewa, str.	K3	Trans-Nzoia	0 53.5 N 34 48 E–0 53 N 34 52 E	Tweedie
Machinga, t.a.	T8	Kilwa/Lindi	c.9 20 39 25 E	
Machinga Nymbu	T2	Mbulu	prob. c.3 26 S 35 48 E	Greenway
Machipi, popl.	T6	Ulanga	8 03 S 36 41 E	Haerdi
Macho Afas, str.	Ethiopia *not* K1		4 43 N 36 35 E–4 41 N 36 07 E	Brown
Machole	T8	prob. Lindi	Not traced	Gillman
Machui, popl.	T3	Tanga	5 10 S 39 05 E	Milne-Redhead & Taylor, Perdue & Kibuwa, Faulkner
Mackinder Valley	K4	N. Nyeri	c.0 09 S 37 18 E–0 05 S 37 15 E	J.G. Williams
Mackinnon Road, popl.	K7	Kwale	3 44 S 39 03 E	Numerous
Madaba, popl., airstrip	T8	Kilwa	8 40 S 37 47 E	Schlieben, Vollesen
[Madabira] see Madibira	T7	Mbeya	8 12 S 34 47.5 E	Carmichael
[Madahani] see Madehani	T7	Njombe	9 21 S 34 02 E	Stolz
Madala, popl.	T3	Lushoto	4 29 S 38 18 E	Parry
Madanda, for.	Mozambique *not* Uganda			Dawe
Madang, hill	U1	Karamoja	3 09 N 34 12	J. Wilson
Madanga, popl., area	T3	Pangani	5 21 38 59 E	Tanner
[Madangi] see Mudangi				
[Maddo Delvek] see Mado Delbek	U3	Mbale	1 10 N 34 26 E	Liebenberg, Eggeling
[Maddo Gaschi] see Mado Gashi	K1	Northern Frontier	1 34 N 38 20 E	
	K1	Northern Frontier	0 43.5 N 39 10.5 E	Bally
Maddu, popl.	U4	Mengo	0 13 N 31 40 E	Eggeling
Madehani, mssn.	T7	Njombe	9 21 S 34 02 E	Wigg, Stolz
Madembu, hill	T7	Iringa	near 7 50 S 36 20 E	Carmichael
Maderoma, mt.	T6	Kilosa	6 31 S 36 59 E	
Madhireni, plantation	U3	Busoga	c.0 37 N 33 14 E	Karani
Madi, distr.	U1	West Nile	c.3 25 N 31 45 E	Grant, Dawe
		(considered as part of West Nile District for F.T.E.A.)		
Madi, co.	U1	West Nile	c.2 45 N 31 15 E	A.S. Thomas, Eggeling, Grant (Dec. 1862)
Madi, popl.	Sudan		c.3 45 N 31 57 E	Grant (1863)
Madibira, popl., mssn.	T7	Mbeya	8 12 S 34 47.5 E	Anderson, Orr, Richards

[Madibira, R.] see Ndembera	T7		Iringa/Mbeya	8 05 S 35 33 E−8 12 S 34 48 E	Richards
Madi, East see East Madi	U1		West Nile	c.3 15 N 31 45 E	
Madimola, admin. post	T6		Uzaramo	6 47 S 38 43 E	Stuhlmann
Madi Opei, popl. r.h.	U1		Acholi	c.3 37 N 33 06 E	
Madi Opei, mt., f.r.	U1		Acholi	c.3 41 N 33 06 E	
Madi, West see West Madi	U1		West Nile	c.3 20 N 31 35 E	
Madizini, popl.	T6		Kilosa	7 09 S 36 50.5 E	Grundy
Madjanga-Kwa-Bagaya (Donde area)	T8		Kilwa	prob. c.9 35 S 37 40 E	Busse
Mado Delbek, w.p.	K1		Northern Frontier	1 34 N 38 20 E	
Mado Gashi, p.p.	K1		Northern Frontier	0 43.5 N 39 10.5 E	Gillett
[Madonya] see Mdonya	T7		Iringa	c.7 42 S 34 29 E−7 44 S 34 56 E	Richards
Madoroma = Maderoma mt.	T6		Kilosa	6 31 S 36 59 E	Swynnerton
Mado Yaka, Hill, w.h., str./s.	K1		Northern Frontier	c.0 45 N 38 18 E	
[Madschame] see Machame	T2		Moshi		
[Madu] see Maddu	U4		Mengo	0 13 N 31 40 E	Eggeling
Madudu, popl.	U4		Mubende	0 41 N 31 29 E	
[Maduerca] see Maguera	T3		Pare	3 44 S 37 42 E	Peter
Madunda, mssn.	T7		Njombe	9 52 S 34 27 E	Richards, Eggeling, Gilli
[Madunga] see Madungu	P		Pemba	5 14.5 S 39 46.5 E	Vaughan
Madungi Lookout	T7		Iringa	7 43 S 34 51 E	Greenway
[Madungu] see Madungi	T7		Iringa	7 43 S 34 51 E	Greenway
Madungu, popl.	P		Pemba	5 14.5 S 39 46.5 E	
[Maehame] see Machame	T2		Moshi	3 12 S 37 14 E	
Mafi, mt.	T3		Lushoto	4 56 S 38 09 E	Procter
Mafia, I.	T6		Rufiji	c.7 50 S 39 45 E	Numerous
Mafinga, area	T7		Iringa	8 20 S 35 16 E	Kibuwa
Mafinga, popl.	T7		Iringa	8 18 S 35 18 E	
Mafisi, popl.	T6		Uzaramo	6 59 S 38 38 E	Stuhlmann
Mafisini, popl.	K7		Kwale	4 28 S 39 22 E	Drummond & Hemsley, ?Greenway
Maftaha, bay	K7		Kwale	4 25 S 39 31 E	

201

Name	Code	District	Coordinates	Collector
[Magangwei] see Magangwe				
Magara, escarp., est., popl.	T7	Mbeya	7 46 S 34 13.5 E	Greenway
Magari, peak	T2	Mbulu	c.3 50 S 35 42 E	Vesey-FitzGerald
Magarini, popl.	T6	Morogoro	6 56 S 37 39 E	Pócs et al.
[Magaso] see Machazo	K7	Kilifi	3 02 S 40 04 E	St. Barbe Baker
Magdireshu, str.	T4	Kigoma	4 54 S 29 45 E	Peter, Procter
Magena, mkt., sch.	T2	Arusha	c.3 16 S 36 47 E –3 23 S 36 52 E	Willan
Magendi, popl.	K5	South Kavirondo	0 41 S 34 31 E	
Magengwe = Miquengue, popl.	T4	Buha	3 55 S 31 20 E	Bullock
[Magi Chumvi] see Maji ya Chumvi	T4	Tabora	5 08 S 32 45 E	Boehm
Magila, popl.	K7	Kwale/Kilifi	3 48 S 39 23 E	Kaessner
Magila, popl., area	T3	Lushoto	4 48 S 38 20 E	Greenway
[Magi Moto] see Maji Moto	T3	Tanga	5 08 S 38 46 E	Peter, Archbold, Holst
[Magi Moto] see Maji Moto	T2	Mbulu	3 40 S 35 44 E	Richards
Magina, t.c.	T4	Mpanda	7 14.5 S 31 24 E	Richards
[Magina] see Magena	K4	Kiambu	0 56 S 36 37.5 E	Greenway
Magius's Estate = Kinja Nurseries	K5	South Kavirondo	0 41 S 34 31 E	Opiko
[Magize] see Mazige	K3	Naivasha	0 48 S 36 16 E	Hooper & Townsend
[Mago] see Magu	U2	Mengo	0 22.5 N 32 49 E	Dummer
Magofu, ln.	T1	Kwimba	c.2 35 S 33 20 E	Fischer
Magogoni, sch., disp.	T3	Handeni	c.5 27 S 37 47 E	Busse
Magogoni, popl., pt.	T6	Morogoro	7 13 S 37 59 E	Peter, Haarer
[Magoje] see Magoye	T6	Uzaramo	6 50 S 39 20.5 E	Vaughan
[Magoma] see Hemagoma	T7	Njombe	9 00 S 33 59 E	
Magoma = Mangoma	T3	Lushoto	4 53 S 38 35 E	Drummond & Hemsley, Gillman, Braun
Magoma, L., sw.	T4	Pare	prob. c.4 19 S 38 00 E	Greenway
Magoma, mt.	T7	Buha	c.4 55 S 31 05 E	Vesey-FitzGerald
Magoma (Elton Plateau), area	T7	Chunya	7 40 S 33 47 E	Richards
Magombera, f.r.	T7	Njombe	9 10 S 34 00 E	Semsei, Vollesen
Magomeni, L.	T6	Ulanga	c.7 55 S 37 03 E	Peter, Vaughan, Semsei
	T6	Uzaramo	6 48 S 39 14.5 E	

Name		District	Coordinates	Collector
Magoro, popl.	U3	Teso	1 44 N 34 06 E	J. Wilson
Magosi, popl., hills	U1	Karamoja	2 55.5 N 34 31 E	
[Magoya] see Magoye				
Magoye, mssn.	T7	Njombe	9 00 S 33 59 E	
[Magroto] see Magrotto				
Magrotto, est., area	T3	Tanga	5 07 S 38 45 E	Faulkner, Geilinger, Peter
Magu, t.a.	T3	Tanga	5 07 S 38 45 E	Fischer
Magu, popl.	T1	Kwimba	c.2 35 S 33 20 E	Tanner
Maguera, popl.	T1	Kwimba	2 36 S 33 26.5 E	Peter
Magugu, popl., r.h.	T3	Pare	3 44 S 37 42 E	Polhill & Paulo, Welch
Magunga, popl.	T2	Mbulu	4 00 S 35 47 E	
Magunga, small popl.	T3	Lushoto	4 36.5 S 38 26.5 E	
Magunga, old pl.	T3	Lushoto	4 55 S 38 31 E	Peter
Magunga, popl.	T3	Lushoto	4 58 S 38 38 E	
Magunga, small popl.	T3	Lushoto	5 02 S 38 26 E	Peter
Magunga, sisal est.	T3	Lushoto	5 06.5 S 38 27.5 E	Numerous
Magunga, small popl.	T3	Tanga	5 08 S 38 33 E	
Magyo, popl., mssn.	T3	Pangani	5 17.5 S 38 39 E	Eggeling
Mahali, mts.	U4	Mengo	0 10 N 33 17.5 E	Oxf. Univ.
Mahali pa Nyati, plain	T4	Mpanda	6 12 S 29 50 E	Richards, Greenway & Kanuri
Mahanje, mssn.	T2	Mbulu	3 23 S 35 50 E	Milne-Redhead & Taylor, Schlieben
[Mahanji] see Mahanje	T8	Songea	9 55 S 35 20 E	
Mahara, peak	T8	Songea	9 55 S 35 20 E	
[Mahari] see Mahali	T6	Kilosa	6 26 S 36 51 E	
[Mahatan] see Uhanyana	T4	Mpanda	6 12 S 29 50 E	
Mahenge, popl., mssn.	T7	Njombe	8 56 S 34 52 E	Numerous
[Mahenge] see Mahanje	T6	Ulanga	8 41 S 36 43 E	Milne-Redhead & Taylor
Mahenge Scarp, f.r.	T8	Songea	9 55 S 35 20 E	Cribb et al.
Mahenye, popl.	T6	Ulanga	8 37.5 S 36 43 E	Eggeling
[Mahera] see Mahara	T7	Njombe	9 32 S 34 45 E	Busse
	T6	Kilosa	6 26 S 36 51 E	

Mahiwa, popl., exper. fm., str.	T8	Lindi	10 21 S 39 16 E	Milne-Redhead & Taylor
Mahoma, L.	U2	Toro	0 21 N 29 58 E	Dawe
Mahoma, ln.	U2	Toro	0 22 N 29 59.5 E	
Mahoma, R.	U2	Toro	c.0 37 N 30 08 E—c.0 03 N 30 16 E	Osmaston
Mahomadenga, pond	T5	Dodoma	6 18 S 35 32 E	Busse
[Mahomedenya] see Mahomadenga	T5	Dodoma	6 18 S 35 32 E	Busse
[Mahonda] see Mhonda	T6	Morogoro	6 07 S 37 35 E	Brenan & Greenway
Mahonda, popl.	Z	Zanzibar	6 00 S 39 15 E	Faulkner, Vaughan
Mahonda, plain, sw.	Z	Zanzibar	c.6 00 S 39 14 E	Faulkner
Mahurunga, popl.	T8	Mikindani	10 34 S 40 15 E	Mason
Mahuta, popl., admin. post	T8	Newala	10 52 S 39 26.5 E	Gillman
Mahyoro, popl.	U2	Toro	0 42.5 N 30 10 E	Hansford
Maibey swamp	K3	Trans-Nzoia	1 07 N 35 22 E	Tweedie
Maikona, w.h.	K1	Northern Frontier	2 56 N 37 38 E	Bally
Maisome, I.	T1	Mwanza	2 18 S 32 02 E	Carmichael, Procter, Stuhlmann
Maiyuge, popl.	U1	Lango	1 42 N 32 06 E	
Majanji, popl., pier	U3	Mbale	0 16 N 33 59 E	
[Majareni] see Majoreni	K7	Kwale	4 34 S 39 17 E	R.M. Graham
[Maji Chumve] see Maji ya Chumvi	K7	Kwale/Kilifi	3 48 S 39 23 E	Kassner
Maji Juu, peak	T3	Pare	3 51 S 37 48 E	
Maji Mazuri, for.	K3	Ravine	c.0 00 35 40 E	Angus, R.M. Graham
Maji Mazuri, rsta.	K3	Ravine	0 01 S 35 41.5 E	
Maji Mekundu, popl.	Z	Zanzibar	6 01 S 39 12 E	Greenway
[Maji Moto] see Maji ya Moto	K3	Baringo	0 16 N 36 03 E	
Maji Moto, spr., hill	K3	Masai	1 22 S 35 42 E	Muchiri
Maji Moto, str.	K6	Kwale	c.4 24 S 39 18 E	Allan
Maji Moto, vet. stn.	K7	Musoma	1 38 S 34 20 E	Watkins
Maji Moto, escarp., est.	T1	Mbulu	3 40 S 35 44 E	Richards, Greenway
Maji Moto, spr.	T2	Mpanda	7 14.5 S 31 24 E	Richards
Majita, area	T4	Musoma	1 52 S 33 25 E	Tanner
	T1			

Makenera, popl.	T7	Iringa	7 50 S 35 38 E	Goetze
[Makenya] see Makanya	T3	Pare	4 20 S 37 51 E	Bally
Makere, popl., mssn.	T4	Buha	4 17 S 30 25 E	Hornby, Eggeling
Makerere, popl., coll.	U4	Mengo	0 20 N 32 34 E	Chandler & Hancock, Hindorf
Makerere, hill	U4	Mengo	c.0 20.5 N 32 34 E	Drummond & Hemsley
Makete, popl.	T7	Rungwe	9 23 S 33 40 E	Richards
Makigabi, popl. (on Daluni R.)	T3	Lushoto	Not traced	Faulkner
[Makimbi] see Makindi	T7	Mbeya/Iringa	7 45 S 34 35 E	Greenway
Makindi, ln., Springs	T7	Mbeya/Iringa	7 45 S 34 35 E	Thulin, Procter, Richards
Makindu, popl., rsta.	K4	Machakos	2 17 S 37 49.5 E	Numerous
[Makindu] see Kiumbi	K4	Machakos	c.2 11 S 37 53 E–2 18 S 37 48 E	Kassner
Makindu, popl.	T6	Rufiji	7 52 S 38 00 E	Schlieben
[Makindu] see Makindi	T7	Mbeya	7 45 S 34 35 E	Greenway
Makini	K7	Kwale	3 59 S 39 21 E	Kassner
[Makinjumbe] see Makinyumbe	T3	Pangani	5 20 S 38 38 E	Drummond & Hemsley, Scheffler
[Makinnon Road] see Mackinnon Road	K7	Kwale	3 44 S 39 03 E	Greenway
Makinyumbe = Makinyumbi	T3	Pangani		
Makinyumbi, popl., rsta.	T3	Pangani	5 20 S 38 38 E	Drummond & Hemsley, Scheffler
Makinyumbi, sisal est.	T3	Pangani	5 20 S 38 36 E	
Makipenzi, area, dam	K4	Machakos	1 11 S 37 33 E	Bally
Makola, old popl.	T1	Mwanza	c.3 05 S 32 45 E	Stuhlmann
Makole, popl. mssn.	U4	Masaka	0 11 N 31 09 E	A.S. Thomas
Makolo, mssn.	T1	Mwanza	3 04 S 32 45 E	Stuhlmann
Makombe, popl.	T7	Iringa	7 56 S 35 15 E	Goetze
Makomero, sw. (Yaida Valley)	T2	Mbulu	c.4 00 S 35 05 E	Richards
Makonde, Plateau, thicket	T8	Newala/Masasi	c.10 43 S 39 12 E	Busse, Gillman, Schlieben
[Makondeni] ? = Makongeni	P	Pemba		Vaughan
Makondo, popl.	U4	Masaka	0 28 S 31 31 E	?Brasnett
Makondo, popl.	U4	Masaka	0 30 S 31 29 E	?Brasnett
Makonduchi = Makunduchi	Z	Zanzibar	6 25 S 39 33 E	Faulkner

Name		Location	Coordinates	Collector(s)
Makuyuni, popl.	T2	Masai	3 33 S 36 06 E	Welch, Leippert, Verdcourt
Makuyuni, popl., rsta., old pl.	T3	Lushoto	5 02 S 38 20 E	Numerous
Makwai, ln.	T8	Songea	c.11 05 S 34 37 E	Hay
[Mala] see Molo				
Malaba, R.	T4	Ufipa	c.8 05 S 31 50 E	Richards
Malaba, popl., rsta.	U3/K5	Busoga/Mbale/N. Kavirondo	0 41 N 34 17 E–0 40 N 33 50 E	Tweedie
Malaba Forest = Mlaba Forest	K5	North Kavirondo	0 38 N 34 16 E	Tweedie
Malabigambo, f.r.	U4	Masaka	c.0 57 S 31 35 E	Dawkins, Drummond & Hemsley, Dawe
[Malabisi] see Malikisi				
Malagarasi, popl., rsta.	K5	North Kavirondo	0 40.5 N 34 25.5 E	Templer
Malagarasi, R.	T4	Kigoma	5 06 S 30 50 E	Bullock, Peter
Malagarasi, sw.	T4/Burundi	Buha/Kigoma	4 27 S 29 45 E–5 14 S 29 51 E	Numerous
Malaget, sawmill	T4	Kigoma/Buha	c.4 57 S 31 00 E	Dyson
Malakala = Mkalala, popl.	K5	Kisumu-Londiani	0 03 S 35 29 E	Polhill & Paulo
[Malakisi] see Malikisi	T7	Iringa	8 22 S 35 18 E	Templer
[Malale] see Mbale	K5	North Kavirondo	0 40.5 N 34 25.5 E	Greenway
[Malamagambo Forest] see Maramagambo	K5	North Kavirondo	0 05 N 34 43.5 E	Purseglove
Malamagambo, popl.	U2	Kigezi	c.0 32 S 29 53 E	Battiscombe
Malama's, popl.	U2	Ankole	0 16 S 30 07 E	?Peter
Malamba, old pl.	K5	North Kavirondo	0 11 N 34 30 E	?Peter
Malamba, popl., area, old pl.	T3	Lushoto	4 52 S 38 49 E	Cribb & Grey-Wilson
Malambo, popl.	T3	Lushoto	4 51 S 38 48 E	Willan
Malampaka, popl.	T7	Rungwe	9 09 S 33 45 E	Numerous
Malangale, popl.	T1	Maswa	3 08 S 33 32 E	Tanner
Malangali, popl., t.c.	T7	Iringa	8 02 S 35 46 E	Bogdan
[Malangali] see Mlangali	T7	Iringa	8 34 S 34 51 E	Tanner
Malanja = Malanya, peak, t.a.	T7	Njombe	9 46 S 34 31 E	Richards
Malan's Farm (32 km. NE of Eldoret)	T2	Masai	3 10 S 35 27 E	
Malanya, peak, t.a.	K3	Uasin Gishu	? c.0 45 N 35 25 E	
Malasa, I.	T2	Masai	3 10 S 35 27 E	
	T4	Ufipa	8 12.5 S 30 56.5 E	

Name	District	Region	Coordinates	Collector(s)
Malava, t.c., for.	North Kavirondo	K5	0 27 N 34 51 E	Bamps
[Malawa] see Malaba	Busoga/Mbale/N. Kavirondo	U3/K5	0 41 N 34 17 E—0 40 N 33 50 E	G.H.S. Wood, Michelmore
[Malearuguru] see Malialuguru	Maswa	T1	2 54 S 33 30 E	Greenway
[Malela] see Marera	Mbulu	T2	c.3 12 S 35 40 E—3 25.5 S 35 47 E	
Malemba, old popl.	Kitui	K4	2 06.5 S 38 08 E	Rodgers, Ludanga, Vollesen
Malemba Thicket, road, str.	Kilwa	T8	c.8 39 S 38 28 E	Hildebrandt
Malemboa = Malemba, popl.	Kitui	K4	2 06.5 S 38 08 E	Newbould
Malenda, area	Masai	T2	c.2 55 S 35 45 E	Schlieben
Malendula (Mahenge)	Ulanga	T6	Not traced	
[Malenge] see Marenje	Kwale	K7	c.4 30 S 39 12 E	Dale, Drummond & Hemsley
[Malesi] see Mwalesi	Rungwe	T7	9 12 S 33 39 E—9 23 S 33 48 E	R.M. Davies
Malialuguru, popl.	Maswa	T1	2 54 S 33 30 E	
Malibui, popl., mssn.	Lushoto	T3	4 42.5 S 38 23.5 E	
Malibwi = Malibui, popl., mssn.	Lushoto	T3	4 42.5 S 38 23.5 E	Tweedie
Malikisi, popl., t.c.	North Kavirondo	K5	0 40.5 N 34 25.5 E	Stolz
[Malila] see Umalila	Mbeya/Rungwe	T7	c.9 05 S 33 15 E	Brasnett
Malima, popl., r.c.	Busoga	U3	1 20 N 33 04 E	Numerous
Malindi, popl., airstrip	Kilifi	K7	3 13 S 40 07 E	Geilinger, Drummond & Hemsley, Swynnerton, Procter
Malindi, popl., r.h.	Lushoto	T3	4 37.5 S 38 18 E	
[Malingali] see Malangale	Iringa	T7	8 02 S 35 46 E	Carmichael
Maliwe, L.	Kilwa	T8	8 50 S 39 00 E	Busse
[Maliya] see Malya	Shinyanga	T1	c.3 45 S 33 00 E	Grundy
Malka Abafayo, w.h.	Northern Frontier	K1	1 12 N 38 58 E	Riva
Malka Daka, ln.	Northern Frontier	K1	3 57 N 41 33 E	Bally & Smith
Malka Dakacha, ln.	Northern Frontier	K1	3 57 N 41 30.5 E	
Malka Gorbesa, area	Northern Frontier	K1	c.0 13 N 38 20 E	
Malka Koja = Melka Koja, w.h.	Northern Frontier	K1	1 12 N 38 58.5 E	
Malka Korokoro	N. Frontier/Tana R.	K1/7	c.0 10 S 39 20 E	F. Thomas
Malka Mari, p.p.	Northern Frontier	K1	4 16 N 40 46 E	Bally & Smith

[Malka Murri] see Malka Mari	K1	Northern Frontier	4 16 N 40 46 E	J. Adamson, Everard
Malnvane (near Ssamunge)	T2	Masai	near 2 09 S 35 42 E	Bally
[Malola] see Malolo	T6	Kilosa	7 18 S 36 35 E	Eggeling
Malolo, popl.	T6	Kilosa	7 18 S 36 35 E	Eggeling
Malone, ln.	T2	Masai	c.2 08 S 35 39 E	Carmichael
[Malonge] see Malonje	T4	Ufipa	8 02 S 31 43 E	Richards
Malongwe, rsta.	T4	Tabora	5 28 S 33 38 E	Peter, Bally, van Rensburg
Malonje, mt, escarp., est.	T4	Ufipa	8 02 S 31 43 E	Numerous
Malu = Maru, mt.	U1	Karamoja	3 03 N 33 57 E	Paget-Wilkes
Malundwe, hill	T6	Kilosa	7 24 S 37 18 E	Lovett
Malupindi, popl.	T7	Rungwe	9 02 S 33 37 E	
Malwelewele = Miwele, track	T7	Mbeya/Iringa	near 7 53 S 34 45 E	Richards
Malya, region	T1	Shinyanga	c.3 45 S 33 00 E	
Malya, popl., f.r.	T1	Maswa	2 59 S 33 31 E	Yeoman, Kadesha, Eggeling
[Mamandu] see Mumandu	K4	Machakos	1 40 S 37 17 E	Nicholson
Mamarehe, g.p.	T1	Maswa	2 40 S 34 24 E	
Mamba, I. (Smith Sound)	T1	Mwanza	near 2 55 S 32 48 E	Tanner
Mamba, popl.	T3	Pare	4 23 S 37 59.5 E	?Greenway, Haarer
Mamba, popl.	T3	Pare	4 26 S 38 00 E	?Greenway
Mamba, mssn.	T4	Mpanda	7 19 S 31 22 E	Richards
[Mamba] see Momba	T4/7	Ufipa/Mbeya	c.8 47 S 32 28 E–8 10 S 32 28 E	Bullock
Mambasasa = Mambosasa, Game, For. Post	K7	Lamu	2 23 S 40 32 E	Greenway & Rawlins, Verdcourt
Mambi, str., falls	T6	Morogoro	c.6 54.5 S 37 39.5 E	Pócs
Mambi, R.	T7	Mbeya	8 51 S 33 56.5 E–8 35 S 33 53 E	
Mambogo, L. (Uluguru Mts.)	T6	Morogoro	Not traced	Paulo
Mamboneke, popl.	T7	Rungwe	c.9 12 S 33 39 E	
Mambore, popl.	K7	Lamu	1 48 S 41 25.5 E	
Mambosasa, Game, Forest Post	K7	Lamu	2 23 S 40 32 E	Gardner
Mambosasa, popl.	K7	Tana River	1 48 S 40 07 E	Greenway & Rawlins, Dale
Mamboya, f.r.	T6	Kilosa	6 15 S 37 06 E	R.M. Graham

Name	Code	Location	Coordinates	Collector
Mandera, popl., mssn., t.c.	T6	Bagamoyo	6 13 S 38 24 E	Sacleux, Mwakalasi, Alexandre
Mandera Bridge	T6	Bagamoyo	6 14.5 S 38 23 E	Procter
[Mandunda] see Madunda	T7	Njombe	9 52 S 34 27 E	
Maneno Mbangu (near Mhinduro)	T3	Lushoto	prob. c.4 47 S 38 45 E	Peter
Maneromanga, popl., t.a.	T6	Uzaramo	7 12 S 38 47 E	Vaughan
Maneromango = Maneromanga, popl., t.a.	T6	Uzaramo	7 12 S 38 47 E	
Manga, ridge, escarp.	K5	South Kavirondo	c.0 38 S 34 49 E	Buxton
Manga, hill	K7	Teita	3 11 S 38 31 E	Bally
[Manga] see Monga	T3	Lushoto	5 06 S 38 37 E	Zimmermann & Eichinger
Manga, popl.	T4	Mpanda	6 59.5 S 31 10.5 E	Richards
Mangala, ln.	T2	Arusha	c.3 15 S 36 50 E	Greenway
Mangaladasi, popl.	T6	Kilosa	6 41 S 36 46 E	
Manganyema, mt. (near Bulongwa)	T7	Njombe	near 9 20 S 34 03 E	Goetze
[Manga Plains] see Mangati Plains				
Mangapwani, popl.	T2	Mbulu	4 35 S 35 30 E	Richards
Mangapwani, popl.	Z	Zanzibar	5 59 S 39 12 E	Greenway
[Manga See] see Manka, L.	Z	Zanzibar	6 00 S 39 11 E	Numerous
Mangati Plains, area	T3	Lushoto	4 44 S 38 05 E	Peter
Mangea, mt.	T2	Mbulu	c.4 35 S 35 30 E	B.D. Burtt, Peter
[Mangero-Mango] see Maneromango	K7	Kilifi	3 15 S 39 43 E	Moomaw, Jeffery, Luke & Robertson
Mango, popl.	T6	Uzaramo	7 12 S 38 47 E	Goetze
[Mango] see Monga	T1	Shinyanga	3 41 S 33 55 E	B.D. Burtt
Mangola, popl., sch.	T3	Lushoto	c.5 05 S 38 36 E	
Mangola Juu, popl.	T2	Mbulu	3 31.5 S 35 20 E	T. Robson, Umesao
[Mangoloma] see Mongoroma	T2	Mbulu	3 24 S 35 30.5 E	
Mangua, old popl.	T5	Kondoa	4 54 S 35 37 E	B.D. Burtt
Mangubu, popl.	T8	Songea	10 42 S 35 27 E	Busse
Mangula, popl.	T3	Lushoto	5 06 S 38 41 E	Greenway
Mangunga, area	T6	Ulanga	7 50 S 36 53 E	Haerdi, Cribb & Grey-Wilson
	T7	Chunya	prob. c.8 16 S 33 15 E	Richards
[Manimani] see Omanimani	U1	Karamoja	c.2 31 N 34 45 E–2 30 N 34 14 E	Clayton

213

[Manjanga] see Manonga			
[Manjanga] see Manyangu			
[Manjangu] see Manyangu			
[Manjera] see Manchera			
[Manjira] see Manchera			
[Manjonga] see Manonga			
[Manjonjo] see Munyonyo			
Manka, L.	T1/4	Shinyanga/Nzega/Kahama	3 42 S 32 49 E–4 08 S 34 12 E
Manka, popl.	T6	Morogoro	c.6 07 S 37 34 E	Bittkau
Manola = Manolo	T6	Morogoro	c.6 07 S 37 34 E	Tanner
Manolo, R.	T1	Musoma	c.2 02 S 34 42 E
Manolo, popl.	T1	Musoma	c.2 02 S 34 42 E
Manonga, R.	T1/4	Shinyanga/Nzega/Kahama	3 42 S 32 49 E–4 08 S 34 12 E
Manow, mssn.	U4	Mengo	0 14.5 N 32 37 E	Stuhlmann
[Mansa] see Manza	T3	Lushoto	4 44 S 38 05 E	Stuhlmann
Manshumbi Inselberg, hill	T3	Lushoto	4 59 S 38 24.5 E	Numerous
[Mansira] see Musira	K7	Kwale	c.4 12 S 39 22 E	Engler
Manta (E. Gereza)	K7	Kwale	c.4 12 S 39 22 E	Kassner
Mantachien, str.	T3	Lushoto	4 37 S 38 13.5 E	Kassner
Mantaogo = Mantavyo	T1/4	Shinyanga/Nzega/Kahama	3 42 S 32 49 E–4 08 S 34 12 E	Drummond & Hemsley, Peter
Mantare, popl.	T7	Rungwe	9 15 S 33 48 E	Eggeling
Mantavyo, ln.	T3	Tanga	4 50 S 39 09 E	Richards, Fuller
[Mantini] see Mwantine	K7	Tana River	0 12 S 38 37 E	
Manyaga, popl.	T1	Bukoba	1 21 S 31 51 E	Procter
Manyangu, for.	T3	Tanga	near 5 11 S 38 01 E	Semsei
Manyani, popl.	K1	Northern Frontier	1 15 N 37 15 E–1 10.5 N 37 16.5 E
Manyani, River Drift, ln.	K7	Kwale	c.4 18 S 39 24 E	Kassner
Manyara, L.	T1	Kwimba	2 43 S 33 13 E	Gillman
[Manyemes] see Manayene	K7	Kwale	c.4 18 S 39 24 E	Kassner
Manyessa, popl.	T1	Shinyanga		B.D. Burtt
	P	Pemba	5 17 S 39 48 E	Greenway
	T6	Morogoro	c.6 07 S 37 34 E	Numerous
	K7	Teita	3 05 S 38 30 E	Tweedie
	K7	Teita	3 05 S 38 33 E	Greenway
	T2	Masai/Mbulu	c.3 35 S 35 50 E	Greenway & Kanuri
	T7	Mbeya	c.9 15 S 33 07 E	
	T3	Tanga	5 05 S 38 45 E	Peter

Name	Code	Region	Coordinates	Authority
[Manyi] see Maanyi	U4	Mengo	0 16 N 31 58 E	Harker
[Manyigi] see Minyugi	T5	Singida	4 58.5 S 34 32 E	B.D. Burtt
[Manyonga] see Manonga	T1/4	Shinyanga/Nzega/Kahama	3 42 S 32 49 E–4 08 S 34 12 E	B.D. Burtt, Stuhlmann
Manyoni, L. (near Mbirira)	T4	Buha	near 4 21 S 30 10 E	Peter
Manyoni, popl., mssn., rsta., airstrip	T5	Dodoma	5 44.5 S 34 50 E	Numerous
[Manyonjo] see Munyonyo	U4	Mengo	0 14.5 N 32 37 E	
Manyovu, popl.	T4	Buha	4 29 S 29 50 E	Eggeling, Procter
[Manyugi] see Minyugi	T5	Singida	4 58.5 S 34 32 E	B.D. Burtt
Manyungo	U4	Mengo	Not traced	Maitland
[Manyungu] see Manyangu	T6	Morogoro	c.6 07 S 37 34 E	Drummond & Hemsley
Manza, popl.	T3	Tanga	4 50 S 39 09 E	Drummond & Hemsley, Kassner
Mao, popl.	T4	Ufipa	8 09 S 31 26 E	Richards, Vesey-FitzGerald
[Maogoro] see Morogoro	T6	Morogoro	6 49 S 37 40 E	Stuhlmann
Maore, popl.	T3	Pare	c.4 16.5 S 38 02.5 E	Sangiwa
[Mapala] see Mpala	T7	Njombe	9 24 S 34 51 E	Schlieben
[Mapan] see Napau	K2	Turkana	2 25 N 34 57 E	Padwa
Mapapu Mbuga, sw.	T5	Kondoa	4 50 S 35 33 E	B.D. Burtt
Mapinga, popl.	T6	Bagamcyo	6 36 S 39 04 E	Peter
Mapiringa, popl.	T4	Tabora	4 49 S 33 32 E	Vesey-FitzGerald
Mapiringa, R.	T4	Tabora	c.4 41 S 33 56 E	
Mapogoro = Ipogoro, popl.	T7	Mbeya	8 19 S 34 42 E	B.D. Burtt
Mapopwe, popl.	Z	Zanzibar	6 14 S 39 23.5 E	Vaughan
[Mapuka] see Mubuku	U2	Toro	c.0 21 N 29 54 E–0 05 N 30 13 E	
[Mar.] see Maramba	T3	Lushoto		Zimmerman
Mara, R.	K4	Embu	0 17 S 37 45 E–0 21 S 37 53 E	Fries
Mara, R.	K6/T1	Masai/N. Mara/Musoma	1 02 S 35 15 E–1 31 S 33 56 E	Numerous
Mara, bdg.	K6	Masai	1 13 S 35 02 E	Glover et al.
Maraagwa, R.	K4	S. Nyeri/Fort Hall	0 39.5 S 36 43 E–0 47 S 37 16 E	
Mara Bridge, popl.	K6	Masai	1 09 S 35 05 E	Glover et al.
[Maraca] see Maracha	U1	West Nile	c.3 15 N 30 54 E	Alonzie

Name	Code	Region	Coordinates	Collector
Marach, ln.	K5	North Kavirondo	0 20 N 34 19 E	Templer
Maracha, co.	U1	West Nile	c.3 15 N 30 54 E	Hazel, A.S. Thomas, Eggeling
Maracha, popl., t.c.	U1	West Nile	3 17 N 30 58 E	Chancellor
Marafa, popl.	K7	Kilifi	3 02 S 39 58 E	Polhill & Paulo, Rawlins, Robertson, Beentje
Maragoli, for.	K5	North Kavirondo	0 00 34 40 E	
Maragua = Maraagwa, R.	K4	S. Nyeri/Fort Hall	0 39.5 S 36 43 E–0 47 S 37 16 E	?Balbo
Maragua, rsta., t.c.	K4	Fort Hall	0 47.5 S 37 08 E	
[Maragwa] see Maraagwa	K4	S. Nyeri/Fort Hall	0 39.5 S 36 43 E–0 47 S 37 16 E	
Marahubi, popl.	Z	Zanzibar	6 09 S 39 13 E	Vaughan, R.O. Williams, Barney
Marahubi Mlangoni = Marahubi, popl.	Z	Zanzibar	6 09 S 39 13 E	
Marakwet = Kapsowar, popl.	K3	Elgeyo	0 59 N 35 33.5 E	Brodhurst-Hill, Dale, Lindsay
Maralal, popl., govt. stn.	K1	Northern Frontier	1 06 N 36 42 E	Numerous
Maramagambo, f.r.	U2	Kigezi	c.0 32 S 29 53 E	Eggeling, Purseglove
Mara Mara = Mar Mar, str.	K5	Kericho	c.0 25 S 35 24 E–0 29 S 35 13 E	Kerfoot
Mara Mara, est.	K5	Kericho	0 30 S 35 14 E	Kerfoot
Mara Masai, g.r.	K6	Masai	c.1 30 S 35 00 E	Bally, Kirrika
Maramba, popl., t.c.	T3	Lushoto	5 03 S 38 37.5 E	Greenway, Bogner
Maramba = Malamba, for.	T3	Lushoto	4 51 S 38 48 E	Peter, Faulkner
[Marambo] ? see Mulamba	U2	Kigezi	0 57 S 29 50 E	Purseglove
Mara, Middle, R.	K4	Embu/Meru	0 15 S 37 39 E–0 16 S 37 45 E	
Mara, N., R.	K4	Embu/Meru	0 11 S 37 32 E–0 17 S 37 47 E	
Maranda, popl.	K5	Central Kavirondo	0 05 S 34 13 E	Gillett
Marang, f.r.	T2	Mbulu	c.3 42 S 35 40 E	Richmond, Carmichael, Procter
Marang, old area	T2	Moshi	c.3 18 S 37 31 E	
Marangu, popl.	T2	Moshi	3 17 S 37 31 E	Numerous
Marangu, L.	T6	Kilosa	c.6 21 S 36 59 E	Cribb & Grey-Wilson
Marani, mkt., sch.	K5	South Kavirondo	0 35 S 34 48 E	Greenway
Marania, popl.	K4	North Nyeri	0 05.5 N 37 28 E	Bally
Marania, E., R.	K1/4	N. Frontier/N. Nyeri/Meru	0 03 S 37 22 E–0 17 N 37 33 E	Fries

Name	Region		Coordinates	Collector
Marania, W., R.	N. Frontier/N. Nyeri	K1/4	c.0 03.5 S 37 20.5 E – 0 20 N 37 32 E	Fries, Schelpe
Marara	Lushoto	T3	prob. c.4 37 S 38 00 E	Fischer
Mararani, popl.	Lamu	K7	1 42 S 41 18 E	?Gillespie, J. Adamson
Mararani, w.h.	Northern Frontier	K1	1 35.5 S 41 14 E	?Gillespie
Mara River Guard Post, ln.	Musoma	T1	1 34 S 34 41 E	Greenway
Mara, S., R.	Embu	K4	0 11 S 37 30 E – 0 16.5 S 37 47 E	
[Marashon] see Marashoni	Nakuru	K3	0 22 S 35 49 E	Dale
Marashoni, popl., t.c.	Nakuru	K3	0 22 S 35 49 E	Bally
Marasi, hill	Teita	K7	3 34 S 38 45 E	Volkens
Mareales, popl.	Moshi	T2	3 17 S 37 30 E	Bogdan
[Marech] see Marich	West Suk	K2	1 31.5 N 35 26.5 E	Dale, Drummond & Hemsley, Verdcourt
Marenge = Marenje, for.	Kwale	K7	c.4 30 S 39 12 E	Greenway
[Marenji] see Marenje	Kwale	K7	c.4 30 S 39 12 E	Pirozynski
Marera, R.	Mbulu	T2	c.3 12 S 35 40 E – 3 25.5 S 35 47 E	Magogo & Glover
Marera, mssn.	Buha	T4	c.4 35 S 30 06 E	R.M. Graham
Marere, hill, for.	Kwale	K7	4 10 S 39 24.5 E	J. Adamson
Marereni, popl.	Kilifi	K7	2 52 S 40 09 E	Greenway
[Marereni] see Mararani	Lamu	K7	1 42 S 41 18 E	
Marero, creek	Musoma	T1	c.2 10 S 33 50 E	Eggeling
Mareyo (Mt. Debasien)	Karamoja	U1	c.1 45 N 34 43 E	Bally, Verdcourt
Margaret, Mount, hill	Naivasha	K3	1 00.5 S 36 33 E	Richards
[Margo Mito] see Maji Moto	Mbulu	T2	3 40 S 35 44 E	
[Marhubi] see Marahubi	Zanzibar	Z	6 09 S 39 13 E	
Mari, area	Northern Frontier	K1	c.4 12 N 40 42 E	Faulkner
Mariakani, popl, rsta	Kilifi	K7	3 52 S 39 28 E	Numerous
Mariakebuni = Marikebuni	Kilifi	K7	3 05.5 N 40 07 E	Polhill & Paulo
Mariakebunio = Marikebuni, popl.	Kilifi	K7	3 05.5 N 40 07 E	Polhill & Paulo
[Mariani] see Marania	North Nyeri	K4	0 05.5 N 37 28 E	
[Mariashoni] see Marashoni	Nakuru	K3	0 22 S 35 49 E	J. Bally
Marich, pass	West Suk	K2	1 31.5 N 35 26.5 E	Bogdan, Tweedie, Carter & Stannard

Marienberg, old mssn.	T1	Bukoba	1 14 S 31 48 E	Conrads
Marigat, popl., p.p., t.c.	K3	Baringo	0 28 N 35 59 E	Numerous
[Marikana] see Mariakani	K7	Kilifi	3 52 S 39 28 E	Kassner
Marikebuni, popl.	K7	Kilifi	3 05.5 S 40 07 E	
Marimante, camp, ln.	K4	Meru	0 08.5 S 37 57 E	Scholte
Marimba, for.	K4	Meru	0 02 N 37 32 E	Verdcourt, Polhill
Marimba, f.r.	T3	Lushoto/Tanga	5 02 S 38 43 E	
Marimba, old pl.	T3	Tanga	5 02 S 38 50 E	Peter
[Mariwe] see Maliwe	T8	Kilwa	8 50 S 39 00 E	Busse
Marlborough, former est., road	K4	Nairobi	c.1 14 S 36 47 E	Bally
Marmanet, f.r.	K3	Laikipia	0 09 N 36 19 E	Birch
Mar Mar, str.	K5	Kericho	c.0 25 S 35 24 E–0 29 S 35 13 E	
[Marobus] see Morobus	K2	West Suk	1 23 N 35 17 E	Carter & Stannard
Marongo, popl.	T3	Tanga	5 14 S 39 05 E	Faulkner
[Marrarani] see Mararani	K7	Lamu	1 42 S 41 18 E	Gillespie
[Marrigabo] see Nabugabo	U4	Masaka	0 22 S 31 54 E	Norman
Marsabit, distr.	K1	Northern Frontier	c.2 45 N 37 50 E	
Marsabit, mt., for.	K1	Northern Frontier	c.2 18 N 37 58 E	Numerous
Marsabit, popl., t.c., mssn., r.h., p.p.	K1	Northern Frontier	2 20 N 37 59 E	Numerous
Maru, mt.	U1	Karamoja	3 03 N 33 57 E	
Maruanda's, popl.	T7	Mbeya	9 10 S 33 21 E	
Maruanda's popl.	T7	Mbeya	9 08 S 33 23 E	
[Maruessa] see Maruvessa	K7	Kwale	3 41 S 39 16 E	Hildebrandt
Marui, old area	T6	Uzaramo	c.7 15 S 38 52 E	Stuhlmann
Maruku, popl., area	T1	Bukoba	1 25 S 31 47 E	Haarer, Kayumbo, Panayotis
Marumba, I., popl.	T8	Masasi	11 15 S 38 44 E	
Marume, R. (Shimba Hills)	K7	Kwale	?4 09 S 39 25 E	Magogo & Glover
Marun, R.	K2/3	W. Suk/Elgeyo	1 06 N 35 26 E–1 34 N 35 31 E	Tweedie
Marungu, popl., est.	T3	Tanga	5 12 S 39 01 E	Lindeman
[Marupindi] see Malupindi	T7	Rungwe	9 02 S 33 37 E	

218

Name	Code	Location	Coordinates	Collector
Maruvera, tea est.	T3	Lushoto	5 04 S 38 39 E	Kabuye
Maruvessa, popl.	K7	Kwale	3 41 S 39 16 E	Hildebrandt
Maruzi, co.	U1	Lango	c.1 50 N 32 25 E	
[Maryonyo] see Munyonyo	U4	Mengo	0 14.5 N 32 37 E	A.S. Thomas
[Masaba] see Elgon	U3/K3/5	several	c.1 08 N 34 33 E	J. Adamson
Masabubu, p.p.	K7	Tana River	1 13 S 40 00 E	Schlieben
Masagati, old area	T6/7	Ulanga/Njombe	c.9 10 S 35 30 E	Schlieben
Masagati, popl., f.r.	T6	Ulanga	9 01 S 35 39 E	Greenway
Masagati = Mesangati (Ruaha Nat. Park)	T7	Mbeya/Iringa	Not traced	
Masai, distr.	K6	Masai	c.1 45 S 36 25 E	E. Polhill
Masai, distr.	T2	Masai	c.4 00 S 36 30 E	
Masai Gorge	K3	Naivasha	0 39 S 36 20 E	
Masai-Mara, g.r.	K6	Masai	c.1 30 S 35 05 E	
Masaita, str.	K5	Kisumu-Londiani	0 04 S 35 38 E–0 10 S 35 35 E	Smart
Masaiyika, str. (near Madanga)	T3	Pangani	near 5 21 S 38 59 E	Tanner
Masaka, distr.	U4	Masaka	c.0 20 S 31 45 E	
Masaka, popl., mssn.	U4	Masaka	0 20 S 31 44 E	Numerous
[Masalani] see Bushwhackers	K4	Machakos	2 19 S 38 07 E	
Masalatu = Msalato, popl.	T5	Dodoma	6 05.5 S 35 45 E	
Masama, popl., r.h.	T2	Moshi	3 15 S 37 11.5 E	Bigger
[Masamba] see Msamba	T4	Ufipa	7 51 S 30 47 E	
Masandari (E. Serengeti)	T2	Masai	prob. c.2 42 S 35 26 E	Newbould
[Masangaware's] see Nsangamales	T7	Mbeya	8 52 S 32 45 E	
Masansi, popl.	T2	Ufipa	8 20 S 31 32 E	Bullock
[Masanza] see Massanza	T1	Mwanza		
Masanza, f.r.	T4	Kigoma	5 03 S 30 18 E	Procter
Masanza, popl.	T4	Kigoma	4 59 S 30 18 E	Forcus
Masasi, distr.	T8	Masasi	c.10 50 S 38 35 E	
Masasi, popl.	T8	Masasi	10 43 S 38 47.5 E	Numerous
[Masawa] see Elgon	U3/K3/5	several	c.1 08 N 34 33 E	

219

Name		Region	Coordinates	Collector
[Mascheua] see Mashewa	T3	Lushoto	4 46 S 38 38 E	Holst
[Mascheva] see Mashewa	T3	Lushoto	4 46 S 38 38 E	Holst, Peter
[Maschewa] see Mashewa	T3	Lushoto	4 46 S 38 38 E	Cribb & Grey-Wilson
Masebe, popl.	T7	Rungwe	9 20.5 S 33 39 E	Richards
Masegera, popl. (Kalambo R.)	T4	Ufipa	prob. c.8 35 S 31 15 E	A.S. Thomas
Masekera, pt.	U4	Masaka	0 25 S 32 27.5 E	Thulin & Mhoro
Masenge, area	T6	Kilosa	6 21 S 36 56 E	Abraham, M.D. Graham
Maseno, popl., mssn.	K5	Central Kavirondo	0 00 34 36 E	Greenway
[Maseseni] see Mafisini	K7	Kwale	4 28 S 39 22 E	
Masha, g.r.	U2	Ankole	c.0 40 S 30 49 E	Robertson
Mashamba, popl.	K4	South Nyeri	0 45 S 37 31 E	Numerous
Mashame = Machame, area, foothill	T2	Moshi	c.3 11 S 37 13 E	
[Mashami] see Machame	T2	Moshi		
Mashati, popl.	T2	Moshi	3 08 S 37 36.5 E	Haarer
Mashewa, popl., old pl.	T3	Lushoto	4 46 S 38 38 E	Numerous
Mashewa, sw.	T3	Lushoto	4 48 S 38 38 E	
Mashewa, popl.	T3	Lushoto	5 01 S 38 23 E	Peter
Mashineka	T3	Lushoto	c.4 30 S 38 18 E	Carmichael
[Mashugira] see Mushongero	U2	Kigezi	1 11 S 29 41 E	Gardner, C.G. Rogers
Masigo, Mulele Hills	T4	Mpanda	c.6 47 S 31 45 E	Procter
Masikia (Monduli Mt.)	T2	Masai/Arusha	c.3 15 S 36 29 E	Greenway
[Masinde] see Mazinde	T3	Lushoto	4 48 S 38 13 E	Holst, Busse
Masindi, popl., mssn.	U2	Bunyoro	1 41 N 31 43 E	Numerous
Masindi Port, popl., t.c.	U2	Bunyoro	1 42 N 32 05 E	Eggeling
Masingini, ridge, area	Z	Zanzibar	6 08 S 39 15 E	Greenway, Vaughan, Stuhlmann
[Masinjumbi] see Makinyumbi	T3	Pangani	5 20 S 38 38 E	Scheffler
Masiwa, I.	U3	Busoga	c.0 02 S 33 37 E	
[Masiwe] see Maliwe	T8	Kilwa	8 50 S 39 00 E	
Maskati, mssn.	T6	Morogoro	6 03 S 37 28.5 E	Busse
Masoka, mts.	T5	Dodoma	c.6 38 S 34 47 E	Thulin & Mhoro, Moreau; von Prittwitz

Masoko, popl., L.	T7	Rungwe	9 20 S 33 45 E	Numerous
Masokwa	T4	Kigoma	Not traced	Mahinde
Masol, area, plains	K2	West Suk	c.1 30 N 35 35 E	Bogdan
Masongaleni, popl.	K4	Machakos	2 29.5 S 38 03 E	Kassner, Tweedie
[Masongolene] see Masongaleni	K4	Machakos	2 29.5 S 38 03 E	Kassner
[Masongoleni] see Masongaleni	K4	Machakos	2 29.5 S 38 03 E	
Masote, area	T7	Chunya	c.8 40 S 33 12 E	Goetze
Masowero, str.	T6	Kilosa	c.7 38 S 37 00 E	Semsei
[Massa] see Masabubu	K7	Tana River	1 13 S 40 00 E	Fischer, Denhardt, F. Thomas
[Massagati] see Masagati	T6	Ulanga	9 01 S 35 39 E	Schlieben
Massaini, ? = Masai District	?K5	Masai		Fischer
Massanza I, t.a.	T1	Mwanza	c.2 29 S 33 30 E	
Massanza II, t.a.	T1	Mwanza	c.2 18 S 33 50 E	
[Massassi] see Masasi	T8	Masasi		Schlieben
[Massaua] see Mashewa	T3	Lushoto	4 46 S 38 38 E	
[Massazine] see Mazizini	Z	Zanzibar	6 12 S 39 12.5 E	
Massazini = Mazizini	Z	Zanzibar	6 12 S 39 12.5 E	Faulkner
Massewe, old area	T7	Rungwe	c.9 20 S 33 35 E	Goetze
[Massoko] see Masoko	T7	Rungwe	9 20 S 33 45 E	Stolz
Masuamu = Masuanu, mt.	T7	Njombe	c.9 36 S 34 24 E	
Masukulu, f.r.	T7	Rungwe	9 26 S 33 47 E	
Masukulu, popl.	T7	Rungwe	9 24 S 33 45 E	Stolz
Masumbwe, popl., area	T4	Kaharra	3 38 S 32 11 E	Joseph
Maswa, controlled area	T1	Maswa/Mwanza	c.2 45 S 34 30 E	
Maswa, distr.	T1	Maswa	c.3 05 S 34 15 E	
Maswa, popl.	T1	Maswa	2 41 S 33 58 E	Rounce
Mata, popl.	K7	Teita	3 29.5 S 37 44.5 E	D.C. Edwards
Mataara, ln., factory	K4	Kiambu	0 52 S 36 48 E	Gachathi
Matagoro, hills	T8	Songea	10 45 S 35 39 E	Numerous
Matagoro, popl.	T8	Songea	10 41 S 35 40 E	

221

Name	Code	District	Coordinates	Authority
Matagoro East, hill	T8	Songea	10 28 S 35 56 E	Milne-Redhead & Taylor
Matai, popl.	T4	Ufipa	8 19 S 31 31 E	Wingfield, Hooper & Townsend
Matakas (near Kasembe)	T8	Songea	near 10 14 S 36 25 E	McLoughlin
Matala, str. (Mt. Meru)	T2	Arusha	Not traced	Richards, Greenway
Matalele, popl.	T5	Singida	5 09 S 34 03 E	B.D. Burtt
[Matama] see Matema	T7	Rungwe	9 29 S 34 01 E	
Matamba (E. Mufindi), area	T7	Iringa	8 29 S 35 20 E	Carmichael
Matamba, popl.	T7	Njombe	8 59 S 33 58 E	Numerous
Matamba Pass	T7	Njombe	9 02 S 33 57.5 E	Leedal
Matambere, popl.	T8	Masasi	c.11 17 S 38 41 E	Busse
[Matambo] see Mtamba	T6	Morogoro	7 05 S 37 47 E	
[Matambwa] see Matamba	T7	Njombe	8 59 S 33 58 E	Richards
[Matamondo] see Matomondo	T5	Mpwapwa	c.6 28 S 36 35 E	B.D. Burtt, Hornby
Matanana, plateau, steppe	T7	Iringa	c.8 15 S 35 00 E	Goetze, Polhill & Paulo, von Prittwitz
Matandu, mt.	T6	Kilosa	6 20 S 36 58 E	Thulin & Mhoro
Matandu, popl.	T8	Kilwa	8 45 S 39 17 E	
Matandu, str.	T8	Kilwa	9 26 S 37 32 E–8 42 S 39 22 E	
Matandu, Juu, popl.	T8	Kilwa	9 04 S 37 38 E	
Matanga, popl., hill	T4	Ufipa	8 01 S 31 31 E	Vesey-FitzGerald
Matanga Twani, popl., area	P	Pemba	4 58 S 39 44 E	Vaughan
[Matangi] see Mudangi	U3	Mbale	1 10 N 34 26 E	
Matapwa, popl.	T8	Lindi	9 41 S 39 26 E	Vesey-FitzGerald
Matapwende, R.	T8	Songea	11 24 S 36 44 E–11 33 S 36 38 E	
Matara, mssn., sch.	K4	Kiambu	0 52 S 36 48 E	Bally
Matarawanda, popl.	T3	Lushoto	4 58.5 S 38 32.5 E	
Matarawanda, popl.	T3	Lushoto	4 58.5 S 38 18 E	
Matarawe, str.	T6	Kilosa	c.6 45 S 37 00 E	Pegler, Renvoize
[Matate] see Matete	T2	Masai/Mbulu	3 49 S 36 05 E	Richards
[Matebende] see Matapwende	T8	Songea	11 24 S 36 44 E–11 33 S 36 38 E	Busse
Matele, popl., area	P	Pemba	5 21 S 39 47	Vaughan

Name	Grid	Region	Coordinates	Collector
[Matelele] see Matatele	T5	Singida	5 09 S 34 03 E	B.D. Burtt
Matema, mssn.	T7	Rungwe	9 29 S 34 01 E	Richards
Matembwe, popl., f.r.	T7	Njombe	9 14 S 35 11 E	Schlieben
Matende, popl.	T4	Buha	3 45 S 30 48 E	Bullock
Matengo Highlands	T3	Songea	c.10 55 S 34 53 E	Numerous
Matesse, R.	T7	Rungwe	9 07 S 33 40 E-9 10 S 33 49 E	Brenan & Greenway
Matete, area	T2	Masai/Mbulu	3 49 S 36 05 E	Mwinyjuma
[Mateyo] see Kawanja ya Matheo	T2	Arusha	c.3 15.5 S 36 53 E	
Mathangauta, str./s.	K4	Machakos/Kitui	c.1 09 S 37 33 E-10 55 S 37 32 E	Bally
[Mathenganta] see Mathangauta	K4	Machakos/Kitui	c.1 09 S 37 33 E-10 55 S 37 32 E	
Matheniko, co.	U1	Karamoja	c.2 40 N 34 30 E	Dale, J. Wilson
Mathews Peak, mt.	K1	Northern Frontier	1 18 N 37 18 E	Numerous
Mathews Range, for.	K1	Northern Frontier	c.1 16 N 37 17 E	Numerous
Mathews Range, mts.	K1	Northern Frontier	c.1 15 N 37 15 E	Richards
[Matheyo] see Kawanja ya Matheo	T2	Arusha	c.3 15.5 S 36 53 E	
Mathiniko, str.	U1	Karamoja	c.2 35 N 34 46 E-2 35 N 34 33 E	Dale
Mathioya S., R.	K4	Fort Hall	0 38.5 S 36 45 E-0 42.5 S 37 07.5 E	Gwynne
[Matima] see Matema	T7	Rungwe	9 29 S 34 01 E	Richards
Matiri, popl.	U2	Toro	0 34 N 30 46 E	Stuhlmann
[Matisi] see Mafisi	T6	Uzaramo	6 59 S 38 38 E	
Matofya, area	K7	Kwale	c.4 23 S 39 28 E	Parker
Matolani, popl.	K7	Kilifi	3 05 S 39 39 E	
Matolani, hill	K7	Kilifi	3 03 S 39 33 E	
Matolo, l.s.	U3	Busoga	c.0 06 N 33 52 E	G.H. Wood
Matombo, popl., mssn.	T6	Morogoro	7 03 S 37 46 E	Anatoli, E.M. Bruce, Haarer
Matomondo, area, valley	T5	Mpwapwa	c.6 28 S 36 35 E	van Rensburg
Matondwe, hill	T3	Lushoto	4 45 S 38 19.5 E	Drummond & Hemsley
Matongura = Matunguru, hill	T7	Iringa	8 15 S 35 50 E	Carmichael
[Matovia] see Matofya	K7	Kwale	c.4 23 S 39 28 E	R.M. Graham
[Matschinga] see Machinga	T8	Kilwa/Lindi	c.9 20 S 39 25 E	

Name	Code	Location	Coordinates	Collector
[Mavumba] see Marumba				
Mavumbi, hill	T8	Masasi	11 15 S 38 44 E	Busse
Mawani	T3	Lushoto	4 44 S 38 40 E	Drummond & Hemsley
Mawembe Rock	P	Pemba	5 23 S 39 45 E	Greenway
Mawene, popl.	U3	Busoga	0 48 N 33 20.5 E	?Maitland
Maweni, ln.	T5	Dodoma	5 53 S 35 07 E	Polhill & Paulo
Maweni, popl.	K7	Kwale	c.4 12 S 39 37 E	Bally
Maweni, prison	T3	Tanga	4 44 S 39 04 E	Faulkner
Mawensi = Mawenzi, peak	T3	Tanga	5 07 S 39 01 E	Volkens
Mawenzi, hut	T2	Moshi	3 05 S 37 27.5 E	King
Mawenzi, peak	T2	Moshi	3 05.5 S 37 27 E	Numerous
Mawere, popl.	T2	Moshi	3 05 S 37 27.5 E	Carnochan
[Mawese] see Muhuwesi	T4	Tabora	5 22 S 32 46 E	Milne-Redhead & Taylor
[Mawesi] see Muhuwesi	T8	Tunduru	10 36 S 36 39 E—11 14 S 38 03 E	Milne-Redhead & Taylor, Richards
Maweso, popl.	T8	Tunduru	10 36 S 36 39 E—11 14 S 38 03 E	Milne-Redhead & Taylor, Semsei
Maw Hills ? = Mwau	T8	Songea	9 47 S 35 14 E	B.D. Burtt
Mawie = Mwawe, str.	T5	Singida	?c.5 15 S 34 53 E	Greenway
Mawogola, co.	T2	Moshi	c.3 15 S 37 14 E	Purseglove, Lye
Mawokota, co.	U4	Masaka	c.0 05 S 31 25 E	Numerous
[Mayage] see Mwayange	U4	Mengo	c.0 00 32 10 E	Richards
Mayanga (L. Lutamba—Rondo)	T7	Iringa	7 48 S 34 57 E	Busse
Mayenze, popl.	T8	Lindi	Not traced	
C.H. Mayer's Farm	U3	Mbale	0 57 N 34 16 E	Verdcourt
Mazeras, popl., rsta.	K3/4	Naivasha/Kiambu	1 03.5 S 36 35.5 E	Numerous
Maziba, r.h.	K7	Kilifi	3 58 S 39 33 E	Purseglove, Eggeling
Mazige, popl.	U2	Kigezi	1 19 S 30 06 E	
Mazimasa, popl.	U4	Mengo	0 22.5 N 32 49 E	
Mazinda, area	U3	Mbale	0 57 N 34 02 E	
Mazinde, popl., rsta., old pl.	T1	Mwanza	c.2 45 S 32 33 E	B.D. Burtt
Mazinga, hill, Nat. Pk. H.Q.	T3	Lushoto	4 48 S 38 13 E	Numerous
	K7	Teita	3 21 S 38 35.5 E	Bally, Sheldrick

Name	Region	Place	Coordinates	Collector
[Mazingini, ridge] see Masingini				
Maziwi I.	Z	Zanzibar	6 08 S 39 15 E	Frazier
Mazizini, popl.	T3	Pangani	5 30 S 39 04.5 E	Faulkner
Mazizini, beach	Z	Zanzibar	6 12 S 39 12.5 E	
Mazombe, popl.	Z	Zanzibar	6 12 S 39 12 E	
Mazombe, area	T7	Iringa	7 43 S 36 00 E	Ward
Mazumbai, popl., area, old pl., f.r.	T7	Iringa	c.7 50 S 35 55 E	Numerous
Mazunyungu, R.	T3	Lushoto	4 48 S 38 30 E	Semsei
Mbaani, for.	T6	Kilosa	Not traced	
[Mbaffu] see Mpafu	K4	Machakos	1 27 S 37 29 E	Stuhlmann
Mbaga, mssn.	T6	Uzaramo	7 17 S 39 21 E	Peter, Semsei
[Mbagai] see Mbagi	T3	Pare	4 06 S 37 48 E	Greenway
Mbagala = Mbagalla, mssn., L.	T7	Iringa	7 35 S 34 53 E	Milne-Redhead, Stuhlmann
Mbagalla = Mbagala, mssn.	T6	Uzaramo	6 54 S 39 15 E	Peter
[Mbagara L.] see Mbagala	T6	Uzaramo	6 54 S 39 15 E	Numerous
Mbagathi, popl.	T6	Uzaramo	6 54 S 39 15 E	
Mbagathi, R.	K4	Nairobi	1 23 S 36 45.5 E	Bogdan, Napier, Bally
[Mbagaya] see Bagoyo	K4	Nairobi	c.1 19 S 36 40 E–1 23 S 36 49 E	Greenway
[Mbagayo] see Bagoyo	T2	Mbulu	c.3 30 S 35 47.5 E	Greenway
[Mbagaza] see Mbagala L.	T2	Mbulu	c.3 30 S 35 47.5 E	Peter
[Mbage] see Mbagi	T6	Uzaramo	6 54 S 39 15 E	Richards
Mbagi, camp, Ruaha Nat. Park	T7	Iringa	7 35 S 34 53 E	Richards, Greenway, Procter
[Mbagola] see Mbagala	T7	Iringa	7 39 S 34 55 E	Milne-Redhead & Taylor
Mbaka, popl.	T6	Uzaramo	6 54 S 39 15 E	
Mbaka, R.	T7	Rungwe	9 19 S 33 48 E	Stolz, Goetze
Mbaka Kilambo, for.	T7	Rungwe	c.9 09 S 33 40 E–9 30 S 33 58 E	Stolz
Mbaka Kirambo = Mbaka Kilambo, for.	T7	Rungwe	c.9 32 S 33 57 E	
Mbakana, R.	T6	Morogoro	7 05 S 37 36 E–7 24 S 37 31 E	Goetze, Mgaza
[Mbakano] see Mbakana	T6	Morogoro	7 05 S 37 36 E–7 24 S 37 31 E	
Mbala, hill, popl.	T3	Lushoto	5 01 S 38 14 E	Peter

Name		District	Coordinates	Collector(s)
Mbalageti, R.	T1	Musoma/Maswa	c.2 53 S 34 54 E–2 12 S 33 49 E	J. Adamson, Greenway
[Mbalambal] see Olbalbal	T2	Masai	c.3 05 S 35 25 E	Greenway
Mbalambala, for.	K1	Northern Frontier	c.0 05 S 39 07 E	J. Adamson
Mbalambala, popl.	K1	Northern Frontier	0 02.5 S 39 03.5 E	J. Adamson
Mbalamu, popl.	T3	Lushoto	4 27.5 S 38 20 E	
Mbalangeti = Mbalageti, R.	T1	Musoma/Maswa	c.2 53 S 34 54 E–2 12 S 33 49 E	Greenway
Mbale, distr.	U3	Mbale	c.0 50 N 34 10 E	
(comprising Bukedi, Bugisu and Sebei Districts)				
Mbale, popl.	U3	Mbale	1 05 N 34 10 E	Maitland, Snowden
Mbale, popl.	K5	North Kavirondo	0 05 N 34 43.5 E	Ossent
Mbale, mssn.	K7	Teita	3 23 S 38 23 E	Rees
[Mbali] see Mlahi	T6	Ulanga	8 17.5 S 37 06.5 E	Wingfield
Mbalizi, R.	T7	Mbeya	9 04 S 33 24 E–8 50.5 S 33 10.5 E	Wingfield
Mbalizi, popl., bdg.	T7	Mbeya	8 55.5 S 33 21 E	Bally, Uhlig, Engler
Mbalu, mt.	T3	Lushoto	4 33 S 38 12 E	McLoughlin, Milne-Redhead & Taylor
Mbamba Bay, popl.	T8	Songea	11 17 S 34 46 E	Schlieben
[Mbambaku] see Mkambaku	T6	Morogoro	7 10 S 37 42 E	Eggeling, Schlieben, Rees
Mbangala, popl.	T6	Ulanga	8 36 S 36 44 E	Richards
[Mbangala] see Bangala	T7	Chunya	prob.c.8 22 S 32 55 E	Milne-Redhead & Taylor, Busse
Mbangala, R.	T8	Masasi	c.10 40 S 38 20 E–11 09 S 38 55 E	Busse
[Mbangarandu] see Mbarangandu	T6/8	several	10 40 S 36 35 E–8 57 S 37 24 E	Thorold
Mbara, str.	K2	West Suk	1 39 N 35 23 E–1 38.5 N 35 27 E	
[Mbaraganda] see Mbarangandu	T6/8	several	10 40 S 36 35 E–8 57 S 37 24 E	Busse
Mbaramu = Mbalamu, popl.	T3	Lushoto	4 27.5 S 38 20 E	Holst
Mbarangandu, popl.	T8	Songea	10 12 S 36 48 E	McLoughlin
Mbarangandu, R.	T6/8	several	10 40 S 36 35 E–8 57 S 37 24 E	Busse
Mbarara, popl., mssn.	U2	Ankole	0 37 S 30 39 E	Numerous
Mbarazi, popl.	T4	Buha	4 00 S 31 20 E	Bullock
Mbarika, popl.	T1	Mwanza	2 55 S 32 51 E	Tanner
Mbaruk, rsta.	K3	Nakuru	0 21 S 36 12.5 E	Tweedie

[Mbaruka] see Mbarika			
Mbasa, popl.	T1	Mwanza	2 55 S 32 51 E	Tanner
Mbega, R.	T6	Ulanga	8 08 S 36 44 E	Haerdi
Mbegera, popl.	T4	Ufipa	prob.c.8 10 S 31 25 E	Richards
Mbejera, old area	T7	Njombe	9 32 S 34 57 E	Schlieben
[Mbejera] see Mbegera	T7	Njombe	c.9 30 S 34 50 E	Schlieben
Mbemkuru, R.	T8	Kilwa/Masasi/Lindi	10 22 S 37 42 E – 9 29 S 39 40 E	Nicholson, Milne-Redhead & Taylor, Schlieben
Mbere, str.	K3	Trans-Nzoia	1 08 N 34 40 E – 1 07 N 34 51 E	
[Mbereko] see Bereku	K5	Kondoa	4 27 S 35 44 E	Tweedie
[Mberri] see Mbere	K3	Trans-Nzoia	1 08 N 34 40 E – 1 07 N 34 51 E	Richards
[Mbesi] see Mbisi	T4	Ufipa	7 52 S 31 40 E	Paulo
Mbeta R. (Uluguru Mts.)	T6	Morogoro	Not traced	
Mbeya, distr.	T7	Mbeya	c.8 30 S 33 30 E	
Mbeya, f.r.	T7	Mbeya	c.8 55 S 33 26 E	
Mbeya, mts.	T7	Mbeya	c.8 50 S 33 20 E	Numerous
Mbeya, popl., mssn., r.h., airstrip	T7	Mbeya	8 53.5 S 33 26 E	Numerous
Mbeya Peak, f.r.	T7	Mbeya	c.8 50 S 33 18 E	Kerfoot, Mgaza, Cribb & Grey-Wilson
Mbeya Range, f.r.	T7	Mbeya	c.8 50 S 33 18 E	
Mbeye, popl., hill	T7	Rungwe	9 03 S 33 34 E	St. Clair-Thompson
Mbezi, str.	T6	Uzaramo	c.6 49 S 39 05 E – 6 43 S 39 14 E	Vaughan
Mbezi, str.	T6	Ulanga	c.8 56 S 36 39 E	Cribb et al.
[Mbigiri] see Mbijiri	T7	Iringa	c.7 59 S 35 18 E	Goetze
Mbiji, popl.	Z	Zanzibar	5 57 S 39 15 E	Greenway
Mbijiri, area	T7	Iringa	c.7 59 S 35 18 E	Goetze
Mbimba, agric. stn.	T7	Mbeya	9 04 S 32 56 E	Reakes-Williams, Richards
[Mbimbe] see Mbimba	T7	Mbeya	9 04 S 32 56 E	
Mbinga, popl., mssn.	T8	Songea	10 56 S 35 01 E	Zerny
Mbingo = Mbingu	T6	Ulanga	8 12 S 36 15.5 E	Eggeling
Mbingu, r.h., popl.	T6	Ulanga	8 12 S 36 15.5 E	Carmichael

228

Name	Code	Area	Coordinates	Collector
[Mbinzao] see Mbuinzau	K4	Machakos	2 21 S 37 55 E	Joanna
Mbirira, popl.	T4	Buha	4 21 S 30 10 E	Peter
Mbirizi, popl., mssn.	U4	Masaka	0 23 S 31 28 E	Michelmore, Trapnell
Mbisi, mts., for.	T4	Ufipa	7 52 S 31 40 E	Numerous
Mbisi, str.	T4	Ufipa	c.7 52 S 31 40 E	Richards
Mbita, pt., popl.	K5	South Kavirondo	0 25 S 34 12 E	Gachathi & Opan
Mbiuni, ln.	K4	Machakos	1 15 S 37 24 E	Fliervoet
Mbizi = Mbisi, mts., for.	T4	Ufipa	7 52 S 31 40 E	Whellan, Carmichael
Mboamaji, popl., harbour	T6	Uzaramo	6 52 S 39 25 E	Vaughan
Mboga, area	Zaire *not* Uganda		c.1 10 N 30 10 E	Dawe
Mbogo, mt.	T6	Ulanga	8 56 S 36 40 E	Cribb et al.
Mbogo, popl.	T7	Chunya	7 26 S 33 26 E	Geilinger
Mbogo, mt., plateau	T7	Mbeya	9 10 S 33 18 E	?Goetze, Procter, Leedal
Mbogo, mt.	T7	Rungwe	9 20 S 33 18 E	?Goetze, Cribb & Grey-Wilson
Mbogoli, hill	K4	Embu	0 15.5 S 37 36.5 E	Rammel
[Mbogoria] see Bogoria	K3	Baringo	c.0 15 N 36 06 E	
[Mbogwe] see Mabogwe	T4	Buha	4 05 S 31 10 E	
Mbogwe, area	T4	Kahama	c.3 25 S 32 16 E	Bullock
Mbokwa, hill	T6	Morogoro	6 40 S 37 51 E	Carmichael
Mbola = Mpola, popl.	T7	Rungwe	9 06 S 33 42 E	B.D. Burtt
[Mbolo] see Mbola	T7	Rungwe	9 06 S 33 42 E	Cribb & Grey-Wilson
Mbololo, hill	K7	Teita	3 17 S 38 28 E	Cribb & Grey-Wilson
Mbololo, R.	K7	Teita	3 20 S 38 26 E–3 02.5 S 38 44 E	Numerous
Mbombo (Mikumi Nat. Park)	T6	Kilosa	Not traced	Greenway
[Mbomore] see Bomole	T3	Lushoto	c.5 06 S 38 37 E	Greenway
Mbono, R.	T1	Maswa	c.3 07 S 34 50 E–2 57 S 34 38 E	
Mbooni, hills	K4	Machakos	c.1 40 S 37 27 E	Greenway
Mbora, mt.	T6	Morogoro	6 54 S 37 44 E	Bogdan, Nicholson
[Mboret] see Mboreti	T2	Masai	4 02 S 36 26 E	
Mboreti, area	T2	Masai	4 02 S 36 26 E	Peterson

[Mbori] see Mbozi				
Mbosi = Mbozi, popl., mssn.	T7	Mbeya	9 02 S 32 56 E	Numerous
Mbotiboli (perhaps a vernacular name)	T7	Mbeya	9 02 S 32 56 E	Rigby
Mbowu, R.	T5	Dodoma	Not traced	
[Mboyo Plateau] see Mbogo	T7	Mbeya	c.8 40 S 32 33 E–8 17 S 32 25 E	Goetze
Mbozi, popl., mssn., area	T7	Mbeya	9 10 S 33 18 E	Leedal
Mbozi Circle, road	T7	Mbeya	9 02 S 32 56 E	Jacobsen
Mbuba, popl., mkt.	T7	Mbeya	c.9 02 S 32 56 E	Napper, Richards
Mbudya, I.	T1	Ngara	2 40 S 30 36 E	Tanner
[Mbuera] see Mbwera	T6	Uzaramo	6 39.5 S 39 15 E	Harris
Mbuga ya Larmakau	T8	Kilwa	8 39 S 37 42 E	Schlieben
Mbuga ya Longil, ln.	T2	Mbulu	c.4 10 S 36 10 E	
Mbugwe, area	T2	Arusha	c.3 15.5 S 36 53 E	
Mbugwe, popl.	T2	Mbulu	c.3 55 S 35 50 E	Haarer, Hornby
[Mbui] see Mpui	T4	Kahama	3 17 S 31 18 E	Bullock
Mbuiga, popl.	T4	Ufipa	8 21 S 31 50 E	Richards
Mbuinzau, hill	T6	Kilosa	7 25 S 37 17 E	Grant
Mbujuni, popl.	K4	Machakos	2 21 S 37 55 E	
Mbuka, R., gorge	T2	Masai	3 31 S 36 09 E	Corbett
Mbula, old area	T2	Masai/Mbulu	c.3 01 S 35 58 E	Richards
Mbulamuti, f.r.	T3	Lushoto	4 57 S 38 08 E	Peter
Mbulu, distr.	U3	Busoga	0 51 N 33 03 E	G.H. Wood
Mbulu, popl., mssn.	T2	Mbulu	c.3 50 S 35 35 E	
Mbulumbul, t.a.	T2	Mbulu	3 51 S 35 32 E	Numerous
Mbulumbulu, popl.,t.c.	T2	Mbulu	c.3 15 S 35 40 E	
Mbulu Plateau, area	T2	Mbulu	3 16 S 35 47.5 E	Greenway
[Mbundugi] see Dundugi	T2	Mbulu	c.3 50 S 35 32 E	Bally, B.D. Burtt, Geilinger
Mburahati = Buharati (near Dar Salaam)	T7	Iringa	7 34 S 34 49 E	Greenway
Mburu ? see Mbowu, R.	T6	Uzaramo	near 6 48 S 39 15 E	Peter
Mburumui, hill	?T7	?Mbeya	?8 40 S 32 33 E–8 17 S 32 25 E	Michelmore
	T6	Ulanga	9 44 S 36 41 E	Rees

Name	Code	District	Coordinates	Collector
[Mburununi] see Mburumui	T6	Ulanga	9 44 S 36 41 E	Rees
Mbuya, hill	U4	Mengo	0 20 N 32 38 E	Rwaburindoro
Mbuyuni, popl.	K7	Teita	3 14 S 38 30 E	Scott Elliot
Mbuyuni, area	T2	Moshi	c.3 22 S 37 24 E	Corbett
[Mbuyuni] see Buyuni	T3	Pangani	5 57 S 38 47.5 E	Peter
Mbuyuni, popl.	T6	Kilosa	7 28 S 36 31.5 E	Thulin & Mhoro, Bjornstad, Bally & Carter
Mbuyuni, popl.	P	Pemba	5 11 S 39 49 E	Vaughan
Mbuzini, popl., area	P	Pemba	5 12 S 39 47 E	Vaughan
[Mbwamaji] see Mboamaji	T6	Uzaramo	6 52 S 39 25 E	Welch
Mbwawa, R.	T8	Songea	c.11 10 S 34 50 E	Semsei
Mbwemburu = Mbemkuru, R.	T8	Kilwa/Masasi/Lindi	10 22 S 37 42 E–9 29 S 39 40 E	Faulkner, Greenway, Vaughan
Mbweni, popl., beach	Z	Zanzibar	6 12.5 S 39 12 E	Faulkner
Mbwera, popl., area	T8	Kilwa	8 39 S 37 42 E	Bally
Mbwewe, popl., area	T6	Bagamoyo	6 04 S 38 14 E	Tweedie
[Mbwinzau] see Mbuinzau	K4	Machakos	2 21 S 37 55 E	Peter
[McCall's] see McCoys	K3	Trans-Nzoia	1 05.5 N 35 00 E	Vaughan
McCoy's Bridge	K3	Trans-Nzoia	1 05.5 N 35 00 E	Greenway
[Mchaji] see Musosi	T4	Buha	4 48 S 30 01 E	Greenway
Mchanga, Mto, str.	Z	Zanzibar	c.6 00 S 39 14 E	Greenway
Mchanga, Mto wa, str.	T2	Mbulu	c.3 25 S 35 48 E	Vaughan, Greenway
Mchanga, Mto ya, str.	T1	Musoma	1 34 S 34 50 E–1 45 S 34 54 E	Richards, Napper
Mchangani, popl.	T6	Rufiji	7 54.5 S 39 48.5 E	Richards
Mchangani, popl. area	Z	Zanzibar	6 03 S 39 20 E	
Mchata, mt.	T4	Ufipa	8 13 S 31 34 E	
Mchembo, est. (Mbozi area)	T7	Mbeya	c.9 02 S 32 56 E	
Mchinga = Mchinja, bay, popl.	T8	Lindi	9 44 S 39 42 E	Milne-Redhead & Taylor
Mchinja = Mchinga, bay, popl.	T8	Lindi	9 44 S 39 42 E	
Mchinjiri, popl.	T8	Lindi	10 08 S 39 11 E	Numerous
Mchocha (Chwaka Rd. Mile 5 (South))	Z	Zanzibar	c.6 11.5 S 39 15 E	Vaughan
Mchombe, mssn.	T6	Ulanga	8 18.5 S 36 07 E	Haerdi

Melinda = Malenda, area	T2	Masai	2 55 S 35 45 E	Herlocker
[Melka Bulfayo] ? see Malka Abafayo	K1	Northern Frontier	?1 12 N 38 58 E	Pratt
Melka Cumbisu (near Ramu)	K1	Northern Frontier	near 3 56 N 41 13 E	J. Adamson
Melka Koja, w.h.	K1	Northern Frontier	1 12 N 38 58.5 E	J. Adamson
[Melka Murri] see Malka Mari	K1	Northern Frontier	4 16 N 40 46 E	
Melka Rupia	?K7	?Tana River	?0 05 S 38 30 E	J. Adamson
Melsa, for. (Amani)	T3	Lushoto	prob. c.5 06.5 S 38 37 E	Ali Omare
Mendatini, mt.	K6	Masai	2 07 S 37 14 E	
Menengai, crater	K3	Nakuru	c.0 13 S 36 04 E	Numerous
Menengai, hill, for.	K3	Nakuru	0 14 S 36 05.5 E	
Mengo, distr.	U4	Mengo	c.0 45 N 32 30 E	Grant, Stuhlmann
[Mengo] see Mengwe	T2	Moshi	3 15 N 37 36 E	Haarer
Mengwe, area	T2	Moshi	3 15 N 37 36 E	
Mennell's Estate = Korongo Farm	K3	Naivasha	0 45 S 36 17 E	Hooper & Townsend
Meranga, area	K4	S. Nyeri	c.0 30 S 37 15 E	Alluaud
Merara, popl., mssn.	T6	Ulanga	8 38 S 35 57 E	Anderson, Haerdi
Merera, mssn.	T6	Ulanga	8 38 S 35 56.5 E	Haerdi
[Merile] see Merille	K1	Northern Frontier	c.1 24 N 37 40 E—1 38 N 38 20 E	
Merille, R.	K1	Northern Frontier	c.1 24 N 37 40 E—1 38 N 38 20 E	?Shantz, Magogo
Merille, t.a.	K1	Northern Frontier	c.3 20 N 41 15 E	?Shantz
Meris, popl., r.h.	U1	Karamoja	3 22 N 33 46 E	Liebenberg
[Merris] see Meris	U1	Karamoja	3 22 N 33 46 E	Liebenberg
Merkerstein, hill	T2	Masai	2 43 S 36 31 E	Greenway
[Merogai Gelai] see Gelai Meru-goi	T2	Masai	2 39 S 36 06 E	Carmichael
Merti, plat., escarp.	K1	Northern Frontier	c.1 03 N 38 30 E	G. Adamson, Salkeld, Bally & Smith
Meru, distr.	K4	Meru	c.0 00 37 55 E	
Meru, g.r.	K4	Meru	c.0 07.5 N 38 10 E	Numerous
Meru, popl.	K4	Meru	0 03 N 37 39 E	Numerous
Meru, f.r.	T2	Arusha	c.3 15 S 36 48 E	
Meru, mt.	T2	Arusha	3 14 S 36 45 E	Numerous

233

[Merue] see Kwa Mberue	T3	Handeni	5 27 S 38 36 E	Fischer
Mesangati = Masagati (Ruaha Nat. Park)	T7	Mbeya/Iringa	Not traced	Greenway & Kanuri
Mesema = Misima, mt.	T3	Lushoto	prob. c.5 10 S 38 40 E	Greenway
[Mesembe] see Msembe	T7	Iringa	7 39 S 34 55 E	Richards
Meserani, dam, hill	T2	Masai	3 30 S 36 26.5 E	Greenway, Moreau
[Messumbwa] see Mesumba	T6	Morogoro	6 13 S 37 26 E	Schlieben
Mesumba, peak	T6	Morogoro	6 13 S 37 26 E	Schlieben
Metu, popl., r.c.	U1	West Nile	3 40 N 31 47 E	Numerous
Metuli, popl.	U1	West Nile	3 43 N 31 43 E	Numerous
Meturu, area	U1	West Nile	c.3 40 N 31 52 E	Eggeling
Meu = Miu, popl.	K4	Machakos	1 31 S 37 34.5 E	Clayton
Mfangano, I.	K5	South Kavirondo	0 28 S 34 00 E	Napier
Mferi (near Mbingu)	T6	Ulanga	near 8 12 S 36 15 E	Carmichael
Mferu ? = Weru [Veru]	T7	Iringa	?7 51 S 35 30 E	Lynes
Mfimbwa, mt.	T4	Ufipa	8 24 S 31 43 E	Fromm, Muenzner
Mfiyoza = Mfuizi, popl.	T6	Uzaramo	6 58 S 38 52 E	
[Mfuesi] see Mfuizi	T4	Ufipa	7 46 S 31 09 E–7 10 S 31 07 E	
Mfuizi, R.	T4	Ufipa	7 46 S 31 09 E–7 10 S 31 07 E	
Mfumba Steppe, area	T3	Lushoto	4 24 S 38 18 E	Greenway
Mfumbini = Fumbini, beach	K7	Kilifi	3 36 S 39 49 E	Jeffery
[Mfumbwe] see Mfumbwi	Z	Zanzibar	6 09 S 39 25.5 E	Greenway
Mfumbwi, popl., area	Z	Zanzibar	6 09 S 39 25.5 E	
Mfumoni	T3	Pangani	near 5 21 S 38 59 E	Tanner
[Mfwanganu] se Mfangano	K5	South Kavirondo	0 28 S 34 00 E	
Mfyoza = Mfiyoza, popl.	T6	Uzaramo	6 58 S 38 52 E	
[Mga.] see Monga	T3	Lushoto	5 06 S 38 37 E	Ruffo
Mgahinga, mt.	U2/Rwanda	Kigezi	1 23 S 29 39 E	Zimmermann
[Mgaka] see Ngaka (upper reaches)	T8	Songea	10 58 S 34 54 E–10 47 S 35 02 E	Numerous
Mgama, popl.	T7	Iringa	8 03 S 35 36 E	Polhill & Paulo
[Mgambani] see Migombani	Z	Zanzibar	5 59.5 S 39 18 E	Greenway

Name	Region	District	Coordinates	Collector
Mgambo, popl., f.r.	T3	Handeni	5 34 S 38 30 E	Procter, Richards, Shabani
Mgambo, area, old pl.	T3	Lushoto	5 03 S 38 38 E	
[Mgambo] see Ngambo	T3	Lushoto	5 01.5 S 38 36 E	Greenway, Peter
Mgambo, area, popl.	T6	Morogoro	7 05 S 37 43 E	Schlieben
Mgambo, old area	T6	Bagamoyo	c.6 40 S 38 55 E	Stuhlmann
[M'Ganda] see Ganda	K7	Kilifi	3 13 S 40 04 E	Hacker
[Mgane] see Mgange	K7	Teita	3 24 S 38 18.5 E	Bally
Mgange, popl.	K7	Teita	3 24 S 38 18.5 E	
[Mgangwe] see Magangwe	T7	Mbeya	7 48 S 34 12 E	Greenway
[Mgari] see Magari	T6	Morogoro	6 56 S 37 39 E	Pócs et al.
[Mgasi] see Mngazi	T6	Morogoro	c.7 11 S 37 40 E–7 27 S 37 42 E	Goetze
[Mgatij] see Mangati	T2	Mbulu	c.4 35 S 35 30 E	B.D. Burtt
Mgende, popl.	T4	Buha	4 39.5 S 31 11 E	Bullock
Mgera, popl., r.h., t.c.	T3	Handeni	5 23 S 37 32 E	Geilinger, Busse, B.D. Burtt
[Mgeregere] see Migeregere	T8	Kilwa	8 48 S 39 13 E	Busse
Mgeta, popl., mssn., r.h.	T6	Morogoro	7 02 S 37 34 E	Numerous
Mgeta, R.	T6	Morogoro	7 08 S 37 38 E–7 17 S 38 06 E	Numerous
Mgeta, str., t.c.	T6	Ulanga	8 19 S 36 08 E	Haerdi
[Mgigile] see Mugiligili	T3	Pare	3 45 S 37 43 E	Haarer
Mgila, popl.	T3	Lushoto	5 01 S 38 29 E	Peter
[Mgita] see Mgeta	T6	Morogoro		
Mgiwe = Kilondo, str.	T7	Njombe	c.9 35 S 34 25 E	
[Mgohing] see Ngohingo	T6	Uzaramo	6 47 S 38 50 E	Wigg
[Mgole] see Migole	T7	Iringa	7 07 S 35 50 E	Eggeling
Mgololo, mssn.	T6	Morogoro	6 48 S 37 44 E	
Mgololo, area	T7	Iringa	c.8 44 S 35 09 E	Goetze, Pegler, Renvoize
[Mgombezi] see Ngombezi	T3	Lushoto	5 09.5 S 38 25 E	Semsei
Mgongo Thembo, popl.	T5	Dodoma	5 54 S 34 07 E	Grant
Mgongowa Ngamia	Z	Zanzibar	prob. c.6 12 S 39 15 E	Vaughan
Mgori, popl., r.h., t.c.	T5	Singida	4 50 S 34 59 E	Polhill & Paulo

Mgunda, popl.	T6	Kilosa	7 27 S 37 16 E	Goetze
Mgunda, mt.	T7	Njombe	9 13 S 34 05 E	
Mgunda Mkali, area	T5	Dodoma	c.5 50 S 33 45 E–6 15 S 34 45 E	
[Mgungu] see Mugungu	T1/2	Musoma/Masai	c.2 30 S 35 20 E–2 19 S 34 53 E	
Mgwashi, popl., mssn.	T3	Lushoto	4 46.5 S 38 29 E	Greenway
Mgwina, str.	T6	Ulanga	9 22 S 37 07 E–9 11 S 37 15 E	Cribb & Grey-Wilson, L. Tanner
Mgwina, str.	T6	Ulanga	c.9 37 S 36 45 E	
Mhali = Mlahi, ferry	T6	Ulanga	8 17.5 S 37 06.5 E	
[Mhesa] see Mheza	T3	Lushoto	4 50 S 38 02 E	Greenway
[Mheza] see Muheza	T3	Pare	4 34.5 S 37 44 E	Greenway
Mheza, popl.	T3	Lushoto	4 50 S 38 02 E	Peter
[Mheza] see Muheza	T3	Tanga	5 10 S 38 48 E	Cribb & Grey-Wilson
[Mhindulo] see Mhinduro	T3	Lushoto	4 57 S 38 46 E	
Mhinduro, hill	T3	Lushoto	4 57 S 38 46 E	Peter, Bogner
Mhonda, popl., mssn.	T6	Morogoro	6 07 S 37 34.5 E	Sacleux, Semsei
Mhonda, sawmill	T6	Morogoro	6 07 S 37 35 E	Brenan & Greenway, Schlieben
Mhumbo = ? Lohumbo	T1	Shinyanga	?3 50 S 33 05 E	B.D. Burtt
[Mhutwe] see Muhutwe	T1	Bukoba	1 33 S 31 42 E	Haarer
Mhwala, R.	T4	Tabora	c.4 57 S 33 56 E	
Micahani	K7	Kilifi	3 50 S 39 29 E	Kassner
Michamvi, popl., area	Z	Zanzibar	6 09 S 39 30 E	Vaughan
Michamvi, Ras., pt.	Z	Zanzibar	6 07.5 S 39 30 E	
Micheweni, popl.	P	Pemba	1 57.5 S 39 50 E	Ruffo
Mida, creek	K7	Kilifi	3 21 S 39 58 E	Bogdan, Greenway, Tweedie
Mida, popl.	K7	Kilifi	3 19 S 39 58 E	Numerous
Middle Mara, R.	K4	Embu	0 15 S 37 39 E–0 16 S 37 45 E	
Midigo = Jebel Midigo, mt.	U1	West Nile	3 37 N 31 14 E	Eggeling, Mooney, A.S. Thomas
Miesi, str.	T8	Masasi	10 45 S 38 45 E–11 10 S 38 56 E	
Miewi, I.	T6	Rufiji	7 56 S 39 48 E	
Miewi Kubwa, I.	T6	Rufiji	c.7 56 S 39 48 E	Greenway

Name	Region	District	Coordinates	Collector
Migeregere, popl.	T8	Kilwa	8 48 S 39 14 E	Rose Innes & Magogo
Migole, popl.	T7	Iringa	7 07 S 35 50 E	
Migombani, ln.	T3	Tanga	5 05 S 39 04 E	?Greenway
Migombani, ln.	T3	Tanga	5 12 S 39 02.5 E	?Greenway
Migombani, popl.	Z	Zanzibar	5 59.5 S 39 18 E	Vaughan, Faulkner
Migori, R.	K5/6	S. Kavirondo/Masai	c.0 54 S 35 07 E–0 57 S 34 08 E	Glover et al.
Migwani, ln.	K4	Kitui	1 06 S 38 01 E	Napper
Mihama, popl.	T1	Shinyanga	3 56 S 33 59 E	
Mihama, popl.	T4	Tabora	5 15 S 33 57 E	
Mihumu (near Lugufu R.)	T4	Kigoma	prob. c.5 35 S 30 00 E	Kielland
Mihunga, ridge	U2	Toro	0 21.5 N 30 01 E	Osmaston, ?Fishlock & Hancock, ?Loveridge
Mihunga, popl., Ruwenzori	Zaire *not* U2		0 15.5 N 29 45.5 E	?Fishlock & Hancock, ?Loveridge
Mjelengwa, R.	T7	Rungwe	c.9 20 S 33 10 E	Stolz
[Mijusi] see Mujuzi				
Mijusi, str.	U4	Masaka	0 35 S 31 47 E	Osmaston
Mijusi, str.	U2	Toro	c.0 23.5 N 29 57 E	Hedberg
Mikese, popl., rsta., t.c.	T6	Morogoro	6 46 S 37 54 E	Greenway
Mikese, popl., rsta., t.c.	T6	Morogoro	6 46 S 37 54 E	Forest Guard in Eggeling
[Mikesse] see Mikese				
Mikindani, distr.	T8	Mikindani	c.10 30 S 40 00 E	Gillman, Schlieben
Mikindani, popl.	T8	Mikindani	10 17 S 40 07 E	Tanner
Mikinguni, popl.	T3	Pangani	5 31.5 S 38 56 E	Kraenzlin, Tanner, Zimmermann
Mikocheni, popl.	T3	Pangani	5 22 S 38 50 E	
Mikocheni, popl.	T3	Pangani	5 22 S 38 50 E	
[Mikosheni] see Mikocheni				
Mikumi, rsta. t.c.	T6	Kilosa	7 24 S 36 59 E	Procter, W.H. Lewis
Mikumi National Park, g.r.	T6	Kilosa	c.7 15 S 37 05 E	Pegler, Renvoize
[Mikwesi] see Mkwesi				
Milala, popl., dambo	T5	Dodoma	5 38 S 34 48 E	
Milala, popl., dambo	T4	Mpanda	6 19 S 31 02 E	Sanane
Milala, popl., dambo	T3	Tanga	5 08 S 38 52 E	Sandford
Milala, popl., dambo	T2	Masai	3 10 S 35 27 E	Robson
[Milangani] see Mlingano				
[Milanja] see Malanja				
Milanje Forest	T4	Buha	c.4 30 S 30 05 E	Carmichael

Name		Region	Coordinates	Collector/Authority
Milanzi, popl.	T4	Ufipa	7 59 S 31 34 E	Mwasumbi
Mile 46 = Elangata Wuas, rsta.	K6	Masai	1 53.5 S 36 35 E	Glover & Cooper
Milengeza	T6	Morogoro	Not traced	Holtz
Milepa, popl.	T4	Ufipa	8 04 S 31 56 E	Numerous
Milewemi = Mtwemi	T4	Kahama	Not traced	B.D. Burtt
Milhoi Creek	K7	Lamu	c.2 16 S 40 45 E	Rawlins
Miliatata, mt.	T2	Masai	prob. c.2 45 S 36 20 E	Richards
Mililingwa, area, r.h.	T6	Morogoro	c.6 50 S 37 59 E	
Milimani, popl.	K3	Trans-Nzoia	1 02 N 35 00 E	Tweedie
Milimani, w.h./s.	K7	Lamu	1 48 S 40 56 E	J. Adamson
Mill House	K3	Trans-Nzoia	1 09.5 N 34 45 E	Tweedie
Mill Turning	K3	Trans-Nzoia	1 10 N 34 49 E	Tweedie
Milo, mssn.	T7	Njombe	9 53 S 34 38 E	Numerous
Milola, popl., str.	T8	Lindi	9 59 S 39 24 E	Schlieben, Gillman
[Milonyi] see Mironji	T8	Songea	11 10 S 36 13 E–11 30 S 36 02 E	Busse
Milumba, popl., plain	T4	Mpanda	7 06 S 31 04 E	Richards
Minaki, mssn.	T6	Uzaramo	6 54 S 39 06 E	Vaughan, Harris & Walker
Minakulu, popl.	U1	Lango	2 31 N 32 22 E	Eggeling
[Minangya] see Manyangu	T6	Morogoro	c.6 07 S 37 34 E	Greenway & Farquahar
Mindu, mts., f.r.	T6	Morogoro	6 50 S 37 35 E	Pócs, Schlieben
Mingio, popl.	T4	Tabora	5 46 S 32 29 E	
Mingoyo, popl.	T8	Lindi	10 06.5 S 39 38.5 E	Gillman
Minga (90 km. NNW. of Tabora)	T4	Kahama/Tabora	c.4 18 S 32 27 E	Grant
[Minjale] see Minjare	T8	Newala	10 46 S 39 18 E	
Minjare, popl.	T8	Newala	10 46 S 39 18 E	Gillman
Minjingu, hill	T2	Mbulu	3 42.5 S 35 54.5 E	Polhill & Paulo
[Minsiro] see Minziro	U4/T1	Masaka/Bukoba	c.1 00 S 31 50 E	
Minyugi, popl.	T5	Singida	4 58.5 S 34 32 E	B.D. Burtt
[Minzilo] see Minziro	U4/T1	Masaka/Bukoba	c.1 02 S 31 33 E	
Minziro, hill	U4/T1	Masaka/Bukoba	c.1 02 S 31 33 E	Maitland

Name	District	Location	Coordinates	Collector
Minziro Forests, f.r. (includes Malabigambo, Namalala and Tero forests of Uganda as well as Miniziro Forest of Tanzania)	U4/T1	Masaka/Bukoba	c.1 00 S 31 50 E	Numerous
Minziwera (Uzinza area)	T1	Mwanza	Not traced	
Miono, popl.	T6	Bagamoyo	6 07 S 38 24 E	B.D. Burtt
Miotoni, str.	K4	Nairobi	c.1 18 S 36 41 E–1 19 S 36 47 E	Procter
Miquengue = Magengwe, popl.	T4	Tabora	5 08 S 32 45 E	C. van Someren
Mirambi, popl.	U2	Toro	0 37 N 30 02 E	A.S. Thomas
[Miringwa] see Mililingwa	T6	Morogoro	c.6 50 S 37 59 E	Busse
Miritini, popl., rsta.	K7	Mombasa	4 00 S 39 34 E	Thorold
[Mirola] see Milola	T8	Lindi	9 59 S 39 24 E	
Mironji, R.	T8	Songea	11 10 S 36 13 E–11 30 S 36 02 E	
Mirungamo, str.	T3	Kilwa	c.8 38.5 S 39 13.5 E	Busse
Misikhu, ln.	K5	North Kavirondo	0 43 N 34 45 E	G.R. Williams
Misikhu, mkt.	K5	North Kavirondo	0 43 N 34 43 E	
Misikhu, sch.	K5	North Kavirondo	0 42.5 N 34 43 E	
Misima = Mesema, Mt.	T3	Lushoto	prob. c.5 10 S 38 40 E	Greenway
Miskitini, popl.	T6	Rufiji	7 42 S 39 53 E	Greenway
Misolai, area	T3	Lushoto	5 02 S 38 38 E	Peter
Misoswe, popl., mssn., area	T3	Tanga	5 03.5 S 38 47 E	Greenway, Holst
Misozi, popl., mssn.	U4	Masaka	0 53 S 31 45 E	
Missenyi, area	T1	Bukoba	c.1 05 S 31 15 E	Haarer
Missomba, mts.	T7	Rungwe	c.9 35 S 33 30 E	
[Missoswe] see Misoswe	T3	Tanga	5 03.5 S 38 47 E	
Missungwi, popl.	T1	Mwanza	2 51 S 33 05 E	Tanner
Misufini	T3	Handeni	Not traced	Semsei
Misufini, popl.	P	Pemba	5 19 S 39 42 E	Greenway
[Misumhilo] see Msunkomilo	T4	Mpanda	6 20 S 31 02 E	Sanane
[Misumhumilo] see Msunkomilo	T4	Mpanda	6 20 S 31 02 E	Sanane
Misungwi = Missungwi	T1	Mwanza	2 51 S 33 05 E	
Mitala Maria, popl.	U4	Mengo	0 05 N 32 08 E	Geilinger

Mitanga	T8	Kilwa	near 8 45 S 38 02 E	Ludanga
Mitano, gorge	U2	Kigezi	c.0 44 S 29 48 E	Purseglove
[Mitiana] see Mityana	U4	Mengo	0 25 N 32 03 E	Godman
Mitole, popl.	T8	Kilwa	8 48 S 39 01 E	Busse
Mitoma, co.	U2	Ankole	c.0 05 S 30 30 E	Eggeling, Snowden
Mitoma, popl., r.h.	U2	Ankole	0 37 S 30 03 E	Purseglove
Mito Miwele (Mafia I.)	T6	Rufiji	c.7 50 S 39 50 E	Greenway
Mito Miwiii = Mito Miwele	T6	Rufiji	c.7 50 S 39 50 E	Greenway
Mitubiri, rsta.	K4	Fort Hall	0 59.5 S 37 08 E	Archer
Mitumba, area	T4	Buha	c.4 38 S 29 40 E	
Mitumbate, popl.	T8	Kilwa	8 50 S 39 27 E	
Mitumbati, popl.	T8	Kilwa	9 16 S 38 39 E	
Mitumbati, str., popl.	T8	Kilwa	c.9 34 S 37 56 E	Busse
Mitumbati Mdogo, str.	T8	Kilwa	c.9 08 S 38 47 E	
Mitumbati Mkubwa, str.	T8	Kilwa	9 16 S 38 39 E	
Mitunguni, t.c.	K7	Kilifi	3 05 S 40 03.5 E	S.A. Robertson
Mitungu, popl., airstrip	K4	Meru	0 06.5 S 37 47 E	Bogdan
Mityana, popl., r.h., rsta.	U4	Mengo	0 25 N 32 03 E	Dawkins
Miulani, popl.	P	Pemba	5 19 S 39 47 E	Greenway
[Miussi] see Mjusi	U2	Toro	c.0 23.5 N 29 57 E	J. Adamson
Miwani, Mafia I.	T6	Rufiji	Not traced	Greenway
Miwele = Malwelewele, track	T7	Iringa	near 7 53 S 34 45 E	Greenway
[Miyao] see Miyau	T8	Songea	11 01 S 34 56 E	Semsei
Miyau, r.h.	T8	Songea	11 01 S 34 56 E	Milne-Redhead & Taylor
Miyongwe = Kabungu, hill	T4	Mpanda	6 23 S 31 03.5 E	
Mizi Miombe, hill	P	Pemba	5 22.5 S 39 41.5 E	Vaughan
[Mizizi] see Muzizi	U2/4	Toro/Mubende	c.0 34 N 31 22 E–1 03 N 30 33 E	Bagshawe
[Mizozue] see Misoswe	T3	Tanga	5 03.5 S 38 47 E	Holst
[Mjambeli] see Mtambile	P	Pemba	5 23 S 39 42 E	Burtt Davy
[Mjanje] see Majanji	U3	Mbale	0 16 N 33 59 E	

Name		Region	Coordinates	Source
[Mjanji] see Majanji	U3	Mbale	0 16 N 33 59 E	Dale
Mjele, str.	T7	Mbeya/Chunya	c.8 45 S 33 09 E–8 39 S 33 07 E	Michelmore
[Mjere] see Mjele	T7	Mbeya/Chunya	c.8 45 S 33 09 E–8 39 S 33 07 E	Michelmore
Mjesani, sisal est.	T3	Tanga	5 02 S 38 53 E	Walker
[Mjessani] see Mjesani	T3	Tanga	5 02 S 38 53 E	Stuhlmann
Mjesse = Njassa, ln.	T5	Dodoma	5 59 S 35 48 E	Vaughan, Procter, Dransfield
Mjimwema, popl.	T6	Uzaramo	6 51 S 39 22 E	Greenway
Mjonga, R.	T6	Morogoro	5 49 S 37 30 E–6 11 S 37 42.5 E	Rawlins
Mjume, for. (near Utwani)	K7	Lamu	near 2 22 S 40 30 E	
Mjwini, peak	T3	Pare	4 20 S 37 55 E	Procter
Mkadini, salt flats	T6	Bagamoyo	6 22.5 S 38 50 E	Vaughan, Greenway
Mkadini, popl.	Z	Zanzibar	6 02 S 39 13.5 E	Ukiriguru Res. Centre
Mkafanya, ln.	T4	Kigoma	5 25 S 30 04 E	Carmichael
Mkaja, popl.	T6	Ulanga	8 25 S 36 00 E	Milne-Redhead & Taylor
Mkako, R.	T8	Songea	c.10 40 S 35 08 E–10 47 S 35 12 E	Milne-Redhead & Taylor
Mkaku = Mkako, R.	T8	Songea	c.10 40 S 35 08 E–10 47 S 35 12 E	
Mkalalala = Malakala, popl.	T7	Iringa	8 22 S 35 18 E	B.D. Burtt, Jaeger, Savile
Mkalama, popl., r.h., t.c.	T5	Singida	4 07 S 34 38 E	Numerous
Mkalinzi, popl.	T4	Buha	4 37 S 29 44 E	R.M. Davies
Mkama (near Mbozi)	T7	Mbeya	near 9 01 S 32 58 E	Schlieben, Brehmer
Mkambaku, mt.	T6	Morogoro	7 10 S 37 42 E	Procter
[Mkange] see Mkangi	T6	Bagamoyo	6 04 S 38 33 E	Hannington
Mkangi, popl.	T6	Bagamoyo	6 04 S 38 33 E	Richards
[Mkanja] see Mkenja	T7	Njombe	c.9 12 S 34 08 E	Vaughan
Mkanjuni, popl., area	P	Pemba	5 14 S 39 47 E	Procter
[Mkara] see Mkata Drive	T6	Kilosa	c.7 10–7 20 S 37 05 E	Tanner, van Rensburg
Mkaramo (near Mkwaja)	T3	Pangani	near 5 47 S 38 51 E	Fuggles Couchman, Michelmore
Mkata, plain	T6	Kilosa/Morogoro	c.7 00 S 37 15 E	Michelmore
Mkata, R.	T6	Kilosa/Morogoro	7 26 S 37 06 E–6 32 S 37 27 E	Peter, Wingfield, Faulkner
Mkata, rsta.	T6	Morogoro	6 45 S 37 21 E	

Mkata Drive	T6	Kilosa	c.7 10 – 7 20 S 37 05 E	Procter
[Mkatta] see Mkata	T6	Kilosa/Morogoro	7 26 S 37 06 E – 6 32 S 37 27 E	Busse
Mkawa = Mkewe	T7	Iringa	prob. c.8 25 S 34 45 E	Richards
Mkenja, popl., str.	T7	Njombe	c.9 12 S 34 08 E	Richards
Mkenja, R.	T7	Rungwe	c.9 11 S 33 33 E	Paulo
Mkenke, str.	T4	Buha/Kigoma	c.4 40 S 29 35 E	Pirozynski, Parnell
Mkewe, popl. (Sao Hill – Malangali)	T7	Iringa	prob. c.8 25 S 34 45 E	Eggeling
Mkhali = Mlali, popl.	T5	Mpwapwa	6 18 S 36 46 E	Speke & Grant
Mkhoma, popl.	T7	Mbeya	9 03 S 32 29 E	R.M. Davies
Mkiashi, mssn.	T2	Moshi	3 18 S 37 29.5 E	
Mkigo = Mukigo, popl.	T4	Buha	4 30 S 29 46 E	Peter
Mkigwa, area	T4	Tabora	c.5 08 S 33 10 E	
[Mkimbura] see Mkumbara	T3	Lushoto	4 45.5 S 38 11 E	Leippert
[Mkindu] see Makindu	T6	Rufiji	7 52 S 38 00 E	
[Mkinji] ? see Mkenja	T7	Njombe	?c.9 12 S 34 08 E	Semsei
Mkiziga (near Mwera)	T3	Pangani	near 5 29 S 38 54 E	Tanner
[Mknumbi] see Mkunumbi	K7	Lamu	2 18 S 40 42 E	Sampson
Mko, str.	T6	Morogoro	6 45 S 38 04 E – 6 51.5 S 38 01 E	Stuhlmann
Mkoani, popl.	P	Pemba	5 22 S 39 39 E	Bojer, Vaughan
Mkoaranga	T6	Rufiji	Not traced	Braun
Mkobwe, hill	T6	Morogoro	6 13.5 S 37 33.5 E	Drummond & Hemsley
[Mkoe] see Mokowe	K7	Lamu	2 14 S 40 51 E	
Mkoe, popl.	T8	Lindi	9 32 S 39 38 E	Gillman
Mkokoni, popl.	K7	Lamu	1 58 S 41 17.5 E	Rawlins
Mkokotoni, popl., harbour	Z	Zanzibar	5 52.5 S 39 15 E	Faulkner, Barney
Mkola Waterhole (Mikumi Nat. Park)	T6	Kilosa	Not traced	Greenway & Kanuri
Mkoloka = Mokoloka	T4	Mpanda	Not traced	Newbould & Jefford
[Mkoma] see Mkhoma	T7	Mbeya	9 03 S 32 29 E	R.M. Davies
Mkoma, popl.	T8	Newala	10 28 S 39 17 E	Nicholson, Gillman
Mkomadatchi (Makonde Plateau)	T8	Masasi/Newala	Not traced	Busse

Name		Region	Coordinates	Collector
Mkomazi, g.r.	T3	Pare/Lushoto	c.4 10 S 38 10 E	Richards
Mkomazi, popl, rsta., mssn., r.h.	T3	Lushoto	4 38 S 38 04.5 E	Numerous
Mkomazi, str.	T3	Pare/Lushoto	c.4 28 S 38 06 E–5 08 S 38 21.5 E	Numerous
[Mkombako] see Mkambaku	T6	Morogoro	7 10 S 37 42 E	comm. Sacleux
Mkondoni, w.h.	K1	Northern Frontier	1 36 S 41 20.5 E	Kuchar
Mkongani, popl.	K7	Kwale	4 17 S 39 16 E	Kassner
Mkongani North, for.	K7	Kwale	4 17 S 39 19 E	
Mkongani West, for.	K7	Kwale	4 20 S 39 16 E	
[Mkoni] see Mkokoni	K7	Lamu	1 58 S 41 17.5 E	Elliot
Mkonji, popl.	T6	Ulanga	c.8 35 S 37 19 E	Schlieben
Mkonka, R.	T4	Ufipa	prob.8 10 S 31 25 E	Richards
Mkoreha, popl.	T8	Newala	10 50 S 39 45 E	Gillman
[Mkowasi] see Mkomazi	T3	Lushoto	4 38 S 38 04.5 E	Peter
[Mkowe] see Mokowe	K7	Lamu	2 14 S 40 51 E	Greenway
[Mku] see Mkuu	T2	Moshi	3 12 S 37 36 E	Haarer
Mkuanga, hill	T8	Songea	10 40 S 35 09 E	Milne-Redhead & Taylor
Mkujuni, str.	T2	Masai	3 32 S 36 03 E–3 29 S 35 53 E	
[Mkukira] see Mkurira	T8	Songea	c.10 42 S 35 57 E	Milne-Redhead & Taylor
Mkula, R.	T6	Ulanga	c.7 46 S 36 54 E	Anderson
Mkulukulu	T4	Tabora	prob. c.5 00 S 32 00 E	Bally
Mkulumuzi, str.	T3	Tanga	5 07 S 38 45 E–5 04 S 39 04 E	Drummond & Hemsley, Verdcourt et al.
Mkuluzi, R.	T8	Songea	c.10 47 S 35 07 E	Milne-Readhead & Taylor
Mkumba, popl. (Kalambo R.)	T4	Ufipa	prob. c.8 35 S 31 15 E	Richards
[Mkumbako] see Mkambaku	T6	Morogoro	7 10 S 37 42 E	Thulin & Mhoro
[Mkumbala] see Mkumbara	T3	Lushoto	4 45.5 S 38 11 E	Drummond & Hemsley
Mkumbara, rsta., old pl.	T3	Lushoto	4 45.5 S 38 11 E	Peter, Drummond & Hemsley
[Mkumbe] see Mkumbi	K7	Lamu	2 15 S 40 40 E	
[Mkumbe] see Mkumbo	T6	Morogoro	6 43 S 37 48.5 E	B.D. Burtt
Mkumbi, popl.	K7	Lamu	2 15 S 40 40 E	Battiscombe
Mkumbiro, R.	T6/3	Rufiji/Kilwa	8 08 S 38 02 E–7 57 S 37 52 E	Schlieben

Name	Region	Location	Coordinates	Collector
Mkumbo, hill	T6	Morogoro	6 43 S 37 48.5 E	Vaughan
Mkumbuu, Ras, cape	P	Pemba	c.5 13 S 39 40 E	Wallace
[Mkun] see Mkuu	T2	Moshi	3 12 S 37 36 E	Richards
M'Kunda, popl.	T4	Ufipa	prob. c.8 10 S 31 25 E	Bullock
Mkunde, popl.	T4	Ufipa	7 51 S 31 26 E	Gillman
Mkundi, area	T3	Lushoto	c.4 27 S 38 12 E	Schlieben
Mkundi, popl.	T6	Morogoro	6 19 S 37 23 E	
Mkundi Mtae, popl.	T3	Lushoto	4 27 S 38 12 E	
[Mkunduchi] see Makunduchi	Z	Zanzibar	6 25 S 39 33 E	Greenway
Mkune, for., sw.	K7	Lamu	c.2 24 S 40 30 E	Rawlins
Mkungaungo, for. (W. Usambaras)	T3	Lushoto	prob. c.4 45 S 38 20 E	G.R. Williams
Mkungwe, hill	T6	Morogoro	6 52.5 S 37 55 E	Faden, Evans & Pócs
[Mkunumbe] see Mkunumbi	K7	Lamu	2 18 S 40 42 E	Rawlins
Mkunumbi, popl., plains	K7	Lamu	2 18 S 40 42 E	Patterson, Rawlins
[Mkunumbwe] see Mkunumbi	K7	Lamu	2 18 S 40 42 E	Rawlins
Mkunza, popl.	T8	Newala	10 59 S 39 27 E	Hay
Mkurira, R.	T8	Songea	c.10 42 S 35 57 E	Milne-Redhead & Taylor
Mkuru, est.	T2	Arusha	3 08 S 36 49 E	Richards
Mkurue, old area	T4	Ufipa	c.8 30 S 32 15 E	
Mkurumuji, peak	K7	Kwale	c.4 15 S 39 24 E	Glover & Magogo
[Mkurutuni] see Kurutini	T6	Uzaramo	6 57 S 39 09 E	Stuhlmann
[Mkusa] see Mkussu	T3	Lushoto	c.4 46 S 38 18–38 24 E	Richards
Mkusi, old area	T3	Lushoto	4 46 S 38 21 E	
Mkusi, popl.	T3	Lushoto	4 45 S 38 20 E	Geilinger, Drummond & Hemsley
[Mkusi] see Mkussu	T3	Lushoto	c.4 46 S 38 18–38 24 E	Eggeling
[Mkusi] see Mkuzi	T3	Tanga	5 14 S 38 50 E	
Mkussu, for.	T3	Lushoto	c.4 46 S 38 18–38 24 E	Richards, Gillman
Mkusu, str., area	T3	Lushoto	c.4 46 S 38 21 E	Drummond & Hemsley, Greenway
Mkuti = Muguti, str.	T4	Buha/Kigoma	4 53 S 30 02 E–4 50 S 29 48 E	Peter, Procter, Verdcourt
Mkuu, popl., r.h., area	T2	Moshi	3 12 S 37 36 E	Haarer, Volkens

244

Name	Grid	District	Coordinates	Collector(s)
[Mkuyoni] see Mkuyuni	T6	Morogoro	6 57 S 37 49 E	Pócs et al.
Mkuyu, pt.	T4	Kigoma	5 28.5 S 29 46 E	Carmichael
Mkuyu, R.	T8	Songea	c.9 56 S 36 35 E–10 03 S 36 53 E	Rees
Mkuyuni = Mkujuni, str.	T2	Masai	3 32 S 36 03 E–3 29 S 35 53 E	Lamprey
Mkuyuni, popl., mkt.	T6	Morogoro	6 57 S 37 49 E	Greenway, Drummond & Hemsley
Mkuzi = Mkusi, popl.	T3	Lushoto	4 45 S 38 20 E	Tanner
Mkuzi, mssn.	T3	Tanga	5 14 S 38 50 E	Yusufu
Mkuzi Katani, popl.	T3	Pangani	5 20 S 38 56 E	Peter, Tanner, van Rensburg
[Mkuzu] see Mkussu	T3	Lushoto	c.4 46 S 38 18 E–38 24 E	Vaughan
[Mkwabi] see Mkwaki	K4	North Nyeri	0 16 N 37 06 E	Moreau
Mkwaja, popl.	T3	Pangani	5 47 S 38 51 E	Harris et al.
Mkwajuni, popl., area	P	Pemba	5 07 S 39 43 E	Greenway
[Mkwaki] see Engwaki	K4	North Nyeri	0 16 N 37 06 E	Peter
Mkwanga, popl.	T6	Uzaramo	7 07 S 39 12 E	Hay
Mkwawa, spr. (Ruaha Nat. Park)	T7	Mbeya/Iringa	Not traced	B.D. Burtt, Glover
[Mkwaya] see Mkwaja	T3	Pangani	5 47 S 38 51 E	B.D. Burtt
Mkwaya, popl.	T8	Lindi	10 06.5 S 39 40 E	Tweedie, Hooper & Townsend
Mkwedu, popl.	T8	Newala	10 49 S 39 21.5 E	Lowe
Mkweni, area	T4	Kahama	3 44.5 S 32 18.5 E	Rees, Vollesen
Mkwesi, popl., vet. stn.	T5	Dodoma	5 38 S 34 48 E	Greenway
Mlaba, for.	K5	North Kavirondo	c.0 27.5 N 34 51 E	Richards
[Mlagarasi] see Malagarasi	T4	Kigoma/Buha		
Mlahi, str., ferry	T6	Ulanga	c.8 17.5 S 37 06.5 E	
[Mlala] see Ndala	T2	Mbulu	c.3 25 S 35 41 E–3 29.5 S 35 47.5 E	
Mlala, hills	T4	Mpanda	6 47 S 31 45 E	Vaughan
[Mlalakwa] see Mulalakuwa	T6	Uzaramo	c.6 45 S 39 14 E	Leedal
Mlale, popl.	T7	Rungwe	9 26 S 33 08 E	Bally, B.D. Burtt
Mlali, popl., t.c.	T5	Mpwapwa	6 18 S 36 46 E	Semsei, Drummond & Hemsley,
Mlali, popl., mssn., t.c.	T6	Morogoro	6 57.5 S 37 32 E	B.D. Burtt, Holtz

Name	Grid	District	Coordinates	Collector
Mlalila, ln.	T2	Mbulu	c.3 25 S 35 48.5 E	Greenway
Mlalo, popl., r.h.	T3	Lushoto	4 35 S 38 21 E	Numerous
Mlalo, popl.	T3	Lushoto	4 48 S 38 25 E	
Mlalo, popl.	T3	Tanga	4 56 S 38 56 E	
Mlangali, popl.	T7	Njombe	9 46 S 34 31 E	Richards, Thulin & Mhoro
[Mlangani] see Mlangali	T7	Njombe	9 46 S 34 31 E	
Mlango ya Simba = Lango la Simba, bdgs.	K7	Tana River	2 16 S 40 13 E	Greenway & Rawlins
Mlawi, R.	T7	Iringa	7 48 S 36 20 E–7 37 S 36 19 E	Procter
Mlayar, ln.	T2	Masai	c.4 50 S 37 20 E	Jaeger
Mlejeou, popl.	T5	Dodoma	c.6 01 S 35 25 E	Polhill & Paulo
Mlela, f.r.	T4	Kigoma	4 52 S 29 48 E	Procter
Mlele, Controlled Area	T4	Mpanda	c.6 40 S 31 20 E	Richards
Mlembule, popl.	T3	Handeni	5 50 S 37 52.5 E	Peter
Mlembule Kwa-Tsharumbi, popl.	T3	Handeni	5 28 S 37 38 E	Busse
Mleni, popl.	T3	Tanga	5 02 S 39 02 E	Busse
Mlesa, ln.	T3	Lushoto	c.4 48 S 38 30 E	Cribb & Grey-Wilson
Mlewemi	T4	Kahama	Not traced	B.D. Burtt
Mlifu, popl.	T3	Lushoto	4 36 S 38 14 E	Fletcher
Mligaji = Mligasi	T3/6	Handeni/Bagamoyo	5 33 S 37 43 E–5 58 S 38 47.5 E	
Mligasi, R.	T3/6	Handeni/Bagamoyo	5 33 S 37 43 E–5 58 S 38 47.5 E	
Mlinga, peak	T3	Tanga	5 05 S 38 45 E	Semsei
Mlingano, sisal res. stn.	T3	Tanga	5 08 S 38 52 E	Numerous
Mlinguru, popl.	T8	Lindi	10 03 S 39 45 E	Numerous
Mliwaza, str.	T3	Pangani	c.5 31 S 38 51 E–5 29 S 38 54 E	Schlieben
Mloa, popl.	T7	Iringa	7 41 S 35 24 E	Richards, Procter
Mloa, R.	T7	Iringa	7 25 S 36 08 E–7 40 S 36 12 E	
[Mloha] see Mloa	T7	Iringa	7 25 S 36 08 E–7 40 S 36 12 E	
Mlola, popl., r.h.	T3	Lushoto	4 38 S 38 25 E	Goetze
[Mlomu] see Ilomo	T4	Mpanda	c.6 50–6 57 S 30 30 E	Nicholson
Mlowo, popl.	T7	Mbeya	9 00 S 32 59 E	Cribb & Grey-Wilson

Index of place-names (continued). Entries marked "see" are cross-references; others give grid square, administrative district, coordinates and principal collector(s).

Name	Grid	District	Coordinates	Collector(s)
[Mlupa] see Lupa				
Mmemya, mt.	T7	Chunya	8 15 S 31 56 E	Geilinger
Mnacho, popl.	T4	Ufipa	10 16 S 39 01 E	Bullock, Vesey-FitzGerald
[Mnafa] see Muafa				
Mnagei, area	T8	Lindi	4 56 S 38 24 E	Gillman
[Mnalelo] ? see Mnorera				
Mnarani, fy.	T3	Lushoto	c.1 18 N 35 04 E	Buchwald
Mnazi, popl., sisal est.	K2	West Suk	?10 39 S 39 22 E	Gillman
	T8	Newala	3 38.5 S 39 51 E	Faden et al.
	K7	Kilifi	4 26 S 38 18 E	Gillman, Greenway
Mnazi Mmoja, popl.	T3	Lushoto	5 56 S 39 17 E	Faulkner, Vaughan
	Z	Zanzibar	5 56 S 39 17 E	
Mnazi Moja = Mnazi Mmoja				
Mnazini, vet. station, sch.	Z	Zanzibar	1 59 S 40 08 E	Homewood, Gillett
Mnemba, I.	K7	Tana River	5 49 S 39 22.5 E	
	Z	Zanzibar	4 42.5 S 35 52 E	B.D. Burtt
[Mnenia] see Mnenya				
Mnenya, popl.	T5	Kondoa	4 42.5 S 35 52 E	
	T5	Kondoa	3 38.5 S 39 51 E	Bally
[Mnereni] see Mnarani				
Mngazi, R.	K7	Kilifi	c.7 11 S 37 40 E – 7 27 S 37 42 E	
Mnguvia, str.	T6	Morogoro	7 00 S 39 23 E – 6 52 S 39 26 E	Wingfield
[Mnigazi] see Mligasi				
Mnima, popl.	T6	Uzaramo	5 33 S 37 43 E – 5 58 S 38 47.5 E	Peter
Mnima, popl.	T3/6	Handeni/Bagamoyo	10 38 S 39 11 E	Gillman
Mninga, popl., f.r.	T8	Newala	10 29 S 39 43 E	
Mnolela, popl.	T8	Mikindani	c.8 36 S 35 12 E	Richards
Mnorera, popl.	T7	Iringa	10 13 S 39 45 E	
[Mnuaza] ? see Mliwaza				
Mnyera, peak	T8	Lindi	10 39 S 39 22 E	
[Mnyonga] see Manonga				
Mnyusi, popl., rsta.	T8	Newala	c.5 31 S 38 51 E – 5 29 S 38 54 E	Peter
	T3	Pangani	6 22 S 36 56 E	
	T6	Kilosa	3 42 S 32 49 E – 4 08 S 34 12 E	Mabberley et al., Cribb & Grey-Wilson
	T1/4	Shinyanga/Nzega/Kahama	5 13 S 38 35 E	
	T3	Tanga	5 13 S 38 35 E	
[Mnyuzi] see Mnyusi				
Moa, popl.	T3	Tanga	4 46 S 39 10.5 E	Numerous
Moa, sisal est., area	T3	Tanga	4 44 S 39 09 E	Numerous

248

Name		District	Coordinates	Collector
[Mokkarara] see Mukarara	K4	Fort Hall	0 54.5 S 36 56.5 E	Balbo
Moko, Range Exper. Stn.	U2	Ankole	c.0 37 S 30 39 E	Harrington
Mokoloka = Mkoloka	T4	Mpanda	Not traced	Newbould & Jefford
[Mokololo] see Makororo	T8	Kilwa	9 43 S 37 34 E–9 30 S 37 37 E	Busse
Mokowe, popl., airstrip	K7	Lamu	2 14 S 40 51 E	Greenway
Mokoyeti Gorge	K4	Nairobi	1 23 S 36 52 E	Verdcourt, Beentje
Molitar, popl.	U1	Lango	1 38 N 32 50 E	
Molo, popl.	U3	Mbale	0 51 N 34 10 E	
Molo, popl., rsta.	K3	Nakuru	0 15 S 35 44 E	Numerous
Molo, R.	K3	Baringo/Ravine/Nakuru	0 17 S 35 43 E–0 33 N 36 05 E	Bogdan
Molo, popl. (Malonje plateau)	T4	Ufipa	c.8 05 S 31 50 E	Richards, Robinson, Hooper & Townsend
Molomo, est.	T2	Moshi	3 09 S 37 02 E	Owens
Molwe, popl.	T4	Ufipa	8 23 S 31 08 E	Glover
Momandu, for.	K4	Machakos	c.1 42 S 37 18 E	Gibbons
Momandu, hill	K4	Machakos	1 40.5 S 37 17 E	
Momandu, mkt., sch.	K4	Machakos	1 40 S 37 17 E	Nicholson
[Momba] see Mamba	T4	Mpanda	7 19 S 31 22 E	Richards
Momba, R.	T4/7	Ufipa/Mbeya	c.8 47 S 32 28 E–8 10 S 32 28 E	Claxton
Mombasa, popl.	K4	Kitui	1 41 S 37 54.5 E	Trapnell
Mombasa, distr.	K7	Mombasa	c.4 01 S 39 42 E	
Mombasa, popl., I.	K7	Mombasa	4 03 S 39 40 E	Numerous
[Momba Sasa] see Mambosasa	K7	Lamu	2 23 S 40 32 E	
[Mombassasa] see Mambosasa	K7	Lamu	2 23 S 40 32 E	Hooper & Townsend
Mombo, hill	T3	Lushoto	c.5 02 S 38 37 E	Pegler, Renvoize
Mombo, popl, rsta., old pl., f.r.	T3	Lushoto	4 53 S 38 17 E	Numerous
[Mombosasa] see Mambosasa	K7	Tana River	1 48 S 40 07 E	
Momela, farm, est., area	T2	Arusha	3 15 S 36 51 E	Geilinger, Greenway
Momela, game lodge	T2	Arusha	3 14.5 S 36 51.5 E	
Momela, Lakes	T2	Arusha	c.3 13.5 S 36 54 E	Richards, Uhlig
Momela Farm = Trappe Farm	T2	Arusha	3 15 S 36 51 E	Numerous

Name	Region	District	Coordinates	Collector
Momela Gate, ln.	T2	Arusha	3 15 S 36 50.5 E	Greenway & Kanuri, Cribb & Grey-Wilson
Momela kubwa, L.	T2	Arusha	3 13 S 36 54.5 E	
Momela ndogo, L.	T2	Arusha	3 13.5 S 36 53.5 E	
[Momella] see Momela	T2	Arusha		
[Momgongo] see Mwamgongo	T4	Buha	c.4 35 S 29 39 E	Bally, Geilinger, Sanders
[Momwaga] ? see Lomwaga	U1	Acholi	?3 49 N 32 54 E	Pirozynski
Momwere (N. of Wami R.)	T6	Bagamoyo	Not traced	Eggeling
[Mondjo] see Moonjo	T2	Moshi	3 12 S 37 30 E –3 17 S 37 30 E	Holtz
Mondo, popl., r.h., t.c., mssn.	T5	Kondoa	4 58.5 S 35 55 E	Volkens
[Mondul] see Monduli	T2	Masai	3 18 S 36 27 E	B.D. Burtt, Polhill & Paulo
Monduli, coffee est.	T2	Arusha	3 18 S 36 31 E	
Monduli, cone, f.r.	T2	Masai/Arusha	3 15 S 36 29 E	Fischer
Monduli, popl.	T2	Masai	3 18 S 36 27 E	Numerous
Monga, for.	T3	Lushoto	c.5 05 S 38 36 E	Ali Omare, Faulkner
Monga, hill, tea est.	T3	Lushoto	5 06 S 38 37 E	Numerous
[Mangala] see Mangola	T2	Mbulu		
Mongiro, popl.	U2	Toro	0 50 N 30 10 E	Verdcourt
[Mongo] see Monga	T3	Lushoto	c.5 05 S 38 36 E	Paulo
Mongoroma, popl.	T5	Kondoa	4 54 S 35 37 E	
[Mongubu] see Mangubu	T3	Lushoto	5 06 S 38 41 E	
Monune, str.	K4/7	Kitui/Tana River	0 21 S 38 29 E –0 03 S 38 37 E	Greenway
[Moon, mts. of the] see Ruwenzori	U2/Zaire	Toro	c.0 05 –0 50 N 29 45 –30 25 E	Bamps
Moonjo, R.	T2	Moshi	3 12 S 37 30 E –3 17 S 37 30 E	
Mopia Gap, ln.	K4	Kitui	2 59 S 38 41 E	Sheldrick
[Morang] see Marang	T2	Moshi	c.3 18 S 37 31 E	W.E. Taylor
[Moranga] ? see Maragua	K4	Fort Hall	?0 47.5 S 37 08 E	Balbo
Morendat, popl., rsta.	K3	Naivasha	0 40 S 36 23 E	
[Moribus] see Morobus	K2	West Suk	1 23 N 35 17 E	
Morijo, area	K1	Northern Frontier.	1 20 N 36 38 E	
Morijo, area, sch.	K6	Masai	1 42 S 35 49 E	D.C. Edwards, Fayad

Name	Code	Region	Coordinates	Authority
[Morit] see Mureti	K1	Northern Frontier	1 11.5 N 37 22 E	J. Adamson
Morningside, hotel	T6	Morogoro	6 53 S 37 40 E	Numerous
Morobus, hill	K2	West Suk	1 23 N 35 17 E	Napier, Carter & Stannard
Morobus, t.c.	K2	West Suk	1 24 N 35 18 E	
Morogoro, distr.	T6	Morogoro	c.6 55 S 37 50 E	
Morogoro, popl., mssn., airport	T6	Morogoro	6 49 S 37 40 E	Numerous
Morogoro, R.	T6	Morogoro	6 54 S 37 40 E – 6 45.5 S 37 41 E	Wigg
[Morongo] ? see Marungu	T3	Tanga	?5 12 S 39 01 E	Faulkner
Morongole, mt.	U1	Karamoja	3 49 N 34 02 E	J. Wilson, Eggeling
Morongole, f.r.	U1	Karamoja	c.3 48 N 33 55 E	J. Wilson
[Moropus] see Morobus	K2	West Suk	1 23 N 35 17 E	Tweedie
Moroto, co.	U1	Lango	c.2 20 N 33 20 E	
Moroto = Aswa (part) R.	U1/Sudan	Acholi/Lango	2 22 N 33 32 E – 3 44 N 31 55 E	
Moroto, mt.	U1	Karamoja	2 32 N 34 46 E	Numerous
Moroto, popl.	U1	Karamoja	2 32 N 34 39 E	Numerous
Moruangaberu, popl.	U1	Karamoja	1 58 N 34 39 E	J. Wilson, Dyson-Hudson
Moruariwan, hill	U1	Karamoja	2 49 N 34 26 E	
[Moruassigar] see Murua Nysigar	K2	Turkana	c.3 10 N 35 01 E	Newbould
[Moruatukan] see Muruatunokan	U1	Karamoja	2 01 N 34 29 E	A.S. Thomas
[Moruaturukan] see Muruatunokan	U1	Karamoja	2 01 N 34 29 E	
Moruethe = Moruethi	K1	Northern Frontier	2 57 N 35 25.5 E	Newbould
Moruethi, hill	K1	Northern Frontier	2 57 N 35 25.5 E	Bally
Moruita, r.c.	U1	Karamoja	1 54 N 34 45 E	Tweedie, J. Wilson, Eggeling
Moru Kopjes, hills	T1	Maswa	c.2 45 S 34 45 E	Greenway
Morulem, mssn.	U1	Karamoja	2 36 N 33 46 E	J. Wilson
Morulinga, hill	U1	Karamoja	2 25 N 34 28 E	
Morun, R.	K2/3	West Suk/Elgeyo	1 06 N 35 26 E – 1 34 N 35 31.5 E	Tweedie, Carter & Stannard
[Morungule] see Morongole	U1	Karamoja	3 49 N 34 02 E	Numerous
Moruongole = Morongole	U1	Karamoja	3 49 N 34 02 E	Numerous
Moruthipo (Kanyerus R.)	K2	Turkana	c.1 24 N 34 49 E	Dale

Name	Div.	Locality	Coordinates	Collector(s)
Mpala, f.r.	T7	Njombe	9 24 S 34 51 E	Eggeling
Mpala, L. Tanganyika	Zaire *not* Tanzania			De Beerst
Mpalalu, popl., f.r.	T3	Lushoto	4 56.5 S 38 28 E	Peter
Mpalo, popl.	U2	Kigezi	1 10 S 30 02 E	Purseglove, Eggeling
Mpamba, str.	U4	Mubende	0 54 N 30 49 E – 1 02 N 30 49 E	Bagshawe
Mpanda, distr.	T4	Mpanda		
Mpanda, mbuga	T4	Mpanda/Tabora	c.6 40 S 31 10 E	Bullock
Mpanda, mine	T4	Mpanda	c.5 40 S 31 10 E	Greenway
Mpanda, popl., mssn., r.h.	T4	Mpanda	6 22 S 31 01 E	Numerous
Mpanda North East, f.r.	T4	Mpanda	6 22 S 31 02 E	Carmichael
Mpandeni, popl.	T4	Mpanda	c.5 30 S 31 30 E	Zimmerman
Mpanga, R.	T3	Lushoto	5 07.5 S 38 41.5 E	Bagshawe, Dawe, Eggeling, A.S. Thomas
Mpanga, str. (near Mbarara)	U2	Toro	0 39 N 30 08 E – 0 03 N 30 17 E	Eggeling
Mpanga, f.r.	U2	Ankole	c.0 37 S 30 39 E	Pegler, Byabainazi, Katende, Tweedie, Dawkins
Mpanga, f.r.	U4	Mengo	c.0 12.5 N 32 17.5 E	
Mpanga, area	T3	Lushoto	c.4 46 S 38 39 E	Peter
Mpanga, popl.	T7	Iringa	7 12 S 34 36 E	Busse
Mpanga, R.	T7	Iringa	8 33.5 S 35 21 E – 8 59 S 35 54 E	Carmichael
Mpangaduka (near Nampungu)	T8	Tunduru	near 10 57 S 37 04 E	Tanner
Mpangampanga, popl., ridge, str.	T8	Kilwa	c.8 43 S 38 07 E	Ludanga
[Mpangapanga] see Mpangampanga	T8	Kilwa	c.8 43 S 38 07 E	
Mpangwa, R. (Uvinza f.r.)	T4	Kigoma	c.5 10 S 30 20 E	Procter
Mpapa, popl.	T8	Songea	11 05 S 34 54 E	Eggeling, Milne-Redhead & Taylor, Semsei
Mpapule, popl.	T8	Kilwa	8 42 S 38 05 E	Page-Jones
[Mpapwa] see Mpwapwa	T5	Mpwapwa	6 21 S 36 29 E	Busse, Peter
[Mpâsu] see Mpesu	T7	Mbeya	c.9 08 S 33 18 E	Goetze
Mpatila, Plateau	T8	Newala	c.10 30 S 39 18 E	Busse
Mpemba, mt.	T3	Lushoto	5 02 S 38 24 E	Peter
[Mpembe] see Mpemvi	T4	Buha	c.3 45 S 30 36 E – 3 47 S 30 52 E	Bullock
Mpemvi, str.	T4	Buha	c.3 45 S 30 36 E – 3 47 S 30 52 E	Bullock, Shabani, Carmichael

[Mpeon] see Mpesu				
Mpepera, popl.	T7	Mbeya	c.9 08 S 33 18 E	Semsei
Mperembwa ? see Peremehe	T3	Pare	4 07 S 37 52 E	Greenway
Mpesu, mt.	T5	Dodoma	?6 44 S 35 37 E	
Mpeta, ln., str.	T7	Mbeya	c.9 08 S 33 18 E	Carmichael
Mpigi, popl.	T7	Iringa	7 59 S 36 06 E	Dawkins
Mpiji, str.	U4	Mengo	0 13 N 32 20 E	Harris et al.
[Mpika] see Kimarampaka	T6	Uzaramo	6 57 S 38 57.5 E—6 34 S 39 07 E	Richards
[Mpimbue] see Mpimbwe	T8	Songea	c.10 41 S 35 33 E	
Mpimbwe, popl., area	T4	Mpanda	7 14.5 S 31 24.5 E	
Mpizi	T4	Mpanda	7 14.5 S 31 24.5 E	Stuhlmann
Mpogo, popl., hill	T3	Pangani	Not traced	Dummer
Mpogoro = Mapogoro, popl.	U4	Mengo	0 20 N 33 08 E	Hughes
Mpokya, hill	T7	Mbeya	8 19 S 34 42 E	Osmaston
Mpola, popl.	U2	Toro	0 23 N 30 19 E	
Mpololo	T7	Rungwe	9 06 S 33 42 E	Haarer
[Mpoloto] see Poroto	T2	Moshi	Not traced	Stolz
Mpomode, hill	T7	Rungwe/Mbeya/Njombe		G.H. Wood
Mpoponzi R., (Lupembe area)	U3	Busoga	0 35 N 33 09 E	Schlieben
Mpororo, area	T7	Njombe	Not traced	Stuhlmann
Mpororo, hill	U2/T1/Rwanda	Ankole/Kigezi/Bukoba	c.0 45—1 30 S 29 20—30 30 E	Eggeling
[Mporoto] see Poroto, mts.	U2	Ankole	0 42 S 30 50 E	Geilinger, Eggeling
Mporoto = Poroto, sawmill	T7	Rungwe/Njombe/Mbeya	c.9 00 S 33 45 E	St. Clair-Thompson
Mpuguso, popl.	T7	Mbeya	8 59 S 33 38 E	R.M. Davies, Cribb & Grey-Wilson
Mpugwe, mssn.	T7	Rungwe	9 19 S 33 39 E	
Mpui, popl.	U4	Masaka	0 16 S 31 49 E	Bullock, McCallum Webster, Richards, Vesey-FitzGerald
Mpululu, hill, r.p.	T4	Ufipa	8 21 31 50 E	Pegler, Renvoize, Greenway
Mpumbulya = Ipumbuli, popl.	T7	Iringa	7 02 S 34 37 E	Howard
Mpumu, popl.	T4	Nzega	4 19 S 33 04 E	Dummer
	U4	Mengo	0 12 N 32 50 E	

Name		Location	Coordinates	Collector
Mpumu, mssn.	U4	Mengo	0 13 N 32 50 E	
Mpumudde, popl.	U4	Masaka	0 01 S 31 07 E	
[Mpumude] see Mpumudde	U4	Masaka	0 01 S 31 07 E	
Mpunga (Rondo Plat.)	T8	Lindi	Not traced	Busse
[Mpungwe] ? see Mpugwe	U4	Masaka	0 16 S 31 49 E	Snowden
[Mpurakasese] see Mpurukasese	T8	Songea	10 13 S 36 29.5 E	McLoughlin
Mpurukasese, popl.	T8	Songea	10 13 S 36 29.5 E	
Mpwapwa, distr.	T5	Mpwapwa	c.6 30 S 36 30 E	
Mpwapwa, distr.	T5	Mpwapwa	c.6 30 S 36 30 E	
Mpwapwa, popl., airstrip	T5	Mpwapwa	6 21 S 36 29 E	Numerous
Mraru, ridge	T2	Moshi	3 10 S 37 36 E	
Mraru, ridge	K7	Teita	3 19 S 38 28 E	Faden et al., Gillett et al.
[Mrau] see Mrao	T2	Moshi	3 10 S 37 36 E	Haarer
[Mraw] see Mrao	T2	Moshi	3 10 S 37 36 E	Haarer
Mrere, popl.	T2	Moshi	3 09 S 37 36 E	Haarer
Mrima, hill, for.	K7	Kwale	4 29 S 39 16 E	Gardner, Verdcourt, Greenway
[Mrogoro] see Morogoro	T6	Morogoro	6 49 S 37 40 E	
Mroweka, popl.	T8	Lindi	10 05 S 39 38 E	Koerner
Mruazi ? see Mliwaza, str.	T3	Pangani	?5 31 S 38 51 E–5 29 S 38 54 E	Procter
Mrweka = Mroweka, popl.	T8	Lindi	10 05 S 39 38 E	
Msada, area, str.	T6	Kilosa	c.7 13 S 37 09 E	Greenway & Kanuri
Msagali, popl., rsta., hill	T5	Mpwapwa	6 21 S 36 17 E	Peter
Msaginia = Msaginya, R.	T4	Mpanda	6 24 S 31 12 E–6 58 S 31 16 E	Richards
Msakiri	T5	Dodoma	Not traced	Savile
Msala (near Mshihui), hill	T3	Lushoto	near 4 43.5 S 38 38.5 E	Peter
Msala, area	T3	Lushoto	4 38 S 38 42 E	
Msalala, old t.a.	T1	Shinyanga	c.3 28 S 32 55 E	Hannington
[Msalanga] see Salanga	T5	Kondoa		Bally
Msalato = Masalatu, popl.	T5	Dodoma	6 05.5 S 35 45 E	Carmichael
Msale = Msali, ln.	T7	Iringa	c.7 30 S 36 10 E	Carmichael

255

Name		District	Coordinates	Authority
[Msalgali] see Msagali	T5	Mpwapwa	6 21 S 36 17 E	van Rensburg
[Msalla] ? Msala	T3	Lushoto	?4 38 S 38 42 E	Holtz
Msamala, popl.	T8	Tunduru	10 46 S 37 14 E	Tanner
[Msamara] see Msamala	T8	Tunduru	10 46 S 37 14 E	
Msamba, popl.	T4	Ufipa	7 51 S 30 47 E	Van Meel
[Msamba] see Msembe	T7	Iringa	7 44 S 34 57 E	Greenway & Kanuri
Msambia, str.	T4	Ufipa	c.8 42 S 31 37 E –8 26 S 31 51 E	
Msambweni, popl.	K7	Kwale	4 27.5 S 39 29 E	Allan, Drummond & Hemsley, Thorold
[Msamuye] see Samuye	T1	Shinyanga		
[Msamvia] see Msambia	T4	Ufipa	c.8 42 S 31 37 E –8 26 S 31 51 E	Muenzer
[Msamwia] see Msambia	T4	Ufipa	c.8 42 S 31 37 E –8 26 S 31 51 E	Muenzer
Msandama, peak	U2	Toro	0 44 N 30 10.5 E	Maitland
[Msanga] see Msonga	T4	Mpanda	c.6 10 S 31 50 E	Carmichael
Msanga, mssn.	T5	Dodoma	6 02 S 36 01.5 E	
[Msangaware's] ? see Nsangamales				
Msange Mbuga, sw.	T7	Mbeya	8 52 S 32 45 E	Goetze
	T6	Kilosa	7 17 S 37 06 E	Greenway & Kanuri
Msanzi, popl.	T4	Ufipa	8 13 S 31 31 E	Richards, Vesey-FitzGerald, Hooper & Townsend
Msasa, mt.	T3	Lushoto	5 05 S 38 39.5 E	Volkens
Msasa, str.	T2	Mbulu	c.3 26 S 35 48 E	Greenway
[Msasaani] see Msasani	T6	Uzaramo	6 45 S 39 17 E	Vaughan
Msasani, popl., bay	T6	Uzaramo	6 45 S 39 17 E	Vaughan, Batty, Procter
[Msassani] see Msasani	T6	Uzaramo	6 45 S 39 17 E	
Msata, popl.	T6	Bagamoyo	6 20 S 38 23.5 E	Procter
Msau, popl.	K7	Teita	3 25 S 38 24.5 E	Bally
[Msawala] see Sawala	T7	Iringa	8 30.5 S 35 20.5 E	Carmichael
[Msawati] see Ngangata	T8	Songea	c.10 15 S 36 24 E –10 06 S 36 11 E	McLoughlin
Ms Chaiyu	Z	Zanzibar	Not traced	Stuhlmann
Mseko (near Mwera)	T3	Pangani	near 5 29 S 38 57 E	Tanner
Msembe, Ruaha Nat. Park H.Q.	T7	Iringa	7 39 S 34 55 E	Numerous

[Msembi] see Msembe			
[Mseme] see Msembe			
Msengo Kwa Pundugura	T6	Uzaramo	prob. c.6 58 S 38 50 E	Busse
Msera (near Bukoba)	T1	Bukoba	Not traced	Braun
Msera (near Namasina)	T4	Tabora	Not traced	Braun
Msereki (near Angata Salei)	T2	Masai	near 2 40 S 35 40 E	Newbould
Mserere, ln.	T2	Masai	prob. c.3 05 S 36 10 E	Peter
Mshagea ? = Musughaa (Msughaa)	T5	Singida	?5 03 S 35 00 E	B.D. Burtt
[Mshamba] see Nshamba	T1	Bukoba	1 47 S 31 33 E	Haarer
Mshangano, fishponds	T3	Songea	10 39 S 35 40 E	Milne-Redhead & Taylor
Mshaushi (near Mkalama)	T5	Singida	near 4 07 S 34 38 E	B.D. Burtt
Mshihui, popl., mssn.	T3	Lushoto	4 43.5 S 38 38.5 E	Peter
[Mshiweni] see Mjwini	T3	Pare	4 20 S 37 55 E	Leechman
Mshwamba, old area	T3	Lushoto	c.4 28 S 38 22 E	Greenway
Msima, R.	T4	Mpanča	6 45 S 31 45 E–5 49 S 31 28 E	Ruffo, Carmichael
Msima, Farm	T7	Njombe	9 14 S 34 57 E	Emson, Greenway
Msimasi, sw.	T7	Mbeya	8 55 S 33 33 E	Goetze
Msima Stock Farm	T7	Njombe	9 14 S 34 57 E	Emson
Msimbati, pen., pt.	T8	Mikinčani	c.10 20 S 40 28 E	Richards, Wingfield
Msimbazi, bay	T6	Uzaramo	6 48 S 39 18 E	Vaughan
Msimbazi, str.	T6	Uzaramo	c.6 56 S 39 00 E–6 48 S 39 17 E	Procter
Msindi (Amani)	T3	Lushoto	near 5 06 S 38 38 E	Greenway
Msindossi (near Dar es Salaam)	T6	Uzaramo	Not traced	Peter
[Msinga] see Mzinga	T3	Handeni	5 44.5 S 37 53 E	Peter
Msingo	T6	Uzaramo	Not traced	Holtz
Msisi, popl.	T4	Tabora	6 07 S 33 07.5 E	Ukiriguru Res. Stn.
Msiwasi = Msiwazi, area	T7	Iringa	c.8 31 S 35 22 E	Carmichael
Msokero prob. = Mzerekera	T7	Iringa	prob. 7 53 S 36 08 E	Carmichael
[Msolwa] see Msorwa	T6	Uzaramo	7 10.5 S 39 17 E	Procter
Msolwa, popl.	T6	Ulanga	7 43 S 36 52 E	Carmichael

(near Bukoba) — T7 — Iringa — 7 39 S 34 55 E — Greenway
— T7 — Iringa — 7 39 S 34 55 E — Richards

257

Place	Region	District	Coordinates	Collector
Msolwa, R.	T6	Ulanga	7 42.5 S 36 59 E–8 07 S 36 53 E	Carmichael
Msolwa Camp	T6	Ulanga	7 48 S 37 02 E	Vollesen
[Msombas] ? = Missomba Mts.	T7	?Rungwe	?c.9 35 S 33 30 E	R.M. Davies
Msombe, hill	T7	Iringa	c.8 35 S 35 40 E	Carmichael
Msombe, popl.	T7	Iringa	8 29 S 35 44 E	
Msonga (Ugalla f.r.)	T4	Mpanda	c.6 10 S 31 50 E	Carmichael
[Msongezi] = Nsongezi	U2	Ankole	0 59 S 30 45 E	
[Msonza] see Nzonza	T7	Iringa	8 02.5 S 36 02 E–8 07 S 35 55 E	Carmichael
Msorwa, popl.	T6	Uzaramo	7 10.5 S 39 17 E	
[Msosi] see Musosi	T4	Buha	4 48 S 30 01 E	Peter
Msowelo, R.	T6	Kilosa	7 32 S 36 57 E–7 38 S 37 07 E	Semsei
[Msozi] see Misozi	U4	Masaka	0 53 S 31 45 E	
[Mssanga] see Msanga	T5	Dodoma	6 02 S 36 01.5 E	Busse
[Mssawati] see Ngangata	T8	Songea	c.10 15 S 36 24 E–10 06 S 36 11 E	
Msua, popl.	T6	Uzaramo	6 46 S 38 28 E	
Msua, rsta.	T6	Uzaramo	6 49 S 38 26 E	Peter, Cribb & Grey-Wilson, Mwasumbi
Msua, str.	T6	Uzaramo	c.6 45 S 38 30 E	
Msubugwe, f.r.	T3	Pangani	c.5 32 S 38 45 E	Milne-Redhead & Taylor, Semsei, Tanner
[Msuka] see Mkussu	T3	Lushoto	c.4 46 S 38 18 E–38 24 E	Richards
Msul Bani	Z	Zanzibar	Not traced	Stuhlmann
[Msuma] see Musuma	T4	Mpanda	6 13 S 29 49 E–6 14 S 29 45 E	Jefford et al.
Msumarini, popl.	K7	Kilifi	3 56 S 39 48.5 E	
[Msumbi] see Msumbisi	T6	Morogoro	6 53 S 37 51 E	Semsei
Msumbisi, popl.	T6	Morogoro	6 53 S 37 51 E	
Msumbugwe = Msubugwe, f.r.	T3	Pangani	c.5 32 S 38 45 E	
Msunkomilo, popl.	T4	Mpanda	6 20 S 31 02 E	Numerous
Mswaki, popl., r.h.	T3	Handeni	5 28 S 37 46 E	Semsei
Mswega, popl.	T8	Kilwa	8 26 S 37 57 E	Schlieben
[Mtae] see Mtai	T3	Lushoto	4 29 S 38 14 E	Gillman, Carmichael
Mtagata, hot springs	T1	Bukoba	1 14 S 30 51 E	Procter

Mtai, f.r.	T3	Lushoto	c.4 50 S 38 46 E	
Mtai, hill	T3	Lushoto	4 52 S 38 47 E	Drummond & Hemsley, Holst
Mtai, popl., mssn.	T3	Lushoto	4 29 S 38 14 E	Vesey-FitzGerald
Mtakuja, popl.	T4	Ufipa	7 22 S 30 37 E	
Mtama, popl.	T8	Lindi	10 18 S 39 22 E	Hay, Milne-Redhead & Taylor, Busse
Mtama, popl.	T8	Songea	10 53 S 35 02 E	Milne-Redhead & Taylor
Mtamba, popl., mssn.	T6	Morogoro	7 05 S 37 47 E	
[Mtambeli] see Mtambile	P	Pemba	5 23 S 39 42 E	Burtt Davy
Mtambile, popl., area	P	Pemba	5 23 S 39 42 E	Vaughan, Burtt Davy
[Mtambili] see Mtambile	P	Pemba	5 23 S 39 42 E	
Mtambo, R.	T4	Mpanda	5 46 S 31 10 E–6 24 S 31 33 E	Richards
[Mtambo] see Mtamba	T6	Morogoro	7 05 S 37 47 E	
[Mtambwa] see Matamba	T7	Njombe	8 59 S 33 58 E	Richards
Mtambwe Mkuu, I.	P	Pemba	5 04 S 39 42.5 E	Vaughan
Mtanda, str.	T8	Songea	c.10 44 S 33 37 E	
Mtandasi, R.	T8	Songea	c.10 19 S 35 32 E–10 13 S 35 37 E	Milne-Redhead & Taylor
[Mtandazi] see Mtandasi	T8	Songea	c.10 19 S 35 32 E–10 13 S 35 37 E	Milne-Redhead & Taylor
Mtandi, hill	T8	Masasi	10 41 S 38 49 E	Schlieben
Mtandika, popl., t.c.	T7	Iringa	7 32 S 36 26 E	Leach, B.D. Burtt
Mtanga = Mtange, popl.	T8	Lindi	10 00 S 39 39 E	
Mtanga, popl.	T8	Kilwa	8 51 S 39 28 E	
[Mtanga Twani] see Matanga Twani	P	Pemba	4 58 S 39 44 E	Vaughan
Mtange = Mtanga, popl., area	T8	Lindi	10 00 S 39 39 E	Busse, Gillman
[Mtani] see Mtoni	T4	Kahama/Tabora	c.4 09 S 32 24 E–4 25 S 31 58 E	
[Mtansa] see Mtanza	T6	Rufiji	7 52 S 38 25 E	Goetze
Mtanza, popl.	T6	Rufiji	7 52 S 38 25 E	
[Mtapwa] see Matapwa	T8	Lindi	9 41 S 39 26 E	Gillman
Mtarawanda (W. foot W. Usambara Mts.)	T3	Lushoto	Not traced	Greenway
[Mtatanda] see Tatanda	T4	Ufipa	8 32 S 31 29 E	Sanane
Mtawatawa = Tawatawa, ln.	T8	Kilwa	8 38 S 38 35 E	Vollesen

259

260

Mto Mwanakombo, str.	Z	Zanzibar	c.5 57 S 39 13 E	Vaughan
[Mto Ndei] see Mtito Andei				
Mtondia, popl.	K4	Machakos	2 41 S 38 10 E	Scott Elliot
Mtondole (Noto Plateau)	K7	Kilifi	3 34.5 S 39 52 E	Moggridge
Mtondwe valley (Uzinza)	T8	Lindi	c.9 54 S 39 24 E	Schlieben
Mtondwe (in Usagara)	T1	Mwanza	Not traced	B.D. Burtt
	T6	Kilosa	Not traced	Stuhlmann
Mtoni, R.	K4	Machakos/Kitui	c.2 03 S 37 48 E	Kassner
Mtoni, str.	T4	Kahama/Tabora	c.4 09 S 32 24 E–4 25 S 31 58 E	
Mtoni, popl.	T6	Uzaramo	6 52 S 39 16 E	Holtz
Mtoni, popl.	Z	Zanzibar	6 08 S 39 13 E	Vaughan, Faulkner
Mtonto (S. Pare Mts.)	T3	Pare	Not traced	Greenway
Mtonya, popl.	T8	Songea	10 24 S 36 07 E	Tanner
Mtorwi, peak	T7	Njombe	9 04.5 S 34 01 E	Richards, Cribb & Grey-Wilson
[Mtose] see Mtozi	T4	Ufipa	7 39 S 31 21 E–7 10 S 31 07 E	Richards
[Mtoser] see Mtozi	T4	Ufipa	7 39 S 31 21 E–7 10 S 31 07 E	
[Mtotchovu] see Mtotohovu	T3	Tanga	4 42 S 39 10 E	
[Mtoto] see Mtito Andei	K4	Machakos	2 41 S 38 10 E	Kassner
[Mtotohova] see Mtotohovu	T3	Tanga	4 42 S 39 10 E	Kassner
Mtotohovu, popl., sisal est.	T3	Tanga	4 42 S 39 10 E	Brenan, Greenway, Braun
[Mtoto wa Ande] see Mtito Andei	K4	Machakos	2 41 S 38 10 E	Gregory
Mtowabaga, popl.	T2	Masai	2 31 S 35 54 E	
Mto wa Mbu, game controlled area	T2	Masai	c.3 20 S 36 10 E	Numerous
Mto wa Mbu, popl.	T2	Masai	3 22 S 35 52 E	Numerous
Mto wa Mbu, str.	T2	Masai/Mbulu	c.3 17 S 35 49 E–3 26 S 35 52 E	Eggeling
Mto wa Mchanga, str.	T2	Mbulu	c.3 25 S 35 48 E	Greenway
Mto wa Mkindu, str.	T2	Mbulu	c.3 25 S 35 48–49 E	Greenway
[Mto wa Mkundu] see Mto wa Mkindu	T2	Mbulu	c.3 25 S 35 48–49 E	Greenway
Mtowanguuwe, str.	T3	Lushoto	Not traced	Semsei
Mto wa Ole, creek	P	Pemba	5 10.5 S 39 51 E	Vaughan
Mto wa Pwani, popl.	Z	Zanzibar	5 53 S 39 15 E	

261

Mto wa Simba, str.	T2	Masai/Mbulu	3 18 S 35 48.5 E – 3 26 S 35 49.5 E	Greenway
[Mto wa Umbu] see Mto wa Mbu	T2	Masai	3 22 S 35 52 E	Greenway & Kanuri
Mto ya Mchanga, str.	T1	Musoma	1 34 S 34 50 E–1 45 S 34 54 E	Greenway
[Mto ya Umbo] see Mto wa Mbu	T2	Masai	3 22 S 35 52 E	
Mtozi, R.	T4	Ufipa	7 39 S 31 21 E–7 10 S 31 07 E	Richards
[Mtshangni] see Mchangani	T6	Rufiji	7 54.5 S 39 48.5 E	
[Mtshinyiri] see Mchinjiri	T8	Lindi	10 08 S 39 11 E	
Mtua, popl.	T8	Lindi	10 13 S 39 29 E	Braun, Koerner, Busse
Mtulingale, P.W.D. camp	T7	Iringa	8 45 S 34 53 E	Polhill & Paulo
Mtumbati, valley	T8	Kilwa		Crosse-Upcott
			(numerous streams of this name around Kilwa)	
Mtumbi, area, f.r.	T3	Lushoto	4 40 S 38 20 E	Geilinger
[Mtumwambu] see Mto wa Mbu	T2	Masai/Mbulu	c.3 17 S 35 49 E– 3 26 S 35 52 E	Bally
Mtunda Hill, f.r.	T4	Kigoma	4 40 S 29 38 E	Procter
Mtupani	Z	Zanzibar	near 6 22 S 39 28 E	Vaughan
[Mtu wa Mbu] see Mto wa Mbu	T2	Masai	3 22 S 35 52 E	Bally
[Mtuwambu] see Mto wa Mbu	T2	Masai/Mbulu	c.3 17 S 35 49 E– 3 26 S 35 52 E	Bally
Mtwango, R.	T7	Iringa	near 8 00 S 36 00 E	Carmichael
Mtwapa, creek	K7	Kilifi	c.3 57 S 39 43 E	
Mtwapa, popl.	K7	Kilifi	3 56 S 39 45 E	Bogdan, R.M. Graham, Ward
Mtwara	T6	Ulanga	near 8 30 S 36 00 E	Nicholson
Mtwara, distr.	T8	Mikindani	c.10 25 S 40 05 E	
		(considered as part of Mikindani Distr. for F.T.E.A.)		
Mtwara, bay	T8	Mikindani	10 15 S 40 15 E	Eggeling, Richards
Mtwara, est.	T8	Mikindani	10 18 S 40 17 E	B.D. Burtt
Mtwara, popl.	T8	Mikindani	10 16 S 40 12 E	Numerous
[Mtwemi] see Milewemi	T4	Kahama	Not traced	
Mua, hills	K4	Machakos	c.1 27 S 37 12 E	Buchwald, Eick
Muafa, hill	T3	Lushoto	4 56 S 38 24 E	Stolz
[Muaja] see Mwaya	T7	Rungwe	9 33 S 33 57 E	

Name		District	Coordinates	Collector/Author
[Muaka] see Muoka	P	Pemba	5 26.5 S 39 42.5 E	Greenway
[Muakaleli] see Mwakaleli	T7	Rungwe	9 09 S 33 49 E	Goetze
[Muakereri] see Mvvakaleli	T7	Rungwe	9 09 S 33 49 E	Peter
[Muala] see Kamwala	T3	Pare	3 41 S 37 38 E	
Muamba = Umuamba, old distr.	T7	Rungwe	c.9 20 S 33 45 E	
Muanbuga's ? see Mambuga's	T7	Rungwe	c.9 20 S 33 44 E	
[Muangi] see Muwangi	U4	Masaka	0 19 S 31 27 E	Michelmore
[Muani] see Mwani	K4	Machakos	1 59.5 S 37 23 E	Pospischil
[Muansa] see Mwanza	T1	Mwanza	2 31 S 32 54 E	Stuhlmann
Muantipulo's, popl.	T7	Rungwe	9 34.5 S 33 45.5 E	Stolz
Muantipulo's, popl.	T7	Rungwe	9 31.5 S 33 43 E	Geilinger
[Muanza] see Mwanza	T1	Mwanza	2 31 S 32 54 E	Dummer
Mubango, est.	U4	Mengo	0 25 N 33 01 E	
Mubende, distr.	U4	Mubende	c.0 45 N 31 10 E	
(previously extended further east as far as R. Mayanja)				
Mubende, popl., mssn.	U4	Mubende	0 35 N 31 23 E	Numerous
[Mubendi] see Mubende	U4	Mubende	0 35 N 31 23 E	Snowden
Mubuku, str.	U2	Toro	c.0 21 N 29 54 E–0 05 N 30 13 E	Ball, Eggeling, Scott Elliot
Mubunda, popl., t.c.	T1	Bukoba	1 56.5 S 31 31 E	Procter
Muchlur, area	T2	Mbulu	4 02 S 35 16.5 E	Robson, Carmichael
[Muchlur, L.] see Austin	T2	Mbulu	4 02.5 S 35 16.5 E	
[Mucunu] see Mukunu	T1	Mwanza	2 00 S 32 59 E	
Mucwini, popl.	U1	Acholi	3 24 N 33 02 E	Conrads
[Mudagashi] see Mado Gashi	K1	Northern Frontier	0 43.5 N 39 10.5 E	Purseglove
Mudanda Rock	K7	Teita	3 10 S 38 31 E	J. Adamson
Mudangi, r.c.	U3	Mbale	1 10 N 34 26 E	Greenway
[Muddo Gashi] see Mado Gashi	K1	Northern Frontier	0 43.5 N 39 10.5 E	Numerous
[Mudogashi] see Mado Gashi	K1	Northern Frontier	0 43.5 N 39 10.5 E	J. Adamson
Muduma, popl., sawmill	U4	Mengo	0 21 N 32 18 E	Dawkins
Mue, R.	T2	Moshi	3 07.5 S 37 25 E–3 32.5 S 37 29.5 E	Uhlig, Winkler

Name	Region	District	Coordinates	Collector
[Muɛ-bach] see Mue River	T2	Moshi	3 07.5 S 37 25 E – 3 32.5 S 37 29.5 E	Uhlig
Muena, popl.	T5	Dodoma	5 59 S 35 00 E	Busse, Barth
Muengo, str.	T7	Iringa	8 25 S 35 26.5 E –8 44 S 35 39 E	B.D. Burtt
[Muenia] see Mnenya	T5	Kondoa	4 42.5 S 35 52 E	Busse, Schlieben
[Muera, Plateau] see Rondo	T8	Lindi	c.10 10 S 39 15 E	Numerous
Mufindi, area	T7	Iringa	c.8 35 S 35 15 E	
Mufindi Scarp, f.r.	T7	Iringa	c.8 40 S 35 20 E	
[Mufumbiro] see Virunga	U2/Zaire/Rw./Bur.	Kigezi	c.1 20 S 29 30 E	Procter, Pegler, Renvoize
Mufu, Ukwama, ln.	T7	Iringa	c.7 40 S 36 30 E	Ede
Muganza, popl.	T1	Ngara	2 56 S 30 40 E	Tanner
Mugeni, popl., hill	Burundi		4 06 S 30 03 E	Peter
Mughwango, hill	K4	Meru	0 08 N 38 08 E	Mathenge, Ament & Magogo
Mugiligili, area	T3	Pare	3 45 S 37 43 E	
Mugira track (Mikumi Nat. Park)	T6	Kilosa	Not traced	Greenway & Kanuri
[Mugire] ? see Mgiwe	T7	Njombe	c.9 35 S 34 25 E	Goetze
Mugomba, popl.	U4	Mengo	0 16 N 32 44 E	Dummer
[Mugombasi] see Mugombazi	T4	Kigoma	5 50 S 30 13 E	Harley
Mugombazi, popl.	T4	Kigoma	5 50 S 30 13 E	Harley
Muguga, for.	K4	Kiambu	c.1 11 S 36 37.5 E	Clayton, Trapnell, Verdcourt
Muguga, rsta.	K4	Kiambu	1 11 S 36 39 E	
Muguga Juu, popl.	K4	Kiambu	1 10 S 36 38.5 E	
Muguga North, EAVRO	K4	Kiambu	1 12.5 S 36 37 E	Greenway
Muguga South, EAAFRO	K4	Kiambu	1 13.5 S 36 38 E	Numerous
Mugunga ("near Kisii") ? see Mogonga	K5	South Kavirondo		
Mugunga, str.	T4	Buha	4 43 S 30 26 E –4 47 S 30 23 E	Napier
Mugungu, R.	T1/2	Musoma/Masai	c.2 30 S 35 20 E –2 19 S 34 53 E	Peter
Mugurr, area	K1	Northern Frontier	c.2 37 N 36 28 E	Greenway
Muguti, str.	T4	Buha/Kigoma	4 53 S 30 02 E –4 50 S 29 48 E	Mathew
Mugwe, str.	T7	Njombe	c.9 15 S 35 15 E –9 07 S 35 27 E	Schlieben

Name	Code	District	Coordinates	Authority
[Mugwetal] see Mugwe, str.				
Muhaka, for.	T7	Njombe	c.9 15 S 35 15 E – 9 07 S 35 27 E	
Muhaka (near Madanga)	K7	Kwale	4 20 S 39 31.5 E	Brenan et al., Faden
[Muhala] see Mhwala	T3	Pangani	near 5 21 S 38 59 E	Tanner
[Muhaloholo] see Buholoholo	T4	Tabora	c.4 57 S 33 56 E	Stuhlmann
Muhambwe, popl.	T4	Mpanda	c.6 00 S 30 00 E	Carmichael
Muhanga, mssn.	T4	Mpanda	6 19 S 30 22.5 E	Grant
Muhangi, popl., for.	T7	Iringa	8 12 S 36 01 E	Goetze
[Muhango] ? = Mehange	U2	Toro	0 37.5 N 30 29 E	Stuart-Smith
[Muhasi] see Muheza	T3	Lushoto	?c.5 10 S 38 27 E	Stuhlmann
Muhavura, mt.	T3	Pare	4 19 S 37 51 E	Peter
Muhela, popl.	U2/Rwanda	Kigezi	1 23 S 29 41 E	Numerous
[Muhele] see Mhwala	T4	Mpanda	6 12.5 S 29 46 E	Oxford Univ.
[Muhesa] see Muheza	T4	Tabora	c.4 57 S 33 56 E	Stuhlmann
Muheza, popl.	T3	Tanga	5 10 S 38 48 E	Faulkner
Muheza, sw.	T3	Pare	4 19 S 37 51 E	
Muheza, popl., rsta., mssn., old pl.	T3	Pare	4 34.5 S 37 44 E	
Muhinde Steppe	T3	Tanga	5 10 S 38 48 E	Numerous
[Muhinduro] see Mhinduro	T7	Iringa	prob. c.7 45 S 35 50 E	Goetze
[Muhinga] see Mgahinga	T3	Lushoto	4 57 S 38 46 E	
Muhirike, popl.	U2/Rwanda	Kigezi	1 23 S 29 39 E	Snowden
Muhokya, popl.	T6	Kilosa	6 02 S 37 08 E	Busse
Muhonyera, popl.	U2	Toro	0 06 N 30 03 E	Maitland, A.S. Thomas, Michelmore
Muhonyera, popl.	T6	Uzaramo	6 55 S 38 41 E	Grant
Muhoro = Buhoro, popl.	T6	Uzaramo	6 55 S 38 41 E	Grant
Muhoroni, popl., rsta.	T4	Buha	4 25 S 30 10 E	
Muhororo, popl.	K5	Kisumu-Londiani	0 09 S 35 12 E	Battiscombe, Bogdan, Johansen, Bally
[Muhorro] see Muhoro	U4	Mubende	0 55 N 30 46 E	
Muhu, str.	T4	Buha	4 25 S 30 10 E	Peter
Muhulu, f.r.	T7	Iringa	c.8 16 S 35 52 E	
	T6	Ulanga	8 51 S 36 40 E	Schlieben, Cribb & Grey-Wilson

266

Name		Region	Coordinates	Collector
Mukugwa, R.	T2	Buha	c.3 51 S 30 45 E–3 54 S 30 25 E	Eggeling
[Mukugwe] see Mukugwa	T2	Buha	c.3 51 S 30 45 E–3 54 S 30 25 E	Eggeling
Mukuju, popl.	U3	Mbale	0 45 N 34 11.5 E	Volkens
[Mukulumussi] see Mkulumuzi	T3	Tanga	5 07 S 38 45 E–5 04 S 39 04 E	Goetze
[Mukumi] see Mikumi	T6	Kilosa	7 24 S 36 59 E	Moore
Mukungu, str.	T1	Musoma	c.2 20 S 35 15 E–2 17 S 34 50 E	Greenway & Rawlins
Mukunguya, L. = Kenyatta	K7	Lamu	2 25 S 40 41 E	
[Mukunguyu] see Mukunguya	K7	Lamu	2 25 S 40 41 E	
Mukunu, popl.	T1	Mwanza	2 00 S 32 59 E	
Mukura, rsta.	U3	Teso	1 34 N 33 52 E	
[Mukuru] see Mkuru	T2	Arusha	3 08 S 36 49 E	
[Mukuta] see Makutu	U3	Busoga	0 30 N 33 36 E	G.H.S. Wood
[Mukuyu] see Mkuyu	T4	Kigoma	5 28.5 S 29 46 E	Kielland
[Mukuzi] see Mkuzi	T3	Lushoto	4 45 S 38 20 E	Mgaza
Mulagala, R.	T7	Rungwe	c.9 19 S 38 36 E	Stolz
Mulago, popl.	U4	Mengo	0 21 N 32 35 E	Chandler, Snowden
Mulalakuwa, str.	T6	Uzaramo	c.6 45 S 39 14 E	Wingfield
Mulamba, r.h.	U2	Kigezi	0 57 S 29 50 E	
[Mulanga ya Longil] see Longil L.	T2	Arusha	3 15.5 S 36 53 E	Richards
Mulange, hill	U4	Mengo	0 31 N 33 03 E	Dummer, Dawkins
Mulele = Mlala, hills, f.r.	T4	Mpanda	c.6 47 S 31 45 E	Procter
Mulema (60 miles up Kagera valley)	U2	Ankole	c.1 00 S 31 00 E	Bagshawe
Mulinda, for.	T7	Rungwe	c.9 30 S 33 42 E	Stolz
Mulirahangi (near Kilimatinde)	T5	Dodoma	near 5 51 S 34 57.5 E	Claus
Mulole, hill	U2	Kigezi	1 03.5 S 29 37 E	
Mulolosi, L.	T2	Arusha	3 13.5 S 36 57.5 E	Richards
Mumandu, hill, for., popl.	K4	Machakos	1 40 S 37 17 E	
[Mumemmwa] see Mmemya	T4	Ufipa	8 15 S 31 56 E	
Mumias, popl., t.c.	K5	North Kavirondo	0 20 N 34 29.5 E	Richards
Mumoni, for.	K4	Kitui	c.0 32 S 37 59 E	Scott Elliot, Whyte

Name		Region	Area	Coordinates	Collector
Mumoni, mt.	K4	Kitui		0 31 S 38 00.5 E	Gardner
Mumwendo, fy.	T1	Ngara		2 38 S 30 33.5 E	Carmichael, Tanner
[Munaga] see Muengo	T7	Iringa		8 25 S 35 26.5 E–8 44 S 35 39 E	Carmichael
[Mundanyi] see Wundanyi	K7	Teita		3 24 S 38 22 E	
Mundell's Rocks, Swamp	K3	Trans-Nzoia		c.1 08 N 34 49.5 E	Tweedie
[Munekesi] see Mwunekese	T1/4	Biharamulo/Kahama		3 09 S 31 56 E–3 17 S 31 41 E	Procter
Munene, f.r.	T1	Bukoba		1 16 S 31 34 E	Numerous
Munge, popl, w.h.'s	T2	Masai		3 05 S 35 40 E	Greenway
Munge, str.	T2	Masai		c.3 00 S 35 39 E–3 13 S 35 33 E	
[Munge, Kisima cha] ? see Kisima cha Mungu	T2	Masai		5 18 S 36 56 E	Geilinger
[Munge, Kisima cha] ? see Munge	T2	Masai		3 05 S 35 40 E	Geilinger
Mungilo = Mongiro, popl.	U2	Toro		0 50 N 30 10 E	Paulo
Munguni, for.	K4	Embu		0 19 S 37 53 E	
[Muninga] see Mninga	T7	Iringa		c.8 36 S 35 12 E	Richards
Munondo Plains (near Mbale)	U3	Mbale		near 1 05 N 34 10 E	Eggeling
[Muntandi] see Ntandi	U2	Toro		c.0 48 N 30 09 E	Eggeling, Sangster
Munteme, popl.	U2	Bunyoro		1 19 N 31 10 E	
Muntu, popl.	U1	Lango		1 35 N 32 54 E	Langdale-Brown
Mununguna (Yaida Valley)	T2	Mbulu		c.4 00 S 35 05 E	Richards
Munya, str.	T5	Singida		4 40 S 34 38.5 E–4 34 S 34 37 E	Savile
Munyama (Budongo Forest)	U2	Bunyoro		c.1 47 N 31 35 E	Sangster
Munyen, str.	K2	Turkana		c.2 41 N 34 47 E–2 47 N 34 59 E	
Munyonyo, popl.	U4	Mengo		0 14.5 N 32 37 E	Vaughan
[Munyuni] see Muyuni	Z	Zanzibar		6 22 S 39 28 E	Holst, Busse, Peter
[Muoa] see Moa	T3	Tanga		4 44 S 39 09 E	Goetze
Muoi = Muwi, mts.	T7	Iringa		7 45 S 35 53 E	
Muoka, popl.	P	Pemba		5 26.5 S 39 42.5 E	Vaughan
Muongoni, popl, area	Z	Zanzibar		6 18.5 S 39 26 E	Kirk
Mupanda	Z	Zanzibar		Not traced	
Murai, popl.	T2	Mbulu		4 00.5 S 35 33 E	

[Murambo] see Mulamba				
Muramur, dam	K1	1 01 N 36 43 E	Northern Frontier	Nesbit Evans
Murang'a = Fort Hall, popl.	K4	0 43 S 37 11 E	Fort Hall	
[Murca] see Murka	K7	3 25 S 37 56 E	Teita	V.C. Gilbert
[Murchison Falls] see Kabalega	U1/2	2 17 N 31 41 E	Acholi/Bunyoro	Bagshawe, Eggeling, G. Taylor
[Murchison Falls, Nat. Pk.] see Kabalega	U1/2	c.2 15 N 31 50 E	Acholi/Bunyoro	Buechner, Pegler
[Murenga] see Murungu	T4	4 14 S 31 11 E	Buha	Bullock
[Murentat] see Morendat	K3	0 40 S 36 23 E	Naivasha	Fischer
Murera R.	K4	0 19 N 38 05 E–0 05 N 38 20 E	Meru	Ament
Mureti, area	K1	1 11.5 N 37 22 E	Northern Frontier	
Mu Rgwanza, popl., mssn.	T1	2 29 S 30 38 E	Ngara	
Murikamba, R.	T2	c.3 15 S 36 50 E	Arusha	Greenway
Murille = Merille, t.a.	K1	c.3 20 N 41 15 E	Northern Frontier	
Murimba – vernacular name of plant				
Murka, rsta., game lodge	K7	3 25 S 37 56 E	Teita	Gilbert
Muroki Farm (SE. Elgon)	K3	prob. c.1 00 N 34 50 E	Trans-Nzoia	Anderson
[Murole] see Mulole	U2	1 03.5 S 29 37 E	Kigezi	Purseglove
Muronzi, str.	T1	c.2 47 S 30 52 E	Ngara	Tanner
[Murri] see Malka Mari	K1	4 16 N 40 46 E	Northern Frontier	Kirrika
[Muruanisigar] see Murua Nysigar	K2	c.3 10 N 35 01 E	Turkana	A.S. Thomas
[Muruaniwa] see Moruariwan	U1	2 49 N 34 26 E	Karamoja	Champion, Newbould
Murua Nysigar, hills	K2	c.3 10 N 35 01 E	Turkana	
[Muruasigar] see Murua Nysigar	K2	c.3 10 N 35 01 E	Turkana	
[Muruasugar] see Murua Nysigar	K2	c.3 10 N 35 01 E	Turkana	
Muruatunokan, hill	U1	2 01 N 34 29 E	Karamoja	
[Murugwanza] see Mu Rgwanza	T1	2 29 S 30 38 E	Ngara	Tanner
Mu Rukarazo, mssn.	T1	2 27.5 S 30 42 E	Ngara	Tanner
Murungu, popl., area	T4	4 14 S 31 11 E	Buha	Bullock, Mutch
Mu Rusagamba, popl.	T1	2 58 S 30 50 E	Ngara	Tanner
[Muruzi] see Maruzi	U1	c.1 50 N 32 25 E	Lango	

Eggeling

269

270

[Mussini] see Mucwini				
Musughaa, popl.	U1	Acholi	3 24 N 33 02 E	
Musuma, R.	T5	Singida	5 03 S 35 00 E	
[Musumarini] see Msumarini	T4	Mpanda	6 13 S 29 49 E –6 14 S 29 45 E	Newbould & Jefford
[Muta] see Mutha	K7	Kilifi	3 56 S 39 48.5 E	Pinn
Mutai, popl., f.r.	K4	Kitui	1 47 S 38 24 E	Elliot
Mutai, rsta.	U3	Busoga	0 35 N 33 13 E	Kibuko, G.H. Wood
Mutala, popl.	U3	Busoga	0 31.5 N 33 15 E	Eggeling
Mutamayo, est.	T1	Biharamulo	2 43 S 30 56 E	
Mutanda, L.	K3	Trans-Nzoia	1 10 N 34 47 E	Tweedie
[Mutandasi] see Mtandasi	U2	Kigezi	c.1 13 S 29 40 E	Chandler, Eggeling, Purseglove
[Mutandazi] see Mtandasi	T8	Songea	c.10 19 S 35 32 E –10 13 S 35 37 E	
Mutet, popl.	T8	Songea	c.10 19 S 35 32 E –10 13 S 35 37 E	
Mutha, hill	K5	Central Kavirondo	0 01.5 S 34 45 E	Drummond & Hemsley
Mutha, popl., t.c.	K4	Kitui	1 47 S 38 24 E	Bally, Joanna
Muthaiga Club	K4	Kitui	1 48 S 38 25.5 E	
Muthangari (French Mission Land)	K4	Nairobi	1 15.5 S 36 50 E	Napier
Mutito, hill	K4	Nairobi	1 17 S 36 49 E	Napier
Mutolere, mssn.	K4	Kitui	1 14 S 38 09 E	Gardner
[Mutoma] see Mutomo	U2	Kigezi	1 15.5 S 29 43.5 E	Eggeling
Mutomo, hill	K4	Kitui	1 51 S 38 13 E	Bally, Gillett
Mutonga, R.	K4	Kitui	1 51 S 38 13 E	Bally, Kimani
Mutukula, popl.	K4	Meru/Embu	0 06.5 S 37 22 E –0 26 S 37 54 E	D. Davis, Fries
Mutunda, popl., r.h., fy.	U4	Masaka	0 59 S 31 25 E	
Mutungo, hill, popl.	U2	Bunyoro	2 06 N 32 19 E	Purseglove, Eggeling
Mutungo, popl.	U4	Mengo	0 19 N 32 38.5 E	Chandler
[Mutuwambu] see Mto wa Mbu	U4	Mengo	0 12.5 N 32 35 E	Lye
Muva, popl., f.r.	T2	Masai/Mbulu	c.3 17 S 35 49 E –3 26 S 35 52 E	Bally
Muviko – vernacular name of plant	T4	Ufipa	7 54 S 31 36 E	Carmichael
[Muvumoni] see Mvumoni	K7	Kwale		Padwa
	P	Pemba	5 11 S 39 50 E	?Vaughan

271

[Muvumoni] see Mvumoni				
Muwangi, popl.	P	Pemba	5 16 S 39 49 E	?Vaughan
Muwawu, for.	U4	Masaka	0 19 S 31 27 E	
Muwendo, fy.	U4	Masaka	c.0 30 S 31 49 E	
Muwera, popl.	T1	Ngara	2 38 S 30 34 E	Eggeling
Muyembe, popl., r.h., mssn.	T5	Singida	5 08 S 34 11 E	B.D. Burtt
Muyuni, popl., area	U3	Mbale	1 20 N 34 17.5 E	
Muyuni, popl.	Z	Zanzibar	6 22 S 39 28 E	Vaughan
Muzibini = Mujibini	Z	Zanzibar	5 48 S 39 21 E	
Muzizi, R.	T7	Chunya	near 7 39 S 33 24 E	Richards
[Muzombe] see Musombe	U2/4	Toro/Mubende	c.0 34 N 31 22 E–1 03 N 30 33 E	Bagshawe
[Mwakilembi] see Mwakijembe	T7	Iringa	7 18 S 34 30 E	Richards
Mvilingano, popl.	T3	Tanga	4 32 S 38 56 E	Cribb & Grey-Wilson
Mvindeni, popl.	T3	Lushoto	4 43.5 S 38 36 E	Cribb & Grey-Wilson
Mvomero, popl., mssn., r.h., t.c.	K7	Lamu	1 55 S 41 20 E	C.W. Elliot
Mvuji = Mwuhe, str.	T6	Morogoro	6 20 S 37 24.5 E	Eggeling
Mvuleni, popl.	T6	Morogoro	6 08 S 37 33 E–6 14.5 S 37 37 E	
Mvuma = Vuma, hill, area	Z	Zanzibar	5 50 S 39 17.5 E	Vaughan
Mvumi, popl.	T6	Kilosa	7 26 S 37 07 E	Greenway & Kanuri
Mvumi, popl., t.c.	T5	Dodoma	6 21 S 35 51 E	Greenway, Busse, Ruffo
[Mvumie] see Mvumi	T6	Kilosa	6 35 S 37 10 E	
Mvumoi, L.	T5	Dodoma	6 21 S 35 51 E	
Mvumoni, t.a.	T3	Lushoto	c.4 46 S 38 18 E	Renvoize & Abdallah
Mvumoni, popl.	T3	Tanga	c.5 16 S 38 58 E	Tanner
Mvumoni, popl.	P	Pemba	5 16 S 39 49 E	?Vaughan
[Mwa Baya Nyundo] see Nyundo	P	Pemba	5 11 S 39 50 E	?Vaughan
Mwachewatini (Rabai Hills)	K7	Kilifi	3 49 S 39 35 E	
Mwachi, creek	K7	Kilifi	c.3 36 S 39 35 E	W.E. Taylor
Mwachi, for.	K7	Kwale	4 01 S 39 32 E	Numerous
Mwadui, popl.	K7	Kwale	c.4 00 S 39 32 E	Numerous
	T1	Shinyanga	3 33 S 33 36 E	Batty

Name	Region	District	Coordinates	Collector
Mwagao	T4	Kigoma	Not traced	Peter
[Mwagoje] see Magoye	T7	Njombe	9 00 S 33 59 E	Stolz
[Mwagoye] see Magoye	T7	Njombe	9 00 S 33 59 E	Stolz
Mwagusi, str.	T7	Iringa	c.7 30 S 34 35 E–7 35 S 35 01 E	Richards
[Mwagussi] see Mwagusi	T7	Iringa	c.7 30 S 34 35 E–7 35 S 35 01 E	
[Mwaguzi] see Mwagusi	T7	Iringa	c.7 30 S 34 35 E–7 35 S 35 01 E	
Mwai, popl.	T4	Ufipa	7 23 S 31 06 E	Hooper & Townsend
[Mwaja] see Mwaya	T7	Rungwe	9 33 S 33 57 E	Geilinger
[Mwakalele] see Mwakaleli	T7	Rungwe	9 09 S 33 49 E	
Mwakaleli, mssn.	T7	Rungwe	9 09 S 33 49 E	Numerous
[Mwakalila] see Mwakaleli	T7	Rungwe	9 09 S 33 49 E	Stolz
Mwakete, mssn.	T7	Njombe	9 21.5 S 34 14 E	Eggeling, Richards, Stolz
Mwakijembe, popl.	T3	Tanga	4 32 S 38 56 E	Drummond & Hemsley, Kemode
Mwakiki, agric. stn.	K7	Teita	3 23 S 38 23 E	Carter & Stannard
Mwalabila (Bulambia area)	T7	Rungwe	c.9 20 S 33 10 E	Stolz
[Mwale Mdogo] see Mwele Mdogo	K7	Kwale	4 18 S 39 21 E	
Mwalesi, R.	T7	Rungwe	9 12 S 33 39 E–9 23 S 33 48 E	R.M. Davies
[Mwalpindi] ? see Malupindi	T7	Rungwe	?9 02 S 33 37 E	R.M. Davies
Mwaluvanga, popl.	K7	Kwale	4 25 S 39 21 E	Makin
[Mwalwanda] see Maruanda's	T7	Mbeya		Stolz
Mwamakunda, ln.	T1	Maswa	c.3 10 S 34 45 E	Greenway
Mwamara, ln.	T1	Shinyanga	3 49 S 33 56 E	Ukiriguru Res. Centre
Mwambani, popl.	T3	Tanga	5 08 S 39 06 E	Faulkner, Bally & Carter
Mwambirwa, for. stn.	K7	Teita	3 21 S 38 36 E	Faden, Taita Hills Exped.
Mwamboneke = Mamboneke, popl.	T7	Rungwe	c.9 12 S 33 39 E	Stolz
Mwamgongo	T4	Buha	c.4 35 S 29 39 E	
Mwami = Wami, hill	K4	Machakos	1 39 S 37 08 E	Bally
Mwana Kerekwe, popl.	Z	Zanzibar	6 10.5 S 39 13.5 E	Vaughan
Mwanakombo, str.	Z	Zanzibar	6 57 S 39 13 E	Vaughan
Mwana Kwerekwe, popl.	Z	Zanzibar	6 11 S 39 14 E	

273

Name		District	Coordinates	Collector(s)
Mwanambogo Dam	T6	Kilosa	7 09 S 37 10 E	
Mwanamgaru (near Mwera)	T3	Pangani	near 5 29 S 38 57 E	Tanner
Mwananyamara, L.	T6	Uzaramo	6 47 S 39 14.5 E	Vaughan, Mwasumbi
Mwanda, hills, popl.	K7	Teita	3 24.5 S 38 17 E	Gardner, Klungness
Mwanda, popl.	Z	Zanzibar	5 43.5 S 39 18 E	
Mwanda, I., popl.	Z	Zanzibar	5 55 S 39 13 E	Sacleux
Mwandeo, popl.	K7	Kwale	4 26 S 39 21 E	Verdcourt
Mwandi, str.	T8	Songea	prob. c.9 50 S 35 22 E	Milne-Redhead & Taylor
[Mwane Mbogo] see Mwanambogo	T6	Kilosa	7 09 S 37 10 E	
Mwanga, popl.	T4	Kigoma	4 53 S 29 38.5 E	Greenway & Kanuri
Mwangea, hill	K7	Teita	3 23 S 38 33 E	Azuma, Procter, Hooper & Townsend
Mwanghanda (Mbarika)	T1	Mwanza	c.2 55 S 32 51 E	Tanner
Mwangilye Hill, f.r.	T1	Shinyanga	3 53 S 33 10 E	Gane
Mwangoji (prob. Teita Hills or Kasigau Mt.)	K7	Teita	Not traced	Dale
[Mwanhinj] see Mwantine	T4	Nzega	4 23 S 33 09 E	Doggett, Msagamasi
Mwani, hill	K4	Machakos	1 59.5 S 37 23 E	Koritschoner
Mwanihana, f.r.	T7	Iringa	c.7 48 S 36 52 E	Rodgers & Vollesen, Hall
Mwankinja	T7	Rungwe	Not traced	Semsei
Mwantine, area	T1	Shinyanga	c.3 40 S 33 13 E	B.D. Burtt
Mwantine, hills, f.r.	T1	Shinyanga	3 38 S 33 19.5 E	B.D. Burtt
Mwantine, peak	T1	Shinyanga	3 38.5 S 33 19.5 E	
Mwanza, distr.	T1	Mwanza	c.2 50 S 32 30 E	
Mwanza, popl., mssn.	T1	Mwanza	2 31 S 32 54 E	Numerous
[Mwanzye] see Mwazye	T4	Ufipa	8 27 S 31 44 E	
Mwarakaya, popl.	K7	Kilifi	3 47.5 S 39 42 E	Bally & Smith, Faden & Beentje
Mwasangombe, for.	K7	Kwale	c.4 21 S 39 20 E	Drummond & Hemsley
Mwasbweni	K7	?Kilifi	Not traced	Jeffery
[Mwasje] see Mwasye	T4	Ufipa	8 27 S 31 44 E	
[Mwasukulu] see Masukulu	T7	Rungwe	9 24 S 33 45 E	Stolz

Name		Region	Coordinates	Collector
Mwasye = Mwazye	T4	Ufipa	8 27 S 31 44 E	Bullock, Michelmore, Robinson
Mwatate, popl.	K7	Teita	3 30 S 38 23 E	Numerous
Mwatate, rsta.	K7	Teita	3 31 S 38 24 E	Bally
Mwatesi, R.	T7	Rungwe	9 07 S 33 40 E–9 10 S 33 49 E	Cribb & Grey-Wilson, Hepper & Field
Mwau, area, ? hills	T5	Singida	c.5 15 S 34 53 E	
[Mwauguzi] see Mwagusi	T7	Iringa	c.7 30 S 34 35 E–7 36 S 35 02 E	
Mwawe, str.	T2	Moshi	c.3 15 S 37 14 E	
Mwawesa, area	K7	Mombasa	3 57 S 39 37 E	
Mwaya, popl., mssn.	T7	Rungwe	9 33 S 33 57 E	R.M. Davies
Mwayange = Mwayangi, hill, area	T7	Iringa	7 48 S 34 57 E	Greenway
Mwayaya, sacred grove	T4	Buha	c.4 29 S 29 50 E	Procter
Mwayembe, sw.	T7	Iringa	7 38 S 34 46 E	Greenway, Richards
Mwayivi, hills	T7	Mbeya	c.7 48 S 34 37 E	Richards
Mwazye, popl., mssn.	T4	Ufipa	8 27 S 31 44 E	Michelmore
Mwea, R. (E. Mt. Kenya)	K4	South Nyeri/Embu	Not traced	Battiscombe
Mwea, ln.	K4	South Nyeri	0 45 S 37 29 E	
Mwea-Tebere Irrigation Scheme, area	K4	South Nyeri	c.0 42 S 37 22 E	Bogdan, Brunt
Mwega, R.	T6	Kilosa	7 06 S 36 39 E–7 22 S 36 29 E	Greenway & Kanuri
[Mwega Hakai] see Mwoga Hakai	K7	Lamu	2 29 S 40 28 E	Hooper & Townsend
Mweiga, t.c.	K4	North Nyeri	0 19 S 36 54 E	Bally
Mweka, College of African Wildlife Management	T2	Moshi	3 14 S 37 19 E	C.A.W.M., Harris
Mwela, popl.	T4	Ufipa	7 46 S 31 40 E	Michelmore
Mwela, popl.	T7	Mbeya	8 55 S 33 49 E	Cribb & Grey-Wilson
Mwele, area, old pl.	T3	Lushoto	4 51 S 38 49 E	
Mwele, popl.	T3	Lushoto	4 50 S 38 49 E	Peter
Mwele = Mwele Mdogo, hill	K7	Kwale	4 18 S 39 21 E	R.M. Graham
Mwele Mdogo, hill, for.	K7	Kwale	4 18 S 39 21 E	Drummond & Hemsley
Mwele Mkubwa, hills	K7	Kwale	4 13–4 17 S 39 23 E	
[Mwemba] see Mnemba	Z	Zanzibar	5 49 S 39 22.5 E	Stuhlmann

276

Name	Grid	Region	Coordinates	Collector
Mwimbi, area	K4	Embu	c.0 15 S 37 45 E	Rammel
Mwimbi, popl., r.h., sch.	T4	Ufipa	8 39.5 S 31 40 E	Richards
Mwina, val.	K2	West Suk	c.1 23 N 35 26 E	Honoré, Thorold
Mwina, area	K2	West Suk	1 10 N 35 26 E	
Mwingi, hill	K4	Kitui	0 57 S 38 05 E	Greenway, Bally & Smith, Archer
Mwirasandu, mssn.	U2	Ankole	0 59.5 S 30 23 E	Eggeling
Mwiri, hill	U3	Busoga	0 30 N 33 16 E	Kireri, G.H. Wood
Mwisibach = Miesi, str.	T8	Masasi	10 45 S 38 45 E – 11 10 S 38 56 E	Schlieben
Mwiti, mssn.	T8	Masasi	10 41 S 39 03 E	Schlieben
Mwitika, popl.	T7	Rungwe	9 22 S 33 41 E	
[Mwitikera] see Mwitikira	T5	Dodoma	6 31 S 35 39.5 E	Greenway
Mwitikira, popl.	T5	Dodoma	6 31 S 35 39.5 E	
[Mwiyembe] see Muyembe	U3	Mbale	1 20 N 34 17.5 E	Snowden
Mwizi, popl.	U2	Ankole	0 46.5 S 30 36 E	Eggeling
[Mwoga] see Moga	Z	Zanzibar	5 55 S 39 17.5 E	Vaughan
Mwoga Hakai, popl.	K7	Lamu	2 29 S 40 28 E	
Mwuhe = Mvuji, str.	T6	Morogoro	6 08 S 37 33 E –6 14.5 S 37 37 E	Holtz
[Mwuhu] see Mvuhe	T6	Morogoro	6 08 S 37 33 E –6 14.5 S 37 37 E	
Mwunekese, str.	T1/4	Biharamulo/Kahama	3 09 S 31 56 E –3 17 S 31 41 E	
Myangayanga, popl.	T8	Songea	10 56 S 34 58 E	Zerny
Myanzi, r.h.	U4	Mengo	0 26 N 31 55 E	Brockington
Myanzi, rsta.	U4	Mengo	0 25 N 31 55 E	
[Myumi] see Mvumi	T6	Kilosa	6 35 S 37 10 E	Cribb & Grey-Wilson
Mzambaraoni, popl.	P	Pemba	5 02 S 39 44 E	
Mzambaraoni, popl.	P	Pemba	5 05.5 S 39 46 E	
[Mzambaroani] see Mzambaraoni	P	Pemba		
[Mzambe] see Musombe	T7	Iringa	7 18 S 34 30 E	Richards
Mzelezi, f.r.	T6	Ulanga	8 49 S 36 44 E	Cribb et al.
Mzerekera, area	T7	Iringa	7 53 S 36 08 E	
[Mzigwa] see Mkigwa	T4	Tabora	c.5 08 S 33 10 E	

Name	Code	Region	Coordinates	Authority
[Nabau] see Napau				
Nabazega	K2	Turkana	2 25 N 34 57 E	Maitland
Nabelat, spr.	U4	Mengo	Not traced	
Nabere ? = Namabere	K2	Turkana	2 17 N 35 04 E	
Naberera, popl.	U1	Lango	?1 37 N 32 40 E	Mukasa
[Nabieso] see Nabyeso	T2	Masai	4 12 S 36 56 E	Hornby, Trump, Procter
[Nabigiluwa] see Nambigirwa	U1	Lango	1 52 N 32 47.5 E	H.B. Johnston
[Nabilate] see Nabilatuk	U4	Mengo	0 07 N 32 30 E	Freeman
Nabilatuk, popl., t.c.	U1	Karamoja		A.S. Thomas
	U1	Karamoja	2 03 N 34 34 E	Dyson-Hudson, Eggeling, H.D. van Someren
Nabilatuk, R.	U1	Karamoja	c.2 06 N 34 43 E–1 56 N 34 27 E	A.S. Thomas
[Nabingola] see Nabingora	U4	Mubende	0 30 N 31 11 E	
Nabingora, popl., t.c.	U4	Mubende	0 30 N 31 11 E	
[Nabisanke] see Nabusanke	U4	Mengo	0 01 N 32 03 E	
Nabiswa, ln.	U3	Mbale	1 01 N 33 50 E	Norman
Nabiswera, popl., t.c.	U4	Mengo	1 28 N 32 17 E	
[Nabitat] see Nabelat	K2	Turkana	2 17 N 35 04 E	
Nabiwugulu, r.h.	U3	Busoga	0 59 N 33 07 E	Osmaston
Naboa, popl.	U3	Mbale	1 03 N 34 01 E	G.H. Wood
Nabugabo, L.	U4	Masaka	c.0 22 S 31 54 E	Numerous
Nabuganye, fy.	U4	Mengo	0 49 N 33 01.5 E	A.S. Thomas
Nabugulo, f.r.	U4	Mengo	c.0 27 N 32 44.5 E	Dummer
Nabugut, hill	U1	Karamoja	1 37 N 34 36 E	Harker
[Nabukongolo] ? see Nankongolo	U ?3	?Busoga	?0 32 N 33 50 E	A.D. Staff
Nabulongwe Forest (Semliki region, 27 km. W. of Fort Portal)	U2	Toro	prob. c. 0 40 N 30 02 E	Dawkins
Nabusanke, popl., t.c.	U4	Mengo	0 01 N 32 03 E	Langdale-Brown
[Nabuswera] see Nabiswera	U4	Mengo	1 28 N 32 17 E	H.B. Johnston
Nabyeso, popl., t.c., fy.	U1	Lango	1 52 N 32 47.5 E	
[Nachingwa] see Nachingwea	T8	Lindi	10 22.5 S 38 45.5 E	

Nachingwea, distr.	T8	Kilwa/Tunduru/Masasi/Lindi c.8 45 S 38 00 E (considered as parts of the above districts for F.T.E.A.)		
Nachingwea, popl.	T8	Lindi	10 22.5 S 38 45.5 E	Numerous
Nachola, ln.	K1	Northern Frontier	1 47 N 36 43 E	Bally & Carter
Nachor, ln.	K1	Northern Frontier	1 58 N 36 02 E	Mwangangi & Gwynne
[Nadehani] see Madehani	T7	Njombe	9 21 S 34 02 E	Stolz
Nadunget, popl., str.	U1	Karamoja	2 31 N 34 34 E	J. Wilson
Nadungoro, area	T2	Arusha	3 13.5 S 36 41 E	Carmichael
Naeborkaware = Naibor-Kawadie, area	K6	Masai	1 50.5 S 36 35 E	Glover et al.
[Naene Njoget] see Naeningujet	K6	Masai	1 41 S 35 52 E	
Naeningujet, hill	K6	Masai	1 41 S 35 52 E	
Nagalama, popl.	U4	Mengo	0 31 N 32 47 E	
[Naganjwe] see Magangwe	T7	Mbeya	7 46 S 34 13.5 E	Greenway
Nagasseni, mt.	T2	Arusha	3 02 S 36 54 E	Richards
Nagelesa, ln.	K1	Northern Frontier	c.1 56 N 30 01 E	Mwangangi & Gwynne
[Nagoje] see Nagojje	U4	Mengo	c.0 27 N 32 52.5 E	Dummer
Nagojje, est.	U4	Mengo	c.0 27 N 32 52.5 E	Dummer
Nagongera, popl.	U3	Mbale	0 46 N 34 01 E	
[Nagulu] see Ngulu	T4	Tabora	c.5 30 S 33 20 E	Peter
Nagunga, popl.	U4	Mengo	0 13 N 32 50 E	Dummer
Nagungumet, hills	K2	Turkana	4 08 N 35 28 E	Carter & Stannard
Nahomba Valley	T8	Kilwa	8 39.5 S 38.05 E–8 34 S 38 11 E	Vollesen
Nahoro, popl.	T8	Lindi	10 01 S 39 10 E	Milne-Redhead & Taylor
Nai, fm., sw.	K3	Trans-Nzoia	1 08 N 34 52.5 E	Tweedie
Naibardad, hill	T2	Masai	3 07 S 35 06 E	Paulo
Naibor, hill	K6	Masai	2 21 S 36 41 E	Glover & Cooper
[Naibor] see Engare Naibor	T2	Masai	2 24 S 36 24 E	Bally
Naibor-Kawadie = Naeborkaware, area	K6	Masai	1 50.5 S 36 35 E	Glover & Cooper
Naidigidigo = Naidikidiko, escarp.	K6	Masai	2 24 S 36 34 E	Bally
Naigobia, popl.	U3	Busoga	0 49 N 33 21 E	

Name	Region	District	Coordinates	Source
[Naiguru] see Nyaiguru	U2	Kigezi	1 05 S 29 40 E	Dale
Naikara, popl.	K6	Masai	1 36 S 35 38 E	Kuchar
Nainokanoka, popl., r.h.	T2	Masai	3 01 S 35 41 E	Newbould, Eggeling
Nairobi, falls	K4	Kiambu	1 12 S 37 04 E	Gillett
Nairobi, distr.	K4	Nairobi	c.1 18 S 36 55 E	Numerous
Nairobi, popl.	K4	Nairobi	1 17 S 36 49 E	
Nairobi, R.	K4	Kiambu/Nairobi	1 14 S 36 40 E–1 12 S 37 09.5 E	Bogdan
Nairobi, R.	K4	North Nyeri	0 11 S 37 17.5 E–0 24 S 37 01.5 E	Hedberg
Nairobi Golf Course	K4	Nairobi	1 18 S 36 48 E	Bally & Carter
Nairobi National Park	K4	Nairobi	c.1 23 S 36 52 E	Numerous
Naitamaiong, mt.	K2	Turkana	3 16 N 34 54 E	Champion
[Naitamajong] see Naitamaiong	K2	Turkana	3 16 N 34 54 E	Champion
Naivasha, distr.	K3	Naivasha	c.0 40 S 36 25 E	Numerous
Naivasha, L.	K3	Naivasha	c.0 46 S 36 21 E	Numerous
Naivasha, popl., rsta.	K3	Naivasha	0 43 S 36 26 E	Gillman
[Najakato] see Nyakato	T1	Bukoba	1 16 S 31 49 E	Dawkins
Najembe, r.h.	U4	Mengo	0 24 N 33 01 E	Richards
Najuka = Njeku, camp (Mt. Meru)	T2	Arusha	3 15 S 36 47 E	Chandler
Nakabugo ? see Nakivubo Channel	U4	Mengo	?0 19.5 N 32 34.5 E–0 17.5 N 32 39 E	G.H. Wood
Nakalange, hills	U3	Busoga	0 37.5 N 33 17 E	C.G. Rogers, Snowden
[Nakalembe] see Nyakalembe	U2	Kigezi	1 12 S 29 44 E	Crosse-Upcott
Nakalijenge	T8	Kilwa	c.9 28 S 38 06 E	Nicholson
Nakapapa	T8	Kilwa	Not traced	
Nakapiriririt, popl.	U1	Karamoja	1 51 N 34 43 E	J. Wilson
Nakasajja, popl., t.c.	U4	Mengo	0 29 N 32 40.5 E	Lye
Nakasero, hill	U4	Mengo	0 20 N 32 35 E	Rwaburindore
Nakasongola, popl., t.c.	U4	Mengo	1 19 N 32 28 E	Eggeling, Langdale-Brown, Harker
[Nakasozi] ? see Kasozi	U4	Masaka	?0 05 N 31 33 E	Chandler
Nakaswa, popl., mssn.	U4	Mengo	0 36 N 32 51 E	
Nakatimbo, popl.	T6	Ulanga	c.8 03 S 36 40 E	Haerdi

Nakauka, popl.	U4	Mengo	0 11 N 32 28 E	
[Nakavali] see Nakivali	U2	Ankole	c.0 47 S 30 53 E	Kennedy
[Nakawa] see Nakaswa	U4	Mengo	0 36 N 32 51 E	Eggeling
Nakawali, str.	T8	Songea	c.10 42 S 35 10 E	Milne-Redhead & Taylor
[Nakawuka] see Nakauka	U4	Mengo	0 11 N 32 28 E	
Nakibogwe, ln.	U4	Masaka	0 18 S 31 55 E	Lye
Nakifumu, popl.	U4	Mengo	0 32 N 32 47 E	
Nakigalala, for., est.	U4	Mengo	c.0 13 N 32 31 E	A.S. Thomas
Nakigesi, for.	U4	Mubende	c.0 50 N 30 52 E	
Nakilala valley	T8	Kilwa	8 40 S 38 22 E	Vollesen, Rodgers
Nakiloro, popl.	U1	Karamoja	2 38 N 34 44 E	J. Wilson, Lye
Nakiloro, str.	U1	Karamoja	2 27 N 34 44 E–2 22 N 34 40 E	
[Nakipiripirit] see Nakapiririt	U1	Karamoja	1 51 N 34 43 E	J. Wilson
[Nakiriyanyet] see Nakiriyonyet	U1	Karamoja	1 47 N 34 42.5 E–1 49 N 34 28 E	
Nakiriyonyet, str.	U1	Karamoja	1 47 N 34 42.5 E–1 49 N 34 28 E	Eggeling
[Nakitawa] see Nyabitaba	U2	Toro	0 21.5 N 29 59 E	Duke of Abruzzi
Nakitoma, r.h., t.c.	U4	Mengo	1 32 N 32 06 E	
Nakivali, L.	U2	Ankole	c.0 47 S 30 53 E	Kennedy
Nakivubo, chnl.	U4	Mengo	0 19.5 N 32 34.5 E–0 17.5 N 32 39 E	
Nakivubo, popl.	U4	Mengo	0 18.5 N 32 34.5 E	Snowden
Nakiza, for.	U4	Mengo	c.0 08 N 32 59 E	Dawkins
[Nakoroma] see Nyakoromo	T1	Musoma	2 12 S 34 03 E	Greenway
Nakuijit, popl.	K2	Turkana	1 38.5 N 35 08 E	Tweedie
Nakuijit, str.	K2	West Suk	c.1 37 N 35 09 E	
Nakuru, distr.	K3	Nakuru	c.0 15 S 35 55 E	
Nakuru, L.	K3	Nakuru	c.0 21 S 36 05 E	
Nakuru, popl., rsta.	K3	Nakuru	0 17 S 36 04 E	
[Nakwali] see Nakivali	U2	Ankole	c.0 47 S 30 53 E	Engler, Pole Evans & Erens
Nakwapua, str.	U1	Karamoja	c.2 52 N 34 27 E	Numerous
Nakwehe = Kagwalas, str./s.	K2	Turkana	c.3 08 N 35 32 E	

Name	Code	Region	Coordinates	Collector
Nakwero, popl.	U4	Mengo	0 28 N 32 39 E	A.S. Thomas
[Nakwidget] see Nakuijit	K2	Turkana	1 38.5 N 35 08 E	Tweedie
Nala, area	T8	Lindi	10 09 S 39 18 E	Eggeling
[Naleiloro] see Nakiloro	U1	Karamoja	2 27 N 34 44 E–2 22 N 34 40 E	
Nalemoru, area, str.	T2	Moshi	2 58 S 37 28 E	
[Nalwanda] see Maruanda's	T7	Mbeya		Stolz
Namabere, popl.	U1	Lango	1 37 N 32 40 E	Maitland, G.H. Wood
Namaira = Namaiera, hill	U3	Busoga	1 02 N 33 06 E	Richards
Namakambili Rock = Makambale Hill	T3	Tunduru	10 58 S 37 56 E	Fyffe, Norman, Eggeling
Namalala, f.r.	U4	Masaka	0 53 S 31 40 E	?Osmaston
Namalemba, popl.	U3	Busoga	0 55.5 N 33 40 E	?Osmaston, G.H. Wood
Namalemba, popl., sch.	U3	Busoga	0 46 N 33 36 E	Dyson-Hudson
Namalu, R.	U1	Karamoja	1 47.5 N 34 39 E–1 48 N 34 30.5 E	Numerous
Namalu, popl.	U1	Karamoja	1 49 N 34 38 E	
[Namanve] see Namanve	U4	Mengo	c.0 22 N 32 41 E	Trapnell, Trump, Verdcourt
Namanga, str./s.	K6	Masai	2 31 S 36 49 E–2 41 S 37 01 E	Numerous
Namanga, popl., p.p., t.c.	K6/T2	Masai	2 33 S 36 47 E	Anderson, Evans
Namanga, hill	T8	Lindi	10 18 S 38 43 E	Dummer
[Namaniyama] see Namanyama	U4	Mengo	0 28 N 32 56 E	Eggeling, Kato
Namanve, for., sw.	U4	Mengo	c.0 22 N 32 41 E	
Namanve, str.	U4	Mengo	c.0 21 N 32 40–32 41 E	
Namanyama, popl.	U4	Mengo	0 28 N 32 56 E	Dummer, Dawkins
Namanyere, popl., r.h.	T4	Ufipa	7 31 S 31 03 E	Bullock, Procter
Namanyonyi, popl.	U4	Mengo	0 34 N 32 53 E	Dummer
Namanzi ? see Nemazi	T5	Dodoma	?5 42.5 S 35 46.5 E	Ruffo
Namarambi, sch.	K5	North Kavirondo	0 44 N 34 46 E	G.R. Williams
Namarambi, str.	K5	North Kavirondo	c.0 45 N 34 46 E	
Namasagali, popl., t.c., rsta.	U3	Busoga	1 01 N 32 57 E	G.H. Wood, Harker
Namasale, popl.	U1	Lango	1 30 N 32 37 E	For. Dept.
Namasina	T4	Tabora	Not traced	Braun

Name		Region	Coordinates	Source
Namukupa, for.	U4	Mengo	c.0 36 N 32 58 E	
Namulanda, popl.	U4	Mengo	0 08.5 N 32 33 E	Harker
Namulapi (Bundali)	T7	Rungwe	c.9 30 S 33 27 E	Stolz
Namulonge, popl.	U4	Mengo	0 32 N 32 37 E	
[Namunuha] see Nyamunuka	U2	Toro	0 05 S 29 59 E	A.S. Thomas
[Namunyolo] see Namunyoro	U4	Mengo	0 11.5 N 33 17 E	Maitland
Namunyoro, area, mssn.	U4	Mengo	0 11.5 N 33 17 E	Maitland
[Namur] see Nyimur R.	U1	Acholi	c.3 42 N 32 36 E	
Namutumba, popl.	U3	Busoga	0 58.5 N 33 41 E	
[Namwale] see Namwele	T4	Ufipa	7 47 S 31 29.5 E	Bullock
[Namwamba] see Nyamwamba	U2	Toro	0 19 N 29 54 E–0 04 N 30 09 E	G. Taylor
[Namwambe] see Nyamwamba	U2	Toro	0 19 N 29 54 E–0 04 N 30 09 E	G. Taylor
Namwele, popl., mine, sw., escarp.	T4	Ufipa	7 47 S 31 29.5 E	Numerous
Namwendwa, mssn.	U3	Busoga	0 55.5 N 33 16 E	
Namwendwa, rsta.	U3	Busoga	0 55 N 33 20 E	
Namwiwa, popl., r.h., t.c.	U3	Busoga	1 06 N 33 33 E	G.H. Wood
Namyekudo, hill	U2	Bunyoro	1 32 N 31 53 E	
Nandembo, popl.	T8	Kilwa	8 29.5 S 38 53 E	Grass
Nandembo, ln.	T8	Lindi	10 13 S 39 13 E	Milne-Redhead & Taylor
Nandi, distr.	K3	Nandi	c.0 20 N 35 15 E	
Nandi Escarpment	K3/5	Nandi/N. Kavirondo	c.0 10–0 35 N 34 53–34 58 E	Numerous
Nandi Forest, S., for.	K3	Nandi	c.0 05 N 35 00 E	Battiscombe
Nandi Forest, N., for.	K3	Nandi	c.0 20 N 34 59 E	Battiscombe
Nandi Hills, popl., p.p.	K3	Nandi	0 06 N 35 11 E	
Nandida (Longido Mt.)	T2	Masai	c.2 41 S 36 42 E	Carmichael
Nanga, popl.	K5	Kisumu-Londiani	0 08.5 S 34 44 E	A. Turner
[Nanga] see Manda	K7	Lamu	c.2 17 S 40 57 E	
Nanga, hill	T2	Moshi	3 19 S 37 26.5 E	
Nanga, popl.	T2	Moshi	3 23 S 37 25.5 E	Haarer
Nanga, str.	T2	Moshi	c.3 22 S 37 25 E	Bigger

Name		Location	Coordinates	Collector
Nangafana (Donde area)	T8	Kilwa	prob. c.9 45 S 37 50 E	Busse
[Nangangwe] see Magangwe	T7	Mbeya	7 46 S 34 13.5 E	Greenway
Nanganji = Nyanganje	T6	Ulanga	c.8 00 S 36 42 E	Haerdi
Nangaru, popl.	T8	Lindi	9 51 S 39 28 E	Busse
[Nangeya] see Langia	U1	Acholi/Karamoja	c.3 39 N 33 34 E	Carter & Stannard
Nangolemaret, str./s.	K2	Turkana	4 41 N 35 27 E–4 34 N 35 24 E	Haerdi
[Nangonji] see Nanganji	T6	Ulanga	c.8 00 S 36 42 E	Eggeling
Nangua, popl.	T8	Masasi	10 54.5 S 38 15 E	Milne-Redhead & Taylor
Nangurukuru, hill	T8	Songea	10 34 S 35 50 E	Rose Innes & Magogo
Nangurukuru, popl.	T8	Kilwa	8 48 S 39 21 E	Gillman
Nanguruwe, popl.	T8	Mikindani	10 29 S 40 03 E	Greenway, Carmichael
Nangwa, popl.	T2	Mbulu	4 27.5 S 35 26 E	A.S. Thomas
Nangya, popl.	U3	Mbale	0 56.5 N 34 17 E	
Nankongolo, popl.,for.	U3	Busoga	0 32 N 33 50 E	Dawkins
Nansagazi, popl.	U4	Mengo	0 07 N 32 59 E	Katende
Nansana, popl.	U4	Mengo	0 21 N 32 32 E	
Nanya	T8	Lindi	c.10 01 S 39 04 E	Busse
[Nanyuki, distr.] see North Nyeri	K4	North Nyeri	c.0 00 37 15 E	
Nanyuki, popl., mssn.	K4	North Nyeri	0 01 N 37 04 E	Numerous
Napak, mts.	U1	Karamoja	2 05 N 34 18 E	Numerous
Napau, pass, hills	K2	Turkana	2 25 N 34 57 E	Numerous
Napianyenya, popl.	U1	Karamoja	1 52 N 34 35 E	A.S. Thomas
Napienyenya = Napianyenya, popl.	U1	Karamoja	1 52 N 34 35 E	Eggeling
Napoleon Gulf	U3/4	Busoga/Mengo	0 25 N 33 15 E	Greenway
Napolimoru = Napotimoru	K2	Turkana	2 21 N 35 15 E	
Napono, for.	U1	Acholi/Karamoja	2 54 N 33 35 E	J. Wilson
[Napore] see Kaleri	U1	Acholi/Karamoja	3 36 N 33 36 E	
Napotimoru, popl.	K2	Turkana	2 21 N 35 15 E	Eggeling
[Napyenenya] see Napianyenya	U1	Karamoja	1 52 N 34 35 E	
[Napyenya] see Napianyenya	U1	Karamoja	1 52 N 34 35 E	

Naramum			4 44 N 35 10 E	Martin
Naramuru, popl.	K6	Masai	2 45 S 37 44 E	
Narangani, Mt. Kulal	K1	Northern Frontier	c.2 43 N 36 55 E	Oteke
Narasha, L.	K3	Ravine	0 02.5 N 35 32 E	Drummond & Hemsley, Tweedie, Richardson
Narasho Mbuga, ln.	T2	Masai	4 06 S 36 45 E	Procter
[Naremuro] see Narumuru				
Nariam, popl.	K6/T2	Masai/Moshi	3 03 S 37 25 E –2 55 S 37 35 E	Carmichael
[Nariam] see Old Nariam	U3	Teso	1 53 N 34 08.5 E	Eggeling
Narimuru = Nalemoru	U3	Teso	1 57.5 N 34 06 E	Eggeling
Narion prob. = Nariam	T2	Moshi	2 58 S 37 28 E	Harker
Narok, distr.	U3	Teso	prob. 1 53 N 34 08.5 E	
	K5	Masai	c.1 15 S 35 30 E	
(considered as part of Masai Distr. for F.T.E.A.)				
Narok, popl., t.c.	K6	Masai	1 05 S 35 52 E	Numerous
Narok = Ngare Narok, R.	K6	Masai	0 42.5 S 35 44 E –1 12.5 S 35 53 E	
Narok, mt.	T2	Masai	2 34 S 36 26 E	Carmichael
Narok = Engare Narok, str.	T2	Arusha	c.3 15 S 36 45 E –3 25 S 36 40 E	
Naromoru, popl.	K4	North Nyeri	0 10 S 37 01 E	Numerous
Naromoru, R.	K4	North Nyeri	0 11 S 37 05 E –0 03 S 36 54.5 E	Coe, Verdcourt, Schelpe
Naromoru, N., R.	K4	North Nyeri	0 10 S 37 18.5 E –0 11 S 37 05 E	
Naromoru, S., R.	K4	North Nyeri	0 11 S 37 05 E –0 03 S 36 54.5 E	
Naro Moru Track	K4	North Nyeri	0 10 S 37 13.5 E –37 18.5 E	Townsend
Narosura, hill	K6	Masai	1 30 S 35 49 E	
Narosura, hill	K6	Masai	1 33 S 35 48 E	
Narosura, r.h., t.c.	K6	Masai	1 32 S 35 52 E	Glover et al.
Narosura, str.	K6	Masai	c.1 33 S 35 53 E	
Narosura River Bridge	K3	Ravine	0 01.5 N 35 47 E	Tweedie
Narozari, popl.	U4	Masaka	0 26 S 31 46 E	
Narumuru, str.	K6/T2	Masai/Moshi	3 03 S 37 25 E –2 55 S 37 35 E	
Narungombe, str.	T8	Kilwa	9 03 S 38 33 E –8 57.5 S 38 32 E	Lye & Katende

Naruzinna, sw.	U4	Masaka	0 57 S 31 42 E	Lye
[Nasa] see Nyashimo	T1	Mwanza	2 24 S 33 33 E	Rounce, Tanner
Nasampolai, area, str.	K6	Masai	c.0 48.5 S 36 07 E	Greenway
[Nasampulai] see Nasampolai	K6	Masai	c.0 48.5 S 36 07 E	Greenway
Nasarawinding	T2	Masai	c.2 36 S 36 06 E	Carmichael
[Nashorn Hügel] see Kifaru	T3	Pare	3 31 S 37 35 E	Volkens
[Nasilo] see Nasula	T2	Arusha	3 13.5 S 36 48.5 E–3 13.5 S 36 51 E	Richards
Nasokol, mssn., popl.	K2	West Suk	1 16 N 35 05 E	Tweedie
[Nasolo] see Nasula	T2	Arusha	3 13.5 S 36 48.5 E–3 13.5 S 36 51 E	Greenway
Nassa, hills	T1	Mwanza	2 25 S 33 40 E	
Nassa, t.a.	T1	Mwanza	c.2 30 S 33 40 E	Tanner
Nasula, str.	T2	Arusha	3 13.5 S 36 48.5 E–3 13.5 S 36 51 E	Greenway, Richards
[Nasumpulai] see Nasampolai	K6	Masai	c.0 48.5 S 36 07 E	
Nata, popl.	T1	Musoma	2 00 S 34 24 E	Carmichael
Nataparin, popl.	K2	Turkana	2 44.5 N 34 51 E	Champion
Natera Hills, mts.	U1	Karamoja	c.3 47 N 33 53 E	
Natira, R.	U1/K2	Karamoja/Turkana	c.3 47.5 N 34 13 E–3 55 N 34 40 E	Dale
[Natirai Pass] see Natira Pass	K2	Turkana	c.3 40 N 35 04 E	
Natira Pass	K2	Turkana	c.3 40 N 35 04 E	
Natonko, for.	U4	Mengo	c.0 14 N 32 52 E	
Natron, L.	K6/T2	Masai	c.2 25 S 36 00 E	Numerous
Nattle's, fm., dam	K3	Trans-Nzoia	1 10.5 N 34 47 E	Tweedie
Natural Bridge Falls	T7	Rungwe	9 15.5 S 33 32 E	Batty
Nausane, sw. ? see Nansana	U?4	?Mengo	?0 21 N 32 32 E	Liebenberg
[Navugabo] see Nabugabo	U4	Masaka	c.0 22 S 31 54 E	Norman
Nawaikoke, popl., mssn.	U3	Busoga	1 06 N 33 25 E	G.H. Wood, Langdale-Brown
Nawandala, popl.	U3	Busoga	0 50 N 33 24 E	Maitland
Nawanje Thicket	T6	Rufiji	c.8 00 S 38 50 E	Ludanga
Nawendwa, prob. = Namwendwa	U3	Busoga	prob.0 55 N 33 16 E	Maitland
Nawenge, f.r.	T6	Ulanga	8 37 S 36 42 E	Cribb & Grey-Wilson

Name	Region	District	Coordinates	Collector
[Nazi Moio] see Mnazi Mmoja				
Nazi Moja (Lamu I.)	Z	Zanzibar	5 56 N 39 17 E	Kirk
Nchoroko Kangaga (Mafia I.)	K7	Lamu	2 16 S 40 54 E	Greenway & Rawlins
Ndagoni, popl.	T6	Rufiji	Not traced	Braun
Ndabaka Gate, ln.	T6	Rufiji	7 52.5 S 39 45.5 E	Greenway
Ndabaka Plains	T1	Musoma	2 10 S 33 53 E	Greenway
[Ndabasch] see Endabash	T1	Musoma	c.2 07 S 33 55 E	Richards, Greenway
[Ndabash] see Endabash	T2	Mbulu	3 44 S 35 36 E—3 33.5 S 35 47 E	Greenway
Ndabibi, area	T2	Mbulu	3 34.5 S 35 37 E	Greenway
Ndachi, popl.	K3	Naivasha	c.0 43 S 36 16 E	Fischer
Ndaga, popl.	T5	Dodoma	6 21 S 36 07 E	Savile
Ndagaa, popl., area	T7	Rungwe	9 04 S 33 35 E	Cribb & Grey-Wilson
Ndagalu, f.r.	Z	Zanzibar	6 03 S 39 18 E	Faulkner
Ndagara (near Mbingu)	T1	Maswa	c.2 40 S 33 36 E	Carmichael
Ndagoni, popl., area	T6	Ulanga	near 8 12 S 36 15 E	Carmichael
[Ndaha] see Ndoha	P	Pemba	5 13 S 39 43 E	Vaughan
Ndaiga, popl.	T1	Maswa	c.2 30 S 34 25 E	Greenway
Ndala, str.	U4	Mubende	1 01.5 N 30 35 E	Eggeling
Ndala, mssn., t.c.	T2	Mbulu	c.3 25 S 35 41 E—3 29.5 S 35 47.5 E	Richards, Greenway & Kanuri
Ndaleta (Kitumbeine)	T4	Nzega	4 45 S 33 16 E	van Rensburg
Ndama (near Mabira)	T2	Masai	prob. c.2 57 S 36 15 E	Carmichael
Ndanda, mssn.	T1	Bukoba	near 1 14 S 30 57 E	Haarer
	T8	Masasi	10 30 S 39 01 E	?Gerstner, Gillman, McCusker*, Mwasumbi
		(*not Lindi District as cited in 'Combretaceae')		
Ndanda, popl.	T8	Lindi	10 13 S 38 50 E	?Gerstner
[Ndandamaire] see Ndandamire	U2	Bunyoro	2 11 N 31 23 E	
Ndandamire, popl., r.h.	U2	Bunyoro	2 11 N 31 23 E	
Ndandawala, popl.	T8	Kilwa	9 42 S 39 06 E	Gillman
Ndani, popl.	T4	Mpanda	5 59 S 31 36 E	Boehm
Ndapo = Mdapo	T7	Njombe	?c.9 17 S 34 06 E	Semsei
Ndara, hill	K7	Teita	3 31 S 38 36 E	Hildebrandt

289

Name	Code	Region	Coordinates	Collectors
Ndara, rsta.	K7	Teita	3 29 S 38 41 E	
Ndara, popl.	T7	Rungwe	9 09 S 33 46 E	R.M. Davies, Greenway, Cribb & Grey-Wilson
Ndaragwa, hill, popl.	K3	Laikipia	0 04 S 36 31.5 E	Hooper & Townsend
Ndare, for.	K4	North Nyeri	0 10 N 37 21 E	
Ndareda = Dareda, popl., mssn.	T2	Mbulu	4 13 S 35 33 E	Peter, Rowe, Carmichael
Ndarugu, popl.	K4	Kiambu	1 05.5 S 36 32 E	
Ndarugu, R.	K4	Kiambu	c.0 51 S 36 37.5 E—1 07 S 37 10.5 E	C. van Someren
Ndassekera, mt.	T2	Masai	1 58 S 35 42 E	Jaeger
Ndavaya, popl., mssn.	K7	Kwale	4 15 S 39 10 E	J.A. Allan
Ndege, beach	T6	Uzaramo	c.6 38 S 39 12 E	Bjornstad
Ndegere (Utschungwe Mts.)	T7	Iringa	Not traced	Goetze
[Ndeisha] see Ndeizha				
Ndeiya, mkt., grazing scheme	U2	Ankole	0 43.5 S 30 21 E	Snowden
Ndeizha, popl.	K4	Kiambu	1 15 S 36 36 E	Verdcourt & Hemming
Ndeke, popl.	U2	Ankole	0 43.5 S 30 21 E	Eggeling, A.S. Thomas
	U2	Ankole	0 17 S 30 06 E	Eggeling
	T3	Lushoto	5 05 S 38 39 E	
[Ndelema] see Derema				
Ndembera, R.	T7	Mbeya/Iringa	c.8 05 S 35 33 E—8 12 S 34 48 E	?Richards
Ndenga, popl.	T8	Songea	11 03 S 34 54 E	Milne-Redhead & Taylor, Semsei
[Ndengo] see Ndenga				
Ndengu = Ndenga, popl.	T8	Songea	11 03 S 34 54 E	Semsei
	T8	Songea	11 03 S 34 54 E	
[Nderema] see Derema	T3	Lushoto	5 05 S 38 39 E	Drummond & Hemsley, Greenway, Holst, Heinsen, Volkens
Ndgaa = Ndagaa, popl., area	Z	Zanzibar	6 03 S 39 18 E	Faulkner, R.O. Williams
Ndi, mt.	K7	Teita	c.3 18 S 38 28 E	Dale, Gregory, Hildebrandt
Ndi, popl., rsta.	K7	Teita	3 14 S 38 30 E	Hildebrandt
Ndiandasa = Ndiandaza, area	K4	Kitui	c.2 26 S 38 40 E	
Ndigri (Ndoto Mts.) ? = Ndikir	K1	Northern Frontier	?1 38.5 N 37 10 E	Newbould
Ndikir, area	K1	Northern Frontier	1 38.5 N 37 10 E	
[Ndindini] see Dindini	T5	Singida	5 12 S 34 09 E	B.D. Burtt

Name	Region	Area	Coordinates	Collector
Ndiso, hill	T4	Mpanda	6 38 S 31 09 E	Prizov
[Nditima] see Ditima				
Ndiuka, popl.	T7	Njombe	c.9 25 S 35 17 E	Schlieben
Ndiwa Farm	T5	Dodoma	6 13.5 S 35 46 E	
Ndjofu, mt.	K3	Trans-Nzoia	0 59 N 34 57.5 E	Tweedie
[Ndo] see Endo	T6	Morogoro	7 08 S 37 33 E	Goetze
Ndoha Plains	K3	Elgeyo	c.1 16 N 35 39 E	Tweedie
Ndoinet, str.	T1	Maswa	c.2 30 S 34 25 E	Greenway
Ndola, hill	K3/5	Nakuru/Kericho	c.0 21 S 35 44 E–0 26 S 35 31 E	Gillett
Ndole, popl.	T3	Lushoto/Tanga	5 08 S 38 43 E	Verdcourt & Greenway, Faulkner
Ndolezi, est.	T6	Morogoro	6 09 S 37 23.5 E	Schlieben
Ndololo, ln.	T7	Mbeya	c.9 02 S 32 56 E	Richards
Ndolwa, popl.	K7	Teita	3 22 S 38 38 E	Bally, Greenway
Ndolwa, f.r.	U3	Busoga	1 12 N 33 04 E	
Ndoro, area	T3	Lushoto	5 03.5 S 38 14.5 E	Shabani
Ndoroto, fm.	K4	North Nyeri	c.0 13 S 37 07–37 10 E	von Hoehnel
Ndorwa, co.	K3	Nakuru	c.0 07 S 36 14 E	Bally
Ndoto, mts.	U2	Kigezi	c.1 20 S 30 00 E	
Ndugani Kopjes, hills	K1	Northern Frontier	c.1 45 N 37 08 E	Jex-Blake, Kerfoot, Newbould
Ndui Rock Dam	T1	Maswa	3 15 S 34 50 E	Greenway
Nduimet, est., ln.	K4	Kitui	1 46 S 38 16 E	Kimani
Ndula, est.	T2	Moshi	2 55 S 37 06 E	
Ndumbi Forest Reserve = Ndumbi Valley f.r.	K4	Kiambu	1 02 S 37 15 E	Gardner
	T7	Njombe	c.9 02 S 34 02 E	Cribb & Grey-Wilson, Paulo, Procter, Semsei
Ndumbi, R.	T7	Njombe	c.9 04 S 33 52 E–8 55 S 34 07 E	Eggeling, Goetze, Richards
Ndumbi Valley, f.r.	T7	Njombe	c.9 02 S 34 02 E	Paulo, Procter, Semsei
Ndume, popl., beach	T3	Tanga	5 07 S 39 07 E	Faulkner
Ndumi = Ndume, popl., beach	T3	Tanga	5 07 S 39 07 E	
Ndumi, popl.	T6	Bagamoyo	6 03 S 38 43 E	Hannington
[Ndunga] see Ndungu	T3	Pare	4 22 S 38 03 E	Greenway

291

292

Name		District	Coordinates	Collector
[New Langenburg] see Tukuyu	T7	Rungwe	9 15 S 33 38.5 E	
Newman's Camp, Uaso Nyiro	K1	Northern Frontier	0 34.5 N 37 33 E	J. Adamson
New Mill = Suam Mill (1949—)	K3	Trans-Nzoia	1 10 N 34 44 E	Tweedie
[New Moshi] see Moshi	T2	Moshi	3 21 S 37 20 E	
New Picnic Point	K2	West Suk	1 16 N 35 11.5 E	Tweedie
[New Shinyanga] see Shinyanga	T1	Shinyanga	3 40 S 33 26 E	
Ngabalo (Kitumbeine)	T2	Masai	c.2 53 S 36 13 E	Carmichael
Ngabobo, area	T2	Arusha	3 08.5 S 36 55 E	Greenway
Ngabora, str.	T2	Mbulu	c.3 30 S 35 25 E	Peter
[Ngabua] see Nkabwa	U2/Zaire	Kigezi	0 54 S 29 35 E	
Ngagu, popl.	P	Pemba	5 11.5 S 39 39 E	
[Ngaheo] see Ngoheo	T1	Musoma	2 11 S 34 15 E	
Ngaia, for.	K4	Meru	0 22 N 38 02 E	Greenway
Ngaimaronya (Longido Mt.)	T2	Masai	c.2 41 S 36 42 E	Dyson, Wachiori
Ngaka, R.	T8	Songea	c.10 58 S 34 54 E–10 18 S 34 59 E	Carmichael
[Ngalai] see Ngolai	T2	Masai	c.3 15 S 35 25 E–3 21 S 35 19 E	
Ngalaliko, gorge	T2	Arusha	prob. c.3 14 S 36 50 E	Peter
Ngamba, popl.	U2	Toro	0 46.5 N 30 09 E	Richards, Arasululu
[Ngamba] see Engamba	K4	Kitui	1 05.5 S 38 41 E	A.S. Thomas
[Ngamba] see Mgambo	T6	Morogoro	7 05 S 37 43 E	L.C. Edwards
Ngambo, hill	T3	Lushoto	5 08 S 38 25 E	Stuhlmann
Ngambo, Mts.	T6	Morogoro	prob. c.6 10 S 37 32 E	Greenway
Ngambo, peak, pl.	T3	Lushoto	5 01.5 S 38 36 E	Holtz
Ngangao = Ngaongao, for.	K7	Teita	c.3 22 S 38 23 E	Holst, Peter, Scheffler
Ngangata, R.	T8	Songea	c.10 15 S 36 24 E–10 06 S 36 11 E	Drummond & Hemsley, Bally, Gardner, Dale
Ngao, area	K7	Tana River	c.2 24 S 40 12 E	F. Thomas, Sampson
Ngao, mssn.	K7	Tana River	2 24.5 S 40 12 E	Hooper & Townsend
[Ngapikani] see Inkipikoni	K6	Masai	c.1 21 S 36 38 E	
Ngara, distr.	T1	Ngara	c.2 40 S 30 45 E	Bally

Name	Region	Code	Coordinates	Collector
Ngara, popl., r.h., t.c.	Ngara	T1	2 29 S 30 41 E	Dobson, Tanner
Ngara Road Bridge	Nairobi	K4	1 17 S 36 49 E	Kirrika
Ngarani, R. = Kingarane	Masai	T2	c.2 06 S 35 37 E	Carmichael
[Ngaraya] see Garaya	Lushoto	T3	c.5 00–5 02 S 38 26 E	Peter
Ngare Mesheni, R.	Masai	T2	c.2 14 S 35 42.5 E	Carmichael
Ngare Nairobi = Engare Nairobi	Moshi	T2	3 03 S 37 00 E	Fuggles Couchman, Greenway, Sanders, Fischer, Staples
Ngare Nanyuki = Engare Nanyuki, airstrip, spr.	Masai	T2	2 37 S 35 13 E	Greenway & Turner
Ngare Nanyuki = Engare Nanyuki, R.	Arusha/Masai	T2	3 15 S 36 48 E – 2 48 S 36 55 E	Richards
Ngare Nanyuki, popl.	Arusha	T2	3 09 S 36 51.5 E	Peter, Richards
Ngare Narok = Engare Narok, str.	Northern Frontier	K1	1 13 N 36 33 E–0 52 N 36 47 E	Bally, Newbould
Ngare Narok, R.	Masai	K6	0 42.5 S 35 44 E–1 12.5 S 35 53 E	Bally, Gardner, Davidson, J. Bally
Ngare Ndare, R.	N. Frontier/N. Nyeri	K1/4	0 01 N 37 23 E–0 34.5 N 37 23.5 E	
Ngare Ndare Ranch	North Nyeri	K4	0 07 N 37 25 E	
Ngare Njuki, ln.	Masai	T2	c.2 53 S 36 53 E	Fischer
[Ngare Nyiro] see Uaso Nyiro	Masai	K6	c.0 39 S 35 44 E–2 04 S 36 07 E	Bally
[Ngare Olmotoni] see Engare Olmotoni	Arusha	T2		Haarer
[Ngare Suguta] see Ngare Sugutu	Moshi	T2	c.3 19 S 37 07 E	Fischer
Ngare Sugutu, ln.	Moshi	T2	c.3 19 S 37 07 E	Fischer
Ngarika (near Kajiado)	Masai	K6	near 1 51 S 36 47 E	Vesey-FitzGerald
Ngarinaro, R. ? see Ngare Ndare	N. Frontier/N. Nyeri	K4	?0 01 N 37 23 E–0 34.5 N 37 23.5 E	Schelpe
[Ngarisera Furrow] see Ngaserai	Masai	T2	2 48 S 36 54 E	Greenway
Ngarsett (Mt. Kulal)	Northern Frontier	K1	c.2 43 N 36 55 E	J. Adamson
[Ngaruka] see Engaruka	Masai	T2	2 59 S 35 57 E	Bally
Ngarwa, R.	Masai	T2	c.2 06 S 35 37 E	Carmichael
[Ngasama] see Ngasamo	Mwanza	T1	2 29 S 33 53 E	Tanner
Ngasamo, hill, mine	Mwanza	T1	2 29 S 33 53 E	
Ngasasi Lahars, rocks	Arusha	T2	prob. c.3 08 S 36 43 E	Richards
Ngaserai, area, furrow	Masai	T2	2 48 S 36 54 E	Richards, Leippert

Name	Code	Region	Coordinates	Collector
Ngaserai, hill	T2	Masai	2 53.5 S 36 54 E	Mbano & Willy
[Ngasserai] see Ngaserai	T2	Masai	2 53.5 S 36 54 E	Bally, Greenway
Ngatana, popl.	K7	Tana River	c.2 15 S 40 08 E	J.W. Gregory
[Ngau] see Ngao	K7	Tana River	2 24.5 S 40 12 E	
Ngaza (Mafia I.)	T6	Rufiji	c.7 52.5 S 39 45.5 E	Greenway
Ngebai Langata Endul	K6	Masai	c.1 15 S 35 03 E	Glover et al.
Ngejepa, ridge	K6	Masai	1 46 S 36 18.5 E	
Ngeleka, ln.	T1	Mwanza	2 39 S 33 00.5 E	Tanner
Ngelema, popl., area	P	Pemba	5 17.5 S 39 43 E	Vaughan
Ngelesha, for.	K3	Laikipia	c.0 28 N 36 16 E	Napier
Ngelesha N., hill	K3	Laikipia	0 29 N 36 15.5 E	
Ngelesha, S., hill	K3	Laikipia	0 27 N 36 15 E	
[Ngeleshwa] see Ngelesha	K3	Laikipia	c.0 28 N 36 16 E	
Ngeng, str./s.	K1	Northern Frontier	c.1 18 N 37 16 E–1 09 N 37 07 E	J. Bally, Newbould
Ngenge, r.h.	U3	Mbale	1 30 N 34 30 E	Langdale-Brown
Ngengo (Matengo Highlands)	T8	Songea	Not traced	Zerny
[Ngeregere] see Migeregere	T8	Kilwa	8 48 S 39 13 E	Busse
Ngerende, hill	K6	Masai	1 07 S 35 10 E	Glover et al.
[Ngerendei] see Ngerende	K6	Masai	1 07 S 35 10 E	Glover et al.
Ngerenge	T7	?Mbeya	Not traced	Michelmore
Ngerengere, popl., rsta., mssn., r.h.	T6	Morogoro	6 44.5 S 38 07 E	B.D. Burtt, Peter, Welch
Ngerengere, R.	T6	Morogoro/Uzaramo	6 52 37 36.5 E–7 03 S 38 31 E	Bally, Drummond & Hemsley, W.H. Lewis
Ngerenyi, popl., r.h.	K7	Teita	3 26 S 38 20 E	Busse
[Ngerere] see Migerere	T8	Kilwa	8 48 S 39 13 E	
[Ngero] see Ngoro	U2	Ankole	c.0 10 S 30 12 E–0 14 S 30 15 E	Ball & Hamilton
Ngeta, popl., mssn.	U1	Lango	2 18 N 32 55 E	Hill
Ngeta, hill	U1	Lango	2 17 N 32 56 E	
[Ngeta] see Mgeta, R.	T6	Morogoro	7 08 S 37 38 E–7 17 S 38 06 E	Greenway
[Ngetta] see Ngeta	U1	Lango	2 18 N 32 55 E	Eggeling
Ngezi, popl.	P	Pemba	5 20 S 39 44 E	Burtt Davy

295

Ngezi, for.	P	Pemba	c.4 56 S 39 42 E	Vaughan, Greenway, Burtt Davy
[Ngezi Mwitu] see Ngezi Forest	P	Pemba	c.4 56 S 39 42 E	Lyne
[Nghwema] see Nghweme	T6	Morogoro		Stuhlmann
Nghweme, mt.	T6	Morogoro	7 07 S 37 43 E	Stuhlmann
Nghweme, popl.	T6	Morogoro	7 06 S 37 43.5 E	Stuhlmann
[Nghwenn] see Nghweme	T6	Morogoro		Stuhlmann
Ngiga (near Kadimu)	K5	Central Kavirondo	c.0 04 S 34 09 E	Padwa
Ngiminito (Pian County)	U1	Karamoja	c.2 00 N 34 30 E	Harker
[Ngiro] see Nyiru	K1	Northern Frontier	c.2 10 N 36 50 E	
[Ngito] see Engare Engito	K6	Masai	0 55 S 35 35 E–0 56.5 S 35 25.5 E	Glover et al.
[Nglewenu] see Nghweme	T6	Morogoro		Stuhlmann
[Ngoa] see Ngowa	K7	Kwale	4 38 S 39 14 E	
Ngobit, p.p.	K3	Laikipia	0 03 S 36 47 E	R.M. Graham
Ngogwe, popl.	U4	Mengo	0 14 N 32 59 E	Bally, Gardner
Ngoheo, hill	T1	Musoma	2 11 S 34 15 E	Harker
Ngohingo, popl.	T6	Uzaramo	6 47 38 50 E	Greenway
Ngoina, tea est.	K5	Kericho	0 31 S 35 03 E	Perdue & Kibuwa
Ngoitokitok = Leitokitok, spr., ln.	T2	Masai	3 12 S 35 37 E	Greenway & Kanuri
Ngolai, str.	T2	Masai	c.3 15 S 35 25 E–3 21 S 35 19 E	Peter
Ngole ? see Gongolo	T3	Tanga	5 12 S 39 01 E–5 13 S 39 03.5 E	
Ngololo, area	T7	Iringa	c.8 42 S 35 17 E	Greenway
Ngom, str. (near Selengai)	K6	Masai	near 2 11 S 37 10 E	Schlieben
Ngoma, r.h.	U4	Mengo	1 11 N 32 01 E	Langdale-Brown
[Ngomba] see Gomba	T3	Lushoto	5 00 S 38 18 E	Zimmermann
Ngomba, popl.	T7	Chunya	8 23 S 32 53 E	Richards
Ngombani, popl.	K7	Kwale	4 23.5 S 39 07 E	Greenway
Ngombeni, str., popl.	K7	Kilifi	c.3 43 S 39 44 E	Bally & Smith
Ngombeni, est., popl.	T6	Rufiji	7 56.5 S 39 39 E	Schlieben, Batty
Ngombezi, popl., rsta.	T3	Lushoto	5 09.5 S 38 25 E	Peter
Ngomeni, popl., hill	K4	Kitui	0 39 S 38 24 E	Lind & Agnew, Bally

Ngomeni, w.h.	K4	Machakos	2 55 S 38 22 E	Scott Elliot
Ngomeni, popl.	K7	Kilifi	2 59.5 S 40 11 E	Rawlins, MacNaughton
Ngomeni, popl., rsta., area, old pl.	T3	Tanga	5 09 S 38 54 E	Numerous
Ngomero, mt.	T4	Kahama	4 00 S 31 58 E	Bullock
Ngomero, area	T6	Morogoro	c.7 27 S 37 35 E	Grant
Ngondole, hill	T6	Ulanga	8 01 S 36 46.5 E	Haerdi
Ngondole	T7	Rungwe	near 9 17 S 33 39 E	Stolz
Ngong, hills	K6	Masai	c.1 25 S 36 38 E	Numerous
[Ngongare] see Ngongongare	T2	Arusha	c.3 18 S 36 54 E	Greenway
Ngong Boma, popl.	K6	Masai	c.1 22 S 36 39 E	Numerous
Ngongera, ridge	U3	Mbale	c.1 04 N 34 19 E	
Ngongo, hill	T6	Ulanga	c.8 57 S 36 39 E	Schlieben, Cribb & Grey-Wilson
Ngongo, str., popl.	T8	Lindi	c.10 04 S 39 37 E	Gillman
Ngongongare, farm	T2	Arusha	3 17 S 36 52 E	Richards
Ngongongare, t.a.	T2	Arusha	c.3 18 S 36 54 E	Eggeling
Ngongongare Spring, ln.	T2	Arusha	3 17 S 36 54 E	Richards
[Ngongora] see Ngongera	U3	Mbale	c.1 04 N 34 19 E	Dale
[Ngongoro] see Ngongera	U3	Mbale	c.1 04 N 34 19 E	
Ngong Road Forest, for.	K4	Kiambu	c.1 19 S 36 45 E	Bally, Graham, Napier
[Ngongwa] see Nongwa	T7	Njombe	c.9 32 S 35 02 E – 9 42 S 35 01 E	
Ngonho, R.	T1	Shinyanga	c.3 33 S 33 24 E	B.D. Burtt
Ngonja = Gonja, old area	T3	Lushoto	c.4 59 S 38 45 E	Peter
Ngono, R.	T1	Bukoba	c.1 58 S 31 37 E – 1 08 S 31 35 E	Gillman, Stuhlmann, Haarer
Ngono Bridge	T1	Bukoba	1 21 S 31 28 E	Procter
Ngonzini, ln.	K7	Kwale	4 08.5 S 39 24 E	Greenway
Ngora, co.	U3	Teso	c.1 30 N 33 45 E	Philip
Ngora, popl.	U3	Teso	1 27 N 33 46 E	Fiennes
Ngoragaishi, hills	K6	Masai	1 57 S 36 57 E	
Ngore Ngore, hill	K6	Masai	1 01.5 S 35 30.5 E	
Ngore Ngore, storehouse	K6	Masai	1 02 S 35 30 E	Verdcourt, Archer

297

Ngori, prob. = Ngozi	T7	Rungwe/Mbeya	8 59.5 S 33 33 E	Geilinger
[Ngoringaishi] see Ngoragaishi	K6	Masai	1 57 S 36 57 E	Birch & Trapnell
Ngorinit = Ngurunit, w.h.	K1	Northern Frontier	1 45 N 37 18 E	
Ngoro, R.	U2	Ankole	c.0 10 S 30 12 E – 0 14 S 30 15 E	Ball & Hamilton, Synnott
[Ngoro Kalende] see Momela L.				
Ngorongari = Ngongongare, t.a.	T2	Arusha	c.3 13 S 36 54.5 E	Uhlig
Ngorongoro, conservation area	T2	Arusha	c.3 18 S 36 54 E	Procter
Ngorongoro, crater	T2	Masai/Mbulu	c.3 00 S 35 30 E	
Ngorongoro, popl., r.c.	T2	Masai	3 10 S 35 35 E	Numerous
[Ngoronit] see Ngorinit	T2	Masai	3 15 S 35 31 E	Numerous
Ngorono (near Naabi hill)	K1	Northern Frontier	1 45 N 37 18 E	J. Adamson
Ngorotwa, Kalambo R. f.r.	T1/2	Maswa/Masai	near 2 53 S 35 00 E	Newbould & Thesiger
Ngorsett = Ngarsett, Mt. Kulal	T4	Ufipa	c.8 32 S 31 13 E	Carmichael
Ngosi = Ngozi, crater L.	K1	Northern Frontier	c.2 43 N 36 55 E	J. Adamson
Ngosi = Ngozi, mt.	T7	Rungwe	9 00.5 S 33 33 E	B.D. Burtt
Ngosingosi, popl.	T7	Mbeya/Rungwe	8 59.5 S 33 33 E	Geilinger
Ngotsche (near Ngerengere)	T7	Njombe	9 05 S 34 37 E	Goetze
Ngowa, I, creek	T6	Morogoro	Not traced	Peter
Ngowasi, L.	K7	Kwale	4 38 S 39 14 E	
Ngozi, crater L.	T7	Iringa	8 31 S 35 10 E	Polhill & Paulo
Ngozi, peak, for.	T7	Rungwe	9 00.5 S 33 33 E	
Ngua, old pl., f.r.	T7	Rungwe/Mbeya	8 59.5 S 33 33 E	Geilinger, Richards, Cribb & Grey-Wilson
Nguami, popl.	T3	Lushoto/Tanga	5 09 S 38 35 E	Drummond & Hensley, Greenway
[N'guaso Nyiro] see Uaso Nyiro	T6	Kilosa	6 04 S 37 02.5 E	Haarer, Mismahl, Semsei
[N'guaso Nyiro] see Uaso Nyiro	K1/3/4	several	0 19 S 36 39 E – 0 27 N 39 55 E	
[Nguazi] see Ngwazi	K6	Masai	c.0 39 S 35 44 E – 2 04 S 36 07 E	Mearns
Ngubwisi, str.	T7	Iringa	8 31 S 35 10 E	Carmichael
Ngudu, popl.	T7	Rungwe	9 18 S 33 49 E – 9 27 S 33 54 E	Stolz
[Nguelo] see Ngwelo	T1	Kwimba	2 58 S 33 20 E	Lewys Lloyd, Rounce, Staples
Ngugu ? see Ngagu	T3	Lushoto	5 04 S 38 39 E	Scheffler, Holst, Kummer
	P	Pemba	?5 11.5 S 39 39 E	Greenway

Ngugwi, R.	T7	Iringa	c. 8 33 S 35 45 E	Carmichael
[Ngula] see Hangula	T6	Morogoro	7 02 S 37 38.5 E	Greenway
Ngulakula, f.r.	T6	Rufiji	7 50 S 38 54 E	Ngoundai, Shabani
Ngulia, hills	K4/6	Machakos/Masai	c. 3 00 S 38 10 E	
Ngulia, peak	K6	Masai	3 00 S 38 09 E	Bally
Ngulia, w.h.	K6	Masai	2 59.5 S 38 10 E	Greenway
Ngulot, hill	K6	Masai	0 59.5 S 35 22 E	Glover et al.
Ngulu, area	T3	Pare	3 47 S 37 41 E	Haarer, Peter
Ngulu, pass	T3	Pare	4 03 S 37 46 E	
Ngulu, old area	T4	Tabora	c. 5 30 S 33 20 E	Peter
Ngulu (in Bundali Mts.)	T7	Rungwe	c. 9 30 S 33 27 E	Stolz
Ngulugulu, R., popl.	T7	Rungwe	c. 9 25 S 33 31 E	Stolz
Ngulukilo = Ngulukiro, popl.	U2	Kigezi	0 53 S 30 01 E	Snowden
Ngulukiro, popl.	U2	Kigezi	0 53 S 30 01 E	Snowden
Ngulumi, area	T3	Lushoto	4 44 S 38 41 E	Peter
Ngulwe, hill (near Kipembawe)	T7	Chunya	near 7 39 S 33 24 E	Greenway & Hoyle
[Ngumbe] see Kamba Ngombe	T4	Mpanda	7 31 S 31 44 E	Richards
[Ngumbuluni] see Nyumburuni	T6	Rufiji	7 54 S 39 02 E	Ngoundai
[Ngumbwisi] see Ngubwisi	T7	Rungwe	9 18 S 33 49 E –9 27 S 33 54 E	Stolz
Ngundu = Mgunda, mt.	T7	Njombe	9 13 S 34 05 E	Goetze
Ngunga, popl.	T1	Shinyanga	3 41 S 33 34 E	Doggett
Ngungu = Gungu, popl.	T4	Kigoma	4 51.5 S 29 38.5 E	Azuma
Ngungusi, escarp.	T4	Mpanda	prob. c. 7 25 S 31 50 E	Richards
Nguni, popl.	K4	Kitui	0 48 S 38 19 E	Bally & Smith
Ngunja, popl., mssn.	T8	Newala	10 27 S 39 27 E	Gillman
Ngurdoto, crater	T2	Arusha	3 17 S 36 55.5 E	Numerous
[Ngurdoto Crater, Nat. Pk.] see Arusha Nat. Pk.	T2	Arusha	c. 3 15 S 36 53 E	Numerous
Ngureseru Sambu (Loliondo)	T2	Masai	c. 2 04 S 35 36 E	Carmichael
[Nguronit] see Ngorinit	K1	Northern Frontier	1 45 N 37 18 E	Fratkin

Name		District	Coordinates	Collector
[Ngurtoto] see Engurdoto	T2	Masai	c.2 41 S 36 43 E	Carmichael
Nguru, mts., for.	T6	Morogoro	6 00 S 37 30 E	Numerous
[Ngurue] see Hanang	T2	Mbulu	4 26 S 35 24 E	Jaeger
Nguruimi, area	T1	Musoma	c.1 45 S 34 45 E	
Nguruka, rsta.	T4	Kigoma	5 07 S 31 02 E	
Nguruman, escarp., hills	K6	Masai	c.1 37–2 00 S 36 02 E	Bally, Jones, Vesey-FitzGerald
Ngurumbi, mt.	T7	Rungwe/Mbeya	c.9 02 S 33 29 E	Bally, Fischer, Glover & Samuel
Ngurunit = Ngorinit, w.h., mssn.	K1	Northern Frontier	1 45 N 37 18 E	Goetze
Ngurunit, w.h.	K1	Northern Frontier	1 30 N 37 00 E	Synnott
Nguru ya Ndege, hill	T6	Morogoro	6 42 S 37 36 E	
[Ngushai] see Engushai	T2	Moshi	2 58 S 37 12 E–2 50 S 37 07 E	Schlieben, Pócs & Bjornstad
[Ngusi] see Nkusi	U2/4	Bunyoro/Mubende	c.0 39 N 31 24 E–1 07 N 30 40 E	Carmichael
Ngusi, popl.	T7	Njombe	Not traced	Bagshawe
Nguungi, hill	K4	Kitui	1 03.5 S 37 41 E	Eggeling
Ngwambo, popl.	T8	Songea	c.10 55 S 34 56 E	Mabberley & Faden, Napper
Ngwasi = Ngowasi, L.	T7	Iringa	8 31 S 35 10 E	Milne-Redhead & Taylor
Ngwazi = Ngowasi, L.	T7	Iringa	8 31 S 35 10 E	
Ngwelo, peak	T3	Lushoto	5 04 S 38 39 E	Pegler, Renvoize, Procter
Ngwelo, popl.	T3	Lushoto	4 56 S 38 29 E	Scheffler
Ngwelo, popl.	T3	Lushoto	4 46 S 38 28 E	
Ngwelo, popl., area	T3	Lushoto	4 36 S 38 23 E	
Ngwelo, popl.	T3	Tanga	5 06 S 38 46 E	
Ngwena, popl.	K7	Tana River	1 27 S 40 01.5 E	Sampson
Ngweriroi, area	T2	Masai	c.2 14 S 35 42 E	Carmichael
Ngwina = Mgwina, str.	T6	Ulanga		Rees
[N'Honge] see Nhongo	T6	Morogoro/Uzaramo	7 13 S 38 24 E	Stuhlmann
Nhongo, area	T6	Morogoro/Uzaramo	7 13 S 38 24 E	Stuhlmann
[Niabura] ? see Niaburu	T1	Bukoba	?1 33 S 31 54 E	Braun
Niaburu, I.	T1	Bukoba	1 33 S 31 54 E	Braun
[Niamansi] see Nyamanzi	T4	Kigoma/Mpanda	6 08 S 30 41 E–5 38 S 31 06 E	

Niamanzi = Nyamanzi, R.	T4	Kigoma/Mpanda	6 08 S 30 41 E–5 38 S 31 06 E	Hoyle, Procter
Niamba, R.	T7	Mbeya	c.8 52 S 32 40 E–8 25 S 32 37 E	Goetze
Niam Niam, area	T7	Chunya/Mbeya	c.8 15 S 33 40 E	Pitt
Niapea, area	U1	West Nile	c.2 33 N 31 00 E	
[Niawatura] see Nyavatura	U2	Ankole	c.1 00 S 30 40 E	A.S. Thomas
[Nibbi] see Nebbi	U1	West Nile	2 30 N 31 05 E	Bally & Smith
[Niboi] see Neboi	K1	Northern Frontier	3 58 N 41 47 E	Fries
[Nieri] see Nyeri	K4	North Nyeri	0 25 S 36 57 E	
[Nigeregere] see Migeregere	T8	Kilwa	8 48 S 39 13 E	
Nile, R. see under Albert Nile and Victoria Nile				
Nile Bank, f.r.	U3	Busoga	0 38 N 33 03 E	G.H. Wood
[Nimur] see Nyimur R.	U1	Acholi	c.3 42 N 32 36 E	
Nindo, f.r., ta.	T1	Shinyanga	c.3 35 S 33 00 E	B.D. Burtt, Carmichael
[Ninghua] see Ningwa	T1	Shinyanga	3 34 S 33 15 E–3 38 S 33 27.5 E	B.D. Burtt
Ningwa, str.	T1	Shinyanga	3 34 S 33 15 E–3 38 S 33 27.5 E	B.D. Burtt
Niomkolo	Zambia *not* Tanzania		8 46 S 31 07 E	A. Carson
Niororo, I.	T6	Rufiji	7 37 S 39 41 E	
Nirawungu, hill	U2	Ankole	0 50.5 S 30 41 E	Eggeling
[Nisusi] see Kisuzi	T4	Buha	4 12 S 30 05 E	Peter
Ni Swawa la Mamba (Selous Game Reserve)	T8	Kilwa	Not traced	Rees
Nithi, R.	K4	Embu	0 15 S 37 32 E–0 18 S 37 47 E	Coe & Kirrika, Fries
[Niti] see Nithi	K4	Embu	0 15 S 37 32 E–0 18 S 37 47 E	Fries
Njabini, area	K3	Naivasha	c.0 43 S 36 39 E	Gardner
Njabini, fm.	K3	Naivasha	0 43 S 36 40 E	Gardner
Njage = Njagi, popl., falls	T6	Ulanga	8 14 S 36 10.5 E	Carmichael
[Njame] see Nyama	T6/7	Ulanga/Njombe	9 17 S 35 20 E–9 08 S 35 46 E	Schlieben
[Njangau] see Nyangao	T8	Lindi	10 20 S 39 18 E	Schlieben
Njanja, popl.	T7	Chunya	8 04 S 33 18 E	
[Njarasasee] see Eyasi, L.	T1/2	Maswa/Masai/Mbulu	c.3 40 S 35 05 E	Geilinger
[Njaro Lkatendesee] see Momela	T2	Arusha	c.3 13 S 36 54.5 E	Uhlig

301

Name		District	Coordinates	Collector
Njassa = Mjesse, ln.	T5	Dodoma	5 59 S 35 48 E	Stuhlmann
[Njassa See] see Lake Nyasa	T7/8, Moz/Mal	several	c.12 00 S 34 30 E	
Njeku, camp	T2	Arusha	3 15 S 36 47 E	Ross
Njele	T7	Mbeya	Not traced	Delamere, Ossent
Njemps, t.a.	K3	Baringo	c.0 30 N 36 03 E	
Njenje, R.	T8	Kilwa	c.10 20 S 36 52 E–9 05 S 37 26 E	Busse
[Njenye] see Njenje	T8	Kilwa	c.10 20 S 36 52 E–9 05 S 37 26 E	
Njeru, popl.	U4	Mengo	0 27 N 33 11 E	
Njila, popl.	T7	Chunya	8 17 S 32 46 E	
Njinjo, popl.	T8	Kilwa	8 48 S 38 54 E	Rose-Innes & Magogo
[Njiri See] see Amboseli Lake	K6	Masai	2 37 S 37 08 E	Schillings
Njiro, est.	K4	Nairobi	c.1 17.5 S 36 58 E	Bogdan
Njoge, popl.	T2	Masai	5 56 S 36 49 E	Leippert
Njoge Hill, f.r.	T2/5	Masai/Mpwapwa	5 56.5 S 36 42.5 E	Ruffo
Njojka Tank	?U4	?Mengo	Not traced	Harker
Njombe, distr.	T7	Njombe	c.9 35 S 34 25 E	
Njombe, R.	T5/7	several	c.7 58 S 33 53 E–6 56 S 35 06 E	Renvoize
Njombe, popl.	T7	Njombe	9 20 S 34 46 E	Numerous
Njombi R. (Elton Plateau)	T7	Mbeya	Not traced	Richards
[Njonga] see Ngongo	T8	Lindi	10 04 S 39 37 E	Gillman
Njongomeru = Jongomeru, R.	T7	Mbeya	7 46 S 34 20 E–7 55 S 34 35 E	Richards, Greenway
[Njonja] see Ngongo	T8	Lindi	? c.10 04 S 39 37 E	Gillman
[Njonjo] ? Ngongo	T8	Lindi	? c.10 04 S 39 37 E	Gillman
Njoro, popl., rsta.	K3	Nakuru	0 20 S 35 57 E	Numerous
Njoro, hill	T2	Masai	5 17 S 36 29 E	H.F. Elliot
Njoro, str., area	T2	Moshi	c.3 21 S 37 21 E	Milne-Redhead & Taylor
[Njoro-Lkatende See] see Momela L.	T2	Arusha	c.3 13 S 36 54.5 E	Uhlig
Njorowa = Hell's Gate, gorge	K3	Naivasha	c.0 52–0 57 S 36 18.5 E	
[Njowara] see Njorowa	K3	Naivasha	c.0 52–0 57 S 36 18.5 E	Macdonald
Njukini Forest	K4	S. Nyeri/Embu	0 30 S 37 25 E	Brunt

Name		Region	Coordinates	Authority
Njura, settlement	K6	Masai	c.0 56 S 35 40 E	Glover et al.
[Njussi] see Mnyusi	T3	Tanga	5 13 S 38 35 E	Winkler
Nkabwa, mt.	U2/Zaire	Kigezi	0 54 S 29 35 E	
Nkalakasiha, popl., str.	T8	Lindi	9 56.5 S 39 16 E	Busse
[Nkalinzi] see Mkalinzi	T4	Buha	4 37 S 29 44 E	Stauffer
Nkanda, popl.	U2	Kigezi	1 21 S 29 40 E	
Nkanka, str.	T7	Mbeya	c.9 03 S 32 58 E–9 04 S 32 41 E	Tanner
[Nkaramo] see Mkaramo	T3	Pangani	near 5 47 S 38 51 E	
Nkare Narok = Engare Narok, str.	K1	Northern Frontier	1 13 N 36 33 E–0 52 N 36 47 E	
Nkata, I.	U4	Mengo	0 03 S 33 11 E	
Nkenja, popl.	T7	Njombe	9 10 S 34 04 E	Procter
Nkenja, mt.	T7	Njombe	9 10 S 33 59 E	
Nkoko, popl., est.	U4	Mengo	0 36 N 32 53 E	Dummer
Nkoko, popl.	U4	Mubende	0 58 N 31 26 E	Lind
Nkokonjeru, mt.	U3	Mbale	1 01.5 N 34 15 E	Snowden
Nkola Ngola	U3/U4		Not traced	Scott Elliot
Nkolawe, mt.	T6	Morogoro	6 03 S 37 30 E	Thulin & Mhoro
Nkombe, popl.	U3	Busoga	0 39 N 33 39 E	
Nkombola, popl.	T3	Lushoto	4 58.5 S 38 24.5 E	
[Nkombola] see Kombola	T3	Lushoto	4 56.5 S 38 41 E	
Nkondo, popl.	U3	Busoga	1 19 N 33 05.5 E	G.H. Wood
Nkondwe, r.h.	T4	Kigoma	5 52 S 30 52 E	
Nkonge, rsta.	U4	Masaka	0 14 N 31 10 E	
Nkonge, mt.	T3	Lushoto	5 00.5 S 38 27 E	
Nkongoi, area	T3	Lushoto	c.4 46 S 38 31 E	
Nkongwe, for.	U4	Mengo	c.0 08 N 32 55 E	
Nkonje	T6	Uzaramo	Not traced	Stuhlmann
Nkose, I.	U4	Masaka	0 44 S 32 18 E	
Nkulu, mssn.	T7	Chunya	7 42 S 33 25 E	Dawkins
[Nkumbi] see Nkombe	U3	Busoga	0 39 N 33 39 E	Maitland

Name	Code	Region	Coordinates	Collector(s)
North Kavirondo, distr.	K5	North Kavirondo	c.0 30 N 34 30 E	Pegler
North Kinangop, for.	K3	Naivasha	c.0 37 S 36 39 E	
North Kinangop, popl., p.p.	K3	Naivasha	0 36 S 36 34 E	
North Lereko, peak	K3/4	Naivasha/S. Nyeri	0 33.5 S 36 40 E	
North Lupa, f.r.	T7	Chunya	c.8 02 S 33 15 E	
North Mara, distr.	T1	North Mara	c.1 15 S 34 15 E	
North Nandi Forest, for.	K3	Nandi	c.0 20 N 34 59 E	
North Nyanza, distr.	K5	North Kavirondo	c.0 20 N 34 35 E	
(considered as part of North Kavirondo Distr. for F.T.E.A.)				
North Nyeri, distr.	K4	North Nyeri	c.0 00 37 15 E	
North Pare, mts.	T3	Pare	c.3 30–4 05 S 37 33–37 55 E	Semsei
North Uluguru, f.r.	T6	Morogoro	c.6 55 S 37 42 E	Procter, Eggeling
North Usafwa, f.r.	T7	Mbeya	c.8 50 S 33 32 E	Busse, Schlieben
Noto Plateau, area	T8	Lindi	c.9 54 S 39 24 E	Procter
Nou, f.r.	T2	Mbulu	c.4 05 S 35 30 E	J. Wilson
Noyon	U1	Karamoja	Not traced	
[Npitamaiong] see Naitamaiong	K2	Turkana	3 16 N 34 54 E	
[Nquelo] see Ngwelo	T3	Lushoto	5 04 S 38 39 E	Heinsen
Nsanga, hills, for.	T4	Ufipa	c.8 05 S 31 50 E	Richards, McCallum Webster, Robinson
Nsangamales, old popl.	T7	Mbeya	8 52 S 32 45 E	Goetze
Nsangi, popl.	U4	Mengo	0 17 N 32 28 E	
[Nsangu] see Nsanga	T4	Ufipa	c.8 04 S 31 50 E	Richards, Vesey-FitzGerald
[Nsanvia] see Msambia	T4	Ufipa	c.8 42 S 31 37 E–8 26 S 31 51 E	Michelmore
Nsassa, popl.	T5	Dodoma	6 00 S 35 54 E	Busse
Nseko (near Mwera)	T3	Pangani	near 5 29 S 38 54 E	Tanner
[Nseneto] see Senato	T2	Arusha	3 16 S 36 54 E	Greenway
Nshamba, popl., t.c.	T1	Bukoba	1 47 S 31 33 E	Gillman
Nsika, popl., r.h.	U2	Ankole	0 22 S 30 26 E	Lind, Snowden
[Nsimbo] see Simbo	T5	Kondoa	c.5 00 S 35 40 E	B.D. Burtt
Nsinze, rsta.	U3	Busoga	0 49 N 33 36 E	Eggeling

Name	Code	Region	Coordinates	Authority
Nsogiro, Mts.	T5	Mpwapwa	c.7 12 S 36 25 E	Houy
[Nsolo] see Solo	U3	Mbale	c.0 28 N 34 06 E—0 35 N 33 59 E	G.H. Wood
Nsonge, R. = Mahoma (in part)	U2	Toro	c.0 34 N 30 05 E—0 03 N 30 16 E	Lock
Nsongezi, popl., r.h., fy.	U2	Ankole	0 59 S 30 45 E	Brasnett, B.D. Burtt, Eggeling
[Nsongi] see Mahoma	U2	Toro	c.0 37 N 30 08 E—c.0 03 N 30 16 E	Dawe
Nsonza, str.	T7	Iringa	8 02.5 S 36 02 E—8 07 S 35 55 E	
[Nsue] see Nzuhe	T5	Mpwapwa	6 30 S 36 31.5 E	
Nsundas, popl., area	T7	Mbeya	9 12 S 33 09 E	?Goetze
[Nsungma] see Uzungwa	T6/7	Ulanga/Iringa	c.8 15 S 35 50 E	comm. Agric. Dept.
[Ntakafumu] see Ntakafunvu	U4	Mengo	0 16 N 32 47.5 E	Dummer
Ntakafunvu, popl.	U4	Mengo	0 16 N 32 47.5 E	
Ntale = Ntali, str. (below Kungwe Mt.)	T4	Mpanda	c.6 08 S 29 45 E	R.M. Harley
Ntali, str. (Kungwe Mt.)	T4	Mpanda	c.6 08 S 29 45 E	R.M. Harley
[Ntanangozi] see Tanangozi	T7	Iringa	7 55 S 35 35.5 E	Goetze
Ntande, hill	T7	Chunya	8 10 S 33 16 E	Bally & Carter
Ntandi, popl., f.r.	U2	Toro	c.0 48 N 30 09 E	
Ntangamalala	T1	Biharamulo	Not traced	Ford
Ntanzi, popl., t.c.	U4	Mengo	0 13.5 N 32 49 E	
[Ntara] see Mtera	T7	Iringa	7 06 S 35 58 E	
Ntaru (near Mwera)	T3	Pangani	near 5 29 S 38 54 E	Tanner
[Ntatanda] see Tatanda	T4	Ufipa	8 32 S 31 29 E	Sanane
[Ntb] see Entebbe	U4	Mengo	0 03 N 32 29 E	
Ntenjeru, popl.	U4	Mengo	0 12 N 32 48 E	
Nteza, ln.	T4	Mpanda	prob. 7 00 S 31 00 E	Carmichael
Ntima, peak (Pugu Hills)	T6	Uzaramo	c.6 55 S 39 02 E	Peter
Ntinkalu, area	U3	Busoga	c.0 26.5 N 33 19 E	G.H. Wood
Ntomoko = Tomoko, hills	T5	Kondoa	4 47 S 36 03 E	Ruffo
Ntoroko, popl., l.s.	U2	Toro	1 04 N 30 32 E	Lowe, McConnell
[Ntukuju] see Tukuyu	T7	Rungwe	9 15 S 33 38.5 E	
[Ntukuyu] see Tukuyu	T7	Rungwe	9 15 S 33 38.5 E	

Name	Code	Region	Coordinates	Collector
Ntungamo, popl, r.h., t.c.	U2	Ankole	0 53 S 30 16 E	Numerous
Ntungwe = Ntungu, R.	U2	Kigezi	0 43 S 29 57 E–0 26 S 29 43 E	Lock
Ntusi, popl., t.c.	U4	Masaka	0 03 N 31 13 E	Purseglove, Lye
Ntwetwe, popl.	U4	Mengo	0 57 N 31 35 E	Eggeling
Ntwike, popl.	T5	Singida	4 15 S 34 12 E	Michelmore
[Ntyuka] see Ndiuka	T5	Dodoma	6 13.5 S 35 46 E	Ruffo
[Ntyuya] see Ndiuka	T5	Dodoma	6 13.5 S 35 46 E	Ruffo
[Numbanitu] see Nyumbanitu	T7	Iringa	7 48 S 36 21 E	Carmichael
[Numbi] see Ndumbi	T7	Njombe	c.9 04 S 33 52 E–8 55 S 34 07 E	
[Numlij] see Ndumbi	T7	Njombe	c.9 04 S 33 52 E–8 55 S 34 07 E	
Nunda, popl.	T3	Pangani	5 22 S 38 59 E	Tanner
Nundoto, str.	K1	Northern Frontier	1 35 N 36 36 E–0 55.5 N 36 44.5 E	Kerfoot
Nundwe, ln.	T7	Iringa	8 17 S 35 30 E	Paget-Wilkes
Nunga = Nungu, Thicket	T8	Kilwa	8 37.5 S 38 03 E	Vollesen, Rodgers
Nungi, popl.	T5	Dodoma	5 47 S 35 03 E	B.D. Burtt
Nungu Thicket	T8	Kilwa	8 37.5 S 38 05 E	Vollesen, Ludanga
Nunguni, hill	K4	Machakos	1 46 S 37 21 E	Mwangangi
Nunguni, mkt.	K4	Machakos	1 48 S 37 22 E	
Nungwe, bay	T1	Mwanza	2 42 S 31 59 E	
Nungwe, popl.	T1	Mwanza	2 46 S 32 01 E	Morgan
[Nuronzi] see Muronzi	T1	Ngara	c.2 47 S 30 52 E	Tanner
Nuu, for., hill	K4	Kitui	1 02 S 38 19 E	Bally
[Nwarakaya] see Mwarakaya	K7	Kilifi	3 48 S 39 41.5 E	Brenan, Gillett et al.
[Nwela] see Mwela	T7	Mbeya	8 58 S 33 48.5 E	Cribb & Grey-Wilson
[Nyabarogo] see Nyaburogo	U2	Toro	0 47.5 N 30 12 E–0 57 N 30 16 E	Eggeling
[Nyabarongo] see Nyaburogo	U2	Toro	0 47.5 N 30 12 E–0 57 N 30 16 E	Eggeling
Nyabekwabi (Bukwaya)	T1	Musoma	c.1 34 S 33 46 E	Tanner
Nyabeya, hill, for. sch.	U2	Bunyoro	1 41 N 31 33 E	
Nyabibuye, ln.	T4	Buha	2 57 S 30 52 E	Tanner
Nyabikere, L.	U2	Toro	0 30 N 30 19.5 E	

307

Name		Region	Coordinates	Collector
Nyabirongo, popl.	U2	Toro	0 04 N 29 53 E	Katende
Nyabisabu, R.	U2	Bunyoro	1 44 N 31 32 E	Perdue & Kibuwa, Ball, Osmaston
Nyabitaba = Nyinabitaba, hut	U2	Toro	0 21.5 N 29 59 E	Greenway
[Nyaboro] see Nyaroboro	T1	Musoma	c.2 35 S 34 43 E	Purseglove
Nyabubale, popl.	U2	Ankole	0 32 S 30 05 E	Tanner
Nyabugombe, popl.	T1	Biharamulo	2 36 S 30 59 E	
[Nyabukere] see Nyabikere	U2	Toro	0 30 N 30 19.5 E	
Nyabuoba, hill	T4	Buha	4 40 S 31 15 E	Bullock
Nyaburogo, str.	U2	Toro	0 47.5 N 30 12 E–0 57 N 30 16 E	Eggeling, Mukasa
[Nyaburongo] see Nyaburogo	U2	Toro	0 47.5 N 30 12 E–0 57 N 30 16 E	Mukasa
Nyabushozi, co.	U2	Ankole	c.0 30 S 30 50 E	
Nyabuyonza, popl.	T1	Bukoba	1 42 S 30 59 E	Procter
[Nyabyeya] see Nyabeya	U2	Bunyoro	1 41 N 31 33 E	Osmaston
Nyagak, str.	U1	West Nile	c.2 39 N 31 06 E	Eggeling
Nyagoma, str.	U2	Ankole	0 16 S 30 27 E–0 22.5 S 30 28 E	Rwaburindore
Nyahua, rsta.	T4	Tabora	5 24 S 33 19 E	Peter
Nyahururu, popl., falls, rsta.	K3	Nakuru	0 02.5 N 36 22 E	
Nyaiguru, ridge	U2	Kigezi	1 05 S 29 40 E	
Nyaishozi, popl., area	T1	Bukoba	1 47 S 31 07 E	Haarer
Nyakafunjo (near Busingiro)	U2	Bunyoro	near 1 44 N 31 28 E	Dawkins, Eggeling, Synnott
Nyakageme = Nyakagyeme	U2	Kigezi	0 48 S 29 53 E	Purseglove
[Nyakagene] see Nyakagyeme	U2	Kigezi	0 48 S 29 53 E	Purseglove
Nyakagera, str.	U2	Toro	0 34 N 30 21.5 E–0 31 N 30 21.5 E	Katende
Nyakagyeme, popl., r.c.	U2	Kigezi	0 48 S 29 53 E	
[Nyakagzeme] see Nyakagyeme	U2	Kigezi	0 48 S 29 53 E	
Nyakahanga, popl., hill	T1	Bukoba	1 37 S 31 09 E	Haarer
Nyakahura, mssn.	T1	Biharamulo	2 48 S 31 04 E	Ford
[Nyakahuru] see Nyakahura	T1	Biharamulo	2 48 S 31 04 E	
Nyakalembe, popl.	U2	Kigezi	1 12 S 29 44 E	Snowden
Nyakalengija, popl.	U2	Toro	0 20 N 30 02 E	W.H. Lewis

Name		Region	Coordinates	Collector
[Nyakalengiji] see Nyakalengija	U2	Toro	0 20 N 30 02 E	W.H. Lewis
Nyakalilo, popl.	T1	Mwanza	2 27 S 32 26 E	Carmichael
[Nyakamazi] see Nyakanazi	T1	Biharamulo	2 58 S 31 12 E	Carmichael
[Nyakanasi] see Nyakanyasi	T1	Bukoba	1 11 S 31 13 E	Procter
Nyakanazi, popl.	T1	Biharamulo	3 04 S 31 13 E	Carmichael, ?Eggeling
Nyakanazi, ln.	T1	Biharamulo	2 58 S 31 12 E	?Eggeling
[Nyakangasi] see Nyakanyasi	T1	Bukoba	1 11 S 31 13 E	Carmichael
Nyakanyasi = Kagera Port, popl.	T1	Bukoba	1 11 S 31 13 E	Gillman
Nyakasura, popl., coll.	U2	Toro	0 40 N 30 14 E	Hancock, Shillito
[Nyakasuru] see Nyakasura	U2	Toro	0 40 N 30 14 E	
Nyakato, popl.	T1	Bukoba	1 16 S 31 49 E	Gillman, Haarer
Nyakatonzi, exper. fm.	U2	Toro	0 02 S 29 50 E	
Nyakatowo, R.	T7	Iringa	7 41 S 34 32 E – 7 40.5 S 34 48 E	Greenway
Nyakaziba	U2	Ankole	prob. c.0 19 S 30 27 E	Snowden
Nyakinoni, popl.	U2	Kigezi	0 45.5 S 29 45 E	Purseglove
Nyakisanju, popl.	U2	Kigezi	0 35 S 29 39 E	
Nyakisasa, popl.	T1	Ngara	2 42 S 30 42 E	Tanner, Eggeling
Nyakitonto, popl.	T4	Buha	4 28 S 30 14 E	Procter
Nyakizumba, str.	U2	Kigezi	1 19 S 30 05 E – 1 15 S 30 00 E	Norman
Nyakoromo, ln.	T1	Musoma	2 12 S 34 03 E	Greenway
[Nyakromo] see Nyakoromo	T1	Musoma	2 12 S 34 03 E	Greenway
Nyakumu Swamp, sw.	K4	Kiambu	1 15.5 S 36 38 E	Monou
Nyalawa, R.	T7	Iringa	c.8 42 S 35 11 E	Polhill & Paulo, Paget Wilkes
Nyalenda, popl.	K5	Central Kavirondo	0 09 S 35 00 E	Bjornstad
Nyali, bdg.	K7	Mombasa	4 03 S 39 41 E	Drummond & Hemsley
Nyali, beach	K7	Mombasa	4 03 S 39 43 E	Numerous
[Nyalusangi] see Nyalushanje	U2	Kigezi	1 00 S 29 59 E	
Nyalushanje, popl., r.c.	U2	Kigezi	1 00 S 29 59 E	Harris, A.S. Thomas
Nyalwela, popl.	T7	Mbeya	8 57 S 33 44 E	Leedal
Nyama, R.	T6/7	Ulanga/Njombe	9 17 S 35 20 E – 9 08 S 35 46 E	

Name		Region	Coordinates	Collector
Nyamikoma, popl.	T1	Mwanza	2 20 S 33 41 E	Tanner
Nyamirembe, popl.	T1	Biharamulo	2 32 S 31 42 E	Procter
Nyamiruma, str.	T1	Ngara	2 39 S 30 51 E–2 49 S 31 03 E	Eggeling
Nyamiyaga, popl.	T1	Ngara	2 28 S 30 40.5 E	
Nyamkachowe, str., plains	T1	Musoma	c.2 27 S 34 33 E–2 10S 34 31 E	Greenway
[Nyamnsi] see Nyamanzi	T4	Kigoma/Mpanda	6 08 S 30 41 E–5 38 S 31 06 E	
Nyamonge, area	T1	Mwanza	2 49 S 32 08 E	Willan, Makwilo, Carmichael
Nyamtukusa, hill, popl.	T1	Mwanza	3 02 S 32 45 E	
Nyamtumbo, popl.	T8	Songea	10 30 S 36 06 E	Milne-Redhead & Taylor
[Nyamugoye] see Nyamigoye	U2	Kigezi	0 50 S 29 41 E	
Nyamuleju, camp	U2	Toro	0 22 N 29 57.5 E	Lye, Noble
Nyamuma, guard post	T1	Musoma	c.2 26 S 34 27 E	Greenway
[Nyamumuka] see Nyamunuka	U2	Toro	0 05 S 29 59 E	Greenway & Eggeling
Nyamunuka, L.	U2	Toro	0 05 S 29 59 E	Greenway & Eggeling
Nyamwamba, str.	U2	Toro	0 19 N 29 54 E–0 04 N 30 09 E	G. Taylor, Scott Elliot
Nyamweru, for.	K4	Kiambu	c.1 00 S 36 40 E	Trapnell
Nyamyaga, popl.	T1	Ngara	2 28 S 30 40.5 E	Tanner
[Nyamyekudo] see Namyekudo	U2	Bunyoro	1 32 N 31 53 E	
Nyandarua, distr.	K3	Naivasha/Nakuru	c.0 35 S 36 30 E	
[Nyangai] see Mnagei	K2	West Suk	c.1 18 N 35 04 E	Verdcourt
Nyangala, hill	K7	Teita	3 38 S 38 44 E	Bally
Nyangalamila, hill	T1	Mwanza	2 46 S 32 03 E	Carmichael
Nyangallo, popl.	T5	Dodoma	6 02 S 36 07 E	Busse
Nyangamara, popl.	T8	Lindi	10 21 S 39 38 E	Gillman
Nyanganje = Nanganji, popl., f.r.	T6	Ulanga	c.8 00 S 36 42 E	Haerdi
Nyangao, popl.	T8	Lindi	10 20 S 39 18 E	Milne-Redhead & Taylor, Busse, Schlieben
Nyangao, str.	T8	Lindi	c.10 19 S 39 18 E	Milne-Redhead & Taylor
[Nyangaya] ? see Nangya	U3	Mbale	0 56.5 N 34 17 E	A.S. Thomas
[Nyanga-yanga] see Myangayanga	T8	Songea	10 56 S 34 58 E	
[Nyangedi] see Nyengedi	T8	Lindi	10 17 S 39 24 E	Eggeling

311

Name	Prov.	District	Coordinates	Collector(s)
Nyashimo, popl.	T1	Mwanza	2 24 S 33 33 E	Haarer
[Nyashozi] see Nyaishozi	T1	Bukoba	1 47 S 31 07 E	Drummond & Hemsley
Nyassa Bridge	T3	Lushoto	4 50.5 S 38 20.5 E	Mildbraed
Nyavatura, popl.	U2	Ankole	c.1 00 S 30 40 E	Procter
Nyaviyumbu, popl.	T4	Buha	3 40 S 30 48.5 E	Eggeling
Nyebeya (near Kasatora)	U2	Kigezi	near 1 04 S 29 39.5 E	Fundi
Nyeburu, popl.	T6	Uzaramo	6 58.5 S 39 05 E	Leippert
[Nyegese] see Nyegezi	T1	Mwanza	2 35 S 32 53 E	Doggett, Rounce, Carmichael
Nyegezi, popl.	T1	Mwanza	2 35 S 32 53 E	Tanner
Nyegezi, t.a.	T1	Mwanza	c.2 37 S 32 55 E	Braun
Nyembe	T4	Tabora	Not traced	Hammerstein
Nyembwe Bolungwa	T4	Tabora	Not traced	
[Nyemeku] see Nyamakuyu	T7	Iringa	7 45 S 34 52 E	Greenway
Nyenea, popl.	T8	Lindi	prob. c.10 10 S 39 12 E	Milne-Redhead & Taylor
Nyenga, rsta.	U4	Mengo	0 22 N 33 09 E	A.S. Thomas
Nyengedi, popl.	T8	Lindi	10 17 S 39 24 E	Gillman, Schlieben, Magogo & Rose Innes
Nyeri, popl.	K4	North Nyeri	0 25 S 36 57 E	Numerous
Nyeri, rsta. = Kiganjo	K4	South Nyeri	0 22.5 S 37 00 E	
[Nyero] see Nyiru	K1	Northern Frontier	c.2 10 N 36 50 E	
Nyika = Nyika Steppe	T3	Tanga/Lushoto/Pangani	c.4 45 S 39 00 E	
Nyika country, old area	K7	Kilifi/Mombasa/Kwale (coastal area extending from Tanzania border north to the Sabaki River)	c.3 10 – 4 40 S	Wakefield
Nyika Steppe, old area	T3	Lushoto/Tanga/Pangani (bounded by R. Umba, Usambara Mts., R. Pangani & East Coast)	c.4 45 S 39 00 E	Holst
Nyimur, str.	U1	Acholi	c.3 42 N 32 35 E	
[Nyimwur] see Nyimur	U1	Acholi	c.3 42 N 32 36 E	
Nyinabitaba, camp, ridge	U2	Toro	c.0 21.5 N 30 00 E	Eggeling, Fishlock & Hancock, Hedberg
Nyio, f.r.	U1	West Nile	3 01 N 30 48 E	Oakley
Nyiri Desert, desert	K6	Masai	c.2 25 S 37 20 E	Bally
[Nyiro] see Nyiru	K1	Northern Frontier	c.2 10 N 36 50 E	Kerfoot

314

Name		Region/District	Coordinates	Authority
[Nzima] see Mzima Springs				
Nzima, popl., r.c.	K6	Masai	2 59 S 38 01 E	
Nzinjero, hill (near Mbarara)	T1	Shinyanga	3 19 S 32 52 E	Maitland
Nzoia, R.	U2	Ankole	near 0 37 S 30 39 E	Bogdan, Tweedie
	K3/5	Trans-Nzoia/	0 49 N 35 25.5 E–0 04 N 33 56.5 E	
		Uasin Gishu/Kavirondo		
[Nzowi] see Nzaui				
[Nzue] see Nzui	K4	Machakos	1 55 S 37 33 E	Scott Elliot
[Nzue] see Nzuhe	K4	Kitui	0 51 S 38 15 E	Bally
Nzuhe, L.	T5	Mpwapwa	6 30 S 36 31.5 E	Staples
Nzui = Enzui, popl.	T5	Mpwapwa	6 30 S 36 31.5 E	Staples
[Nzui] see Nzuhe	K4	Kitui	0 51 S 38 15 E	Bally, L.C. Edwards
Nzukini, hills	T5	Mpwapwa	6 30 S 36 31.5 E	Hornby
	K4	Fort Hall	1 04.5 S 37 28 E	Bally
Oaklands Hotel	T3	Lushoto	c.4 47.5 S 38 17.5 E	Cribb & Grey-Wilson, Batty
[Oalla] see Wala	T4	Tabora	c.5 00 S 33 10 E–4 46 S 32 03 E	Stuhlmann
Oboa = Kadam, peak	U1	Karamoja	1 45 N 34 43 E	G.H. Wood
Obongi, popl.	U1	West Nile	3 13 N 31 33 E	Eggeling, Leggat, Brooks
Observation Hill, ln.	T6	Uzaramo	6 46 S 39 12 E	Vaughan
Ochodri = Ocodri, r.h.	U1	West Nile	3 04 N 30 05 E	Eggeling
[Ochoro] see Okoro	U1	West Nile	c.2 28 N 31 06 E	Hazel
Ockiuro	T6	Uzaramo	Not traced	Holtz
Oda, w.h.	K1	Northern Frontier	3 30.5 N 39 04 E	Kirrika
Odongo Ekundu (near Magunga)	T3	Lushoto	prob. near 4 58 S 38 38 E	Peter
[Odonyowass] see Oldonyo Was				
Odupi, prob. = Udupi	T2	Masai	3 02 S 35 43 E	Chancellor
[Odzi] see Otzi	U1	West Nile	prob. 3 18 N 31 10 E	Scott
Offaka, popl.	U1	West Nile	3 37 N 31 51 E	
[Offu] see Ofo	U1	West Nile	2 35 N 31 02 E	
[Offude] see Ofude	U1	West Nile	3 22 N 30 57 E	
Ofo, hill	U1	West Nile	3 13 N 30 59 E	
	U1	West Nile	3 22 N 30 57 E	

Entry		Region	Coordinates	Authority
[Ofu] see Ofo	U1	West Nile	3 22 N 30 57 E	Hazel
Ofude, popl.	U1	West Nile	3 13 N 30 59 E	Trapnell
Ogembo, popl.	K5	South Kavirondo	0 48 S 34 43 E	
Ogili, mt., f.r.	U1	Acholi	3 11 N 33 17 E	Lye
Oguen, popl.	U1	Acholi	2 22 N 31 39 E	Eggeling
[Ogujebbe] see Ogujebe	U1	West Nile	3 29 N 31 42 E	
Ogujebe, popl.	U1	West Nile	3 29 N 31 42 E	
Ogunja, mkt.	K5	North/Central Kavirondo	0 11 N 34 18 E	
[Ogur] see Agur	U1	Lango	2 26 N 32 56 E	
[O'Horo Flats] see Buhoro Flats	T7	Mbeya	c.8 30 S 34 30 E	Richards
Oju = Zoka, for.	U1	West Nile	c.3 50 N 31 39 E	A.S. Thomas
[Okame] see Kami	U3	Mbale	c.0 29 N 34 13 E–0 35 N 33 59.5 E	
Okia, popl.	K4	Machakos	1 46 S 37 30 E	Hemming
Okol, hills	U1	Acholi	3 37 N 33 11 E	
Okollo, popl, r.h.	U1	West Nile	2 40 N 31 08 E	Eggeling
[Okolo] see Okollo	U1	West Nile	2 40 N 31 08 E	Eggeling
Okoro, co.	U1	West Nile	c.2 28 N 31 06 E	
Okuza, I.	T8	Kilwa	8 16.5 S 39 35.5 E	Frazier
[Olagasailie] see Olorgesailie	K6	Masai	c.1 35 S 36 26.5 E	Greenway
Olairobi, area	T2	Masai	c.3 13 S 35 27 E	
Olambwe, str./s	K5	South Kavirondo	0 28.5 S 34 18 E–0 43 S 34 12 E	
Olambwe Valley, g.r.	K5	South Kavirondo	c.0 37 S 34 15 E	
Ol Arabel, for.	K3	Laikipia	c.0 19 N 36 14 E	
Ol Arabel, stores	K3	Laikipia	0 18 N 36 16.5 E	E.A. Lewis, Lacey
Ol Arabel, str.	K3	Baringo/Laikipia	c.0 15 N 36 17 E–0 33 N 36 06.5 E	
Ol Ari Nyiro, spr.	K3	Laikipia	0 36 N 36 23 E	Newbould
Olbaata (Olbalbal)	T2	Masai	c.3 05 S 35 25 E	Greenway
[Ol Balambal] see Olbalbal	T2	Masai	c.3 05 S 35 25 E	Greenway
Olbalbal, escarp.	T2	Masai	c.3 05 S 35 25 E	
Olbalbal, ln.	T2	Masai	2 57 S 35 17 E	Paulo, Newbould, Trapnell

Ol Bolossat, for.	K3	Nakuru	c.0 00 36 19 E	Gardner, Meinertzhagen
Ol Bolossat, hill	K3	Nakuru	0 0.5 S 36 23 E	Dowson
Ol Bolossat, L.	K3	Naivasha	c.0 10 S 36 26 E	Glover et al.
Ol Choro Orogwe, ranch	K6	Masai	1 06 S 35 12 E	Greenway, G.R. Williams
Ol Dane Sapuk, str., fm.	K3	Uasin Gishu	0 28 N 35 13 E –0 32 N 35 08.5 E	
[Oldanisabuk] see Ol Dane Sapuk	K3	Uasin Gishu	0 28 N 35 13 E –0 32 N 35 08.5 E	
[Oldani Sapuk] see Ol Dane Sapuk	K3	Uasin Gishu	0 28 N 35 13 E –0 32 N 35 08.5 E	
Oldeani, mt.	T2	Mbulu/Masai	3 16 S 35 26 E	Numerous
Oldeani, popl.	T2	Mbulu	3 21 S 35 34 E	
Ol'debesi-Lemoko	K6	Masai	Not traced	
Old Entebbe, popl.	U4	Mengo	0 02.5 N 32 26.5 E	Glover et al.
Oldiang'arangar, mt.	T2	Masai	c.2 48 S 35 25 E	Harker, Eggeling
Old Kasanga, popl.	T7	Iringa	8 38.5 S 35 09 E	Newbould
[Old Langenburg] see Lumbila	T7	Njombe	9 34.5 S 34 07.5 E	Polhill & Paulo
Old Mill = Suam Sawmill (1937–49)	K3	Trans-Nzoia	1 09 N 34 46 E	Tweedie
Old Moshi, popl., mssn.	T2	Moshi	3 19 S 37 24 E	Haarer, Peter
Old Nariam, popl.	U3	Teso	1 57.5 N 34 06 E	
Ol Doinyo Gol = Gol Kopjes, hills	T2	Masai	c.2 42 S 35 26 E	Greenway, St. Clair-Thompson
[Ol Doinyo Langai] see Ol Doinyo Lengai	T2	Masai	2 45 S 35 54 E	
Ol Doinyo Lengai, mt.	T2	Masai	2 45 S 35 54 E	
[Ol Doinyo Lengeyo] see Mathews Range	K1	Northern Frontier	c.1 15 N 37 15 E	Newbould
[Ol Doinyo Lengio] see Mathews Peak	K1	Northern Frontier	1 18 N 37 18 E	Newbould
[Ol Doinyo Lengiyo] see Mathews Peak	K1	Northern Frontier	1 18 N 37 18 E	
Ol Doinyo Lesatima = Sattimma, peak	K3/4	Laikipia/Naivasha/S. Nyeri	0 19 S 36 37 E	
Ol Doinyo Loibene = Lolbene, mt.	T2	Masai	3 57 S 37 07 E	
Ol Doinyo Loloponi, area	K6	Masai	1 31 S 35 39 E	Kuchar
[Ol Doinyo Losoe] see Mathews Range	K1	Northern Frontier	c.1 15 N 37 15 E	
Ol Doinyo Mara, mt.	K1	Northern Frontier	c.2 15 N 37 02 E	Cockburn
Ol Doinyo Nyuki = Suswa, crater, peak	K3/6	Naivasha/Masai	1 10.5 S 36 21 E	
Ol Doinyo Orok, mt.	K6	Masai	2 28 S 36 46 E	Bally, Napper

317

Ol Doinyo Sabachi = Lolokwi	K1	Northern Frontier	0 50 N 37 32 E	Bally & Smith
Ol Doinyo Sabuk, hill	T2	Masai	3 28 S 35 06 E	Numerous
Ol Doinyo Sambu, mt.	T2	Masai	2 09 S 35 56 E	Numerous
Ol Doinyo Sambu = Ol Donyo Sambu	T2	Arusha	3 10 S 36 39 E	Numerous
Ol Doinyo Sapuk, mt.	K4	Machakos	1 08 S 37 15 E	
[Oldongo Sambu] see Ol Doinyo Sambu	T2	Arusha	3 10 S 36 39 E	Greenway
Oldonyo Labai, hills	T1	Maswa	2 50 S 34 40 E	
Oldonyo Lengai, mt.	T2	Masai	2 45 S 35 54 E	Richards, Fischer
Oldonyo Lengijjawe, hill	T2	Arusha	3 12.5 S 36 36 E	
Oldonyo Lengiyabe, hill	T2	Masai	4 20 S 36 56 E	
Oldonyo Ngailoni, hill	T2	Masai	2 50 S 36 48 E	
Oldonyo Orok = Ol Doinyo Orok	K6	Masai	2 28 S 36 46 E	Bally, H.F. Elliot, Vesey-FitzGerald
[Ol Donyo Sabuk] see Ol Doinyo Sapuk	K4	Machakos	1 08 S 37 15 E	Bally, Napier
Oldonyo Salili, hill	T2	Masai	3 07 S 35 24 E	Greenway
Ol Donyo Sambu, hill	T2	Arusha	3 10 S 36 39 E	
Ol Donyo Sambu, popl., p.p.	T2	Arusha	3 11 S 36 39 E	Greenway
Oldonyo Was, hill	T2	Masai	3 02 S 35 43 E	B.D. Burtt
Oldoro Lolussoi, w.h.	T2	Masai	4 10 S 36 51 E	Procter
[Oldoroto] see Lolderodo	K3	Nakuru	0 07 S 36 14 E	Bally, J. Adamson
[Oldoway] see Olduwai	T2	Masai	c.2 57 S 35 10 E	Eggeling
Old Shinyanga, popl.	T1	Shinyanga	3 33.5 S 33 24.5 E	Welch, Glover
Old Ukuti, popl.	U1	Acholi	c.3 41 N 33 32 E	Eggeling
Olduvai = Olduwai, gorge	T2	Masai	c.2 57 S 35 10 E	Bally, Greenway
Olduwai, gorge	T2	Masai	c.2 57 S 35 10 E	Hyde, Greenway, Richards
Olduwai, t.a.	T2	Masai	c.3 07 S 35 10 E	
Ole, popl., area	P	Pemba	5 11 S 39 48.5 E	Vaughan
Ol'ebolos, ln.	K6	Masai	c.1 08 S 35 09 E	Glover et al.
[Oleibortato] see Oloibortoto	K6	Masai	c.1 48 S 35 58 E—1 51 S 36 06 E	
Oleikaitorror, ln., escarp.	K6	Masai	1 10 S 36 06 E	
[Olekejo-Ngiro] see Ol Keju Nyiro	K6	Masai	c.1 23 S 36 33 E—1 50 S 36 23 E	Glover & Samuel

Olekemonge Plain, plain	K6	Masai	c.1 33 S 36 25 E	Bogdan
Olengejepa, ridge	K6	Masai	1 46 S 36 18.5 E	Glover et al.
[Olenguerone] see Olenguruone	K6	Masai	0 35 S 35 41 E	Glover et al.
Olenguruone, p.p., for. office	K6	Masai	0 35 S 35 41 E	
Olenguruone, Settlement Area	K6	Masai	c.0 37 S 35 38 E	Glover et al.
Olenyamu, ln.	K6	Masai	c.1 33 S 36 34 E	Piers
Oleolondo, rsta.	K3	Naivasha	0 23.5 S 36 22 E	
Olepolos, area	K6	Masai	1 29 S 36 38 E	
Ol Esakut, mt.	K6	Masai	1 30 S 36 33 E	Archer, Milne-Redhead & Taylor, Bally
Ol Esayeti, hill	K6	Masai	1 27.5 S 36 34 E	Beecher
Olgarien, str.	T2	Masai	c.2 36 S 35 22 E–2 31 S 35 39 E	
[Ol Gazi] see Olgos	T2	Masai	2 07 S 35 17 E	Greenway
Olgos, mt.	T2	Masai	2 07 S 35 17 E	
Oliopa, area	K6	Masai	1 09 S 35 17 E	Kuchar
Ol Joro = Olchoro, popl.	K6	Masai	0 49 S 36 00 E	Glover et al.
Ol Joro Nyuki, str.	T2	Masai	c.3 05 S 35 45 E–3 12 S 35 33 E	Greenway
Ol Joro Ole Soyet, for.	K6	Masai	c.1 04 S 35 17 E	Glover & Samuel
Ol Joro Orok, popl., p.p.	K3	Nakuru	0 04.5 S 36 22 E	Numerous
[Olkakola] see Olkokola	T2	Arusha	3 14 S 36 38 E	Hedberg, Carmichael
Ol Kalou, popl.	K3	Naivasha	0 16 S 36 22.5 E	Piers, C. Turner, Bally
Ol Kanjo, mt.	K1	Northern Frontier	0 52 N 37 46 E	
[Ol Kanjou] see Ol Kanjo	K1	Northern Frontier	0 52 N 37 45 E	
Olkanto, ln.	K1	Northern Frontier	1 01 N 37 20 E	J. Adamson
Olkaria = Orgaria, mt.	K3	Naivasha	0 53 S 36 16 E	Hedberg
Olkarien = Olgarien, str.	T2	Masai	c.2 36 S 35 22 E–2 31 S 35 39 E	Newbould
Ol Keju Ado = Kajiado, R.	K6	Masai	1 44 S 36 44 E–2 27 S 37 17 E	
[Ol Keju Nero] see Ol Keju Nyiro	K6	Masai	c.1 23 S 36 33 E–1 50 S 36 23 E	Glover et al.
Ol Keju Nyiro, str./s.	K6	Masai	c.1 23 S 36 33 E–1 50 S 36 23 E	Glover et al.
Ol Keju Rongai, str. (Kitumbeine)	T2	Masai	c.2 53 S 36 13 E	Carmichael
Olkokola, area	T2	Arusha	3 14 S 36 38 E	Huxley, Uhlig

319

Ol Kokwa, I.	K3	Baringo	0 37 N 36 04 E	M.G. Gilbert
[Ol Kongodu] see Orkunodo	T2	Arusha	c.3 10 S 36 55 E	Richards
[Olkuruta] see Olokurta Lukunya	T2	Arusha	3 16.5 S 36 37 E	Carmichael
Ollaioni, R.	K7	Teita	c.3 12 S 37 50 E	W.H. Lewis
Ol Lalang = Heathcote's Farm	K3	Trans-Nzoia	1 02.5 N 34 57 E	Tweedie
Ol'Loigululu Farm	K4	North Nyeri	0 02.5 S 37 03 E	Townsend
Ol Longonot = Longonot Mt.	K3	Naivasha	0 55 S 36 27 E	
Ol Lorgosailic = Olorgesailie	K6	Masai		Numerous
(There are many variations of spelling for both the hill and the National Park)				
Ol Lorgosailie, hill	K6	Masai	1 42 S 36 26 E	
Ol Moisor Ranch	K3	Laikipia	0 27 N 36 37 E	Cameron
[Ol Moloc] see Ol Molog	T2	Moshi	2 52 S 37 07 E	
Ol Molog, area	T2	Masai	c.2 50.5 S 37 07 E	
Ol Molog, popl.	T2	Moshi	2 52 S 37 07 E	Endlich, Fuggles Couchman, Greenway, Carmichael
[Ol Motij] see Ololmoti	T2	Masai	3 00 S 35 38 E	Greenway, Vesey-FitzGerald, Eggeling
Olmotoni = Engare Olmotoni	T2	Arusha	c.3 19 S 36 36 E	Carmichael
[Olmotonyi] see Engare Olmotoni	T2	Arusha	c.3 19 S 36 36 E	Tweedie
Ol Ngatongo Farm	K3	Trans-Nzoia	1 03 N 34 55 E	
Olodaari, str.	K6	Masai	c.1 03 S 36 06 E–1 10 S 35 59 E	Glover et al.
[Olodoari] see Olodaari	K6	Masai	c.1 03 S 36 06 E–1 10 S 35 59 E	Glover & Samuel
Olodungoro, area	K6	Masai	1 49 S 35 53 E	Glover et al.
Oloibortoto, str./s.	K6	Masai	c.1 48 S 35 58 E–1 51 S 36 06 E	Grindlay, Glover & Samuel
Oloitokitok, area	K6	Masai	c.2 53 S 37 37 E	
Oloitokitok = Laitokitok, p.p., r.h., airstrip	K6	Masai	2 56 S 37 30.5 E	Hooper & Townsend
Olokenya, sw.	K6	Masai	2 41 S 37 18 E	
Olokurta Lukunya, crater	T2	Arusha	3 16.5 S 36 37 E	
Olokurto, popl., t.c.	K6	Masai	0 38 S 35 50.5 E	
[Ololkisalie] see Ol Lorgosailie	K6	Masai	1 42 S 36 26 E	Glover et al.
[Ololkisalie, Nat. Pk.] see Olorgesailie	K6	Masai	1 35 S 36 26.5 E	

Ololmoti, crater	T2	Masai	3 00 S 35 38 E	B.D. Burtt
[Ololokwe] see Lolokwi	K1	Northern Frontier	0 50 N 37 32 E	Bally & Smith, Bally & Carter
Olololunga = Olulunga, r.h.	K6	Masai	1 00 S 35 40 E	
Ololongai = Olulunga, r.h.	K6	Masai	1 00 S 35 40 E	
Ololuaa, f.r.	K6	Masai	c.1 22 S 36 42 E	
Olulunga = Olulunga, r.h.	K6	Masai	1 00 S 35 40 E	
[Olomei] see Lolomei	K6	Masai	1 12 S 35 06 E	Glover et al.
[Olomoti] see Ololmoti	T2	Masai	3 00 S 35 38 E	B.D. Burtt
Olongogo (near Olbalbal)	T2	Masai	near 3 00 S 35 20 E	Newbould
[Oloolkisailie] see Olorgesailie	K6	Masai	1 35 S 36 26.5 E	
Olopito (near Rotian)	K6	Masai	near 0 58 S 35 54 E	Mackay
[Olorgasailie] see Ol Lorgosailie	K6	Masai	1 42 S 36 26 E	
[Olorgasaille] see Olorgosailie	K6	Masai	c.1 35 S 36 26.5 E	
[Ol Orgasalic] see Ol Lorgosailic	K6	Masai	1 42 S 36 26 E	
[Ol Orgasalik] see Ol Lorgosailic	K6	Masai	1 42 S 36 26 E	
Olorgesailie = Ol Lorgosailic	K6	Masai		

(There are many variations of spelling for both the hill and the National Park)

Olorgesailie, Nat. Pk.	K6	Masai	c.1 35 S 36 26.5 E	
Olorkesalie = Ol Lorgosailic	K6	Masai		
Oloropil, ln.	K6	Masai	0 43 S 35 59 E	Glover et al.
Olosendo = Olosentu, area	K6	Masai	1 09 S 34 52 E	Glover et al.
Olosirwa, hill	T2	Masai	3 03.5 S 35 46.5 E	Uhlig
Olowa Werikoi, mt.	K1	Northern Frontier	0 49 N 37 26 E	Bally & Carter
Ol Pusimoru, for.	K6	Masai	c.0 37 S 35 45 E	D.C. Edwards, Glover et al.
Ol' Tarakwai (near Narosura)	K6	Masai	c.1 32 S 35 52 E	Glover et al.
Oltepesi, ln.	K6	Masai	1 34 S 36 28 E	Kuchar
Oltiribe, w.h.	T2	Masai	3 14 S 35 25 E	
Ol Tukai, lodge	K6	Masai	2 40 S 37 17 E	Numerous
[Ol Tuki] see Ol Tukai	K6	Masai	2 40 S 37 17 E	
Oluchor, hill	U3	Mbale	0 32 N 34 08 E	G.H. Wood

321

Name		Region	Coordinates	Authority
Olulunga, r.h.	K6	Masai	1 00 S 35 40 E	Glover et al.
Omanimani, R.	U1	Karamoja	c.2 31 N 34 45 E–2 30 N 34 14 E	Purseglove
[Omeya Angima] see Omiya Anyma	U1	Acholi	3 16 N 33 16 E	
Omiya Anyma, popl.	U1	Acholi	3 16 N 33 16 E	
[Omkagkua] see Omukaguha	U2	Kigezi	0 55 S 29 50 E	
Omogo = Omugo, popl.	U1	West Nile	3 16 N 31 07 E	Greenway & Eggeling
Omogo = Omugo, t.a.	U1	West Nile	c.3 15 N 31 10 E	
Omoro, co.	U1	Acholi	c.2 45 N 32 30 E	Chancellor
Omoro, hill	U1	Acholi	2 42.5 N 32 44.5 E	
Omugo = Omogo, popl.	U1	West Nile	3 16 N 31 07 E	Chancellor
Omugo = Omogo, t.a.	U1	West Nile	c.3 15 N 31 10 E	
Omukaguha, popl.	U2	Kigezi	0 55 S 29 50 E	
Omunyal, str., sw.	U3	Teso	2 05 N 33 20 E–1 47 N 33 26 E	Lind
Omurwera, popl.	U2	Toro	0 30.5 N 31 00 E	
Ona, str.	T2	Moshi	c.3 13 S 37 28 E	
Ondiri Swamp, sw.	K4	Kiambu	1 15 S 36 39.5 E	Verdcourt, Bogdan
Ondoni, R.	K4/6	Machakos/Masai	Not traced	Prescott Decie
[Ongalea] see Ngulia	K4/6	Machakos/Masai	c.3 00 S 38 10 E	J.W. Gregory
Onjiko, popl.	K5	Kisumu-Londiani	0 10 S 34 54 E	Opiko
Opit, hill	U1	Acholi/Lango	2 35.5 N 32 28.5 E	Mukasa
Opok, for.	U1	Acholi	c.2 46 N 32 22 E	
Oraba, popl.	U1	West Nile	2 32 N 30 55 E	Hazel
Orangi, R.	T1	Musoma	2 14 S 34 18 E	Greenway
Orchid Place	K5	North Kavirondo	0 33 N 34 48.5 E	Tweedie
Orengitok = Eorengitok, ln.	K6	Masai	0 55 S 35 54 E	Glover et al.
Orero, popl.	T8	Kilwa	8 52 S 39 27 E	Braun
Orgaria, mt.	K3	Naivasha	0 53 S 36 16 E	Greenway
[Orhengnodwo] see Orkunudo	T2	Arusha	c.3 10 S 36 55 E	Richards
Orkunodo, area, lahars	T2	Arusha	c.3 10 S 36 55 E	Richards
Ormutonyi Plains	T2	Arusha	c.3 02 S 36 54 E	Richards

Name	Code	Region	Coordinates	Collector
[Orna Mdogo] see Ona				
Orok, peak	T2	Moshi	c.3 13 S 37 28 E	Volkens
Orom, f.r.	K6	Masai	2 30 S 36 45 E	
Oropoi, air strip	U1	Acholi	c.3 23 N 33 35 E	Dawkins
Oropoi = Natira, R.	K2	Turkana	3 46 N 34 14 E	Newbould
Ortum = Orrtum, popl., p.p.	U1,K2	Karamoja/Turkana	c.3 47.5 N 34 13 E –3 55 N 34 40 E	Dale, Liebenberg
Oru, R.	K2	West Suk	1 26 N 35 21 E	Tweedie, Carter & Stannard
[Oruchor] see Oluchor	U1	West Nile	c.3 19 N 31 03 E –3 14.5 N 31 08.5 E	Chancellor
Oruha, hill	U3	Mbale	0 32 N 34 08 E	G.H. Wood, Synnott
[Oruma] see Orumo	U2	Toro	0 39 N 30 30 E	Sangster, Eggeling
Orumo, popl., r.h.	U1	Lango	2 24 N 33 21 E	Eggeling
Orungo, popl.	U1	Lango	2 24 N 33 21 E	Eggeling
Orus, popl.	U3	Teso	2 02 N 33 27 E	
Oseni, popl.	K3	Baringo	0 59 N 36 19 E	Meyerhoff
[Osi] see Ozi	K7	Lamu	1 57.5 S 41 18 E	Greenway & Rawlins
[Osine] see Oseni	K7	Lamu	2 30 S 40 27 E	Greenway & Rawlins
Ossirwa, Mt.	K7	Lamu	1 57.5 S 41 18 E	Greenway & Rawlins
Ottune Steppe (Mt. Kenya)	T2	Masai	3 03.5 S 35 46.5 E	Oehler
Otuboi, popl., r.h., t.c.	K4		Not traced	Balbo
Otukei, mt., f.r.	U3	Teso	1 55 N 33 18 E	J. Wilson
[Otze] see Otzi	U1	Karamoja	c.2 30 N 33 35 E	Eggeling, E.M. Scott, A.S. Thomas
Otzi, mt.	U1	West Nile	3 37 N 31 51 E	Eggeling, E.M. Scott
[Ourougouro] see Uluguru	U1	West Nile	3 37 N 31 51 E	comm. Sacleux
Ovujo, popl.	T6	Morogoro	c.7 10 S 37 40 E	
Owor Oringenai, gully	U1	West Nile	3 14 N 30 54 E	Glover et al.
Oyam, co.	K6	Masai	c.1 05 S 35 57 E	
Oyster Bay, popl., bay	U1	Lango	c.2 25 N 32 35 E	Vaughan
Oyugis, popl.	T6	Uzaramo	6 47 S 39 17 E	Bogdan
[Ozi] see Tana R. (part)	K5	South Kavirondo	0 30 S 34 43 E	Gregory, Greenway & Rawlins
Ozi, popl.	K7	Tana River/Lamu	c.2 28 S 40 12 E –2 30 S 40 30 E	
	K7	Lamu	2 30 S 40 27 E	

Name	Code	Region/People	Coordinates	Collector
[Paliko] see Patiko	U1	Acholi	3 01 N 32 19 E	Maxwell Forbes
Pallisa, co.	U3	Mbale	c.1 20 N 33 45 E	
Pallisa, popl., mssn., r.h.	U3	Mbale	1 10 N 33 43 E	Eggeling, Dale
Paloga, popl., r.h.	U1	Acholi	3 35 N 32 57 E	Purseglove
[Palora] see Pawe	U1	Acholi	3 06 N 32 09 E	Hancock
[Paluga] see Paloga	U1	Acholi	3 35 N 32 57 E	Eggeling
[Pamota] see Tamota	T3	Lushoto	5 04 S 38 23 E	Liebusch
Pamwa, popl., r.h.	U1	Lango	2 21 N 32 23 E	Eggeling
Panda, ln.	T7	Iringa	c.7 45 S 36 40 E	Carmichael
Pandani, area, popl.	P	Pemba	5 03.5 S 39 46.5 E	Greenway
Pande, popl., area	T3	Tanga	5 03 S 38 56 E	Peter, Geilinger
Pande, hill, f.r.	T6	Uzaramo	c.6 42 S 39 05 E	Leopold, Wingfield
[Pandeni] see Mpandeni	T3	Lushoto	5 07.5 S 38 41.5 E	Drummond & Hemsley, Peter, Zimmermann
[Panga] see Mpanga	U2	Toro	0 39 N 30 08 E–0 03 N 30 17 E	Bagshawe
Panga, popl.	T3	Handeni	5 18 S 38 19 E	Busse
Panga Juu, well	Z	Zanzibar	6 14 S 39 27 E	Vaughan, Faulkner
Pangali, popl.	T4	Ufipa	7 37 S 31 35 E	Bullock
Pangani, ln.	K7	Kilifi	3 51.5 S 39 40 E	Faden
Pangani, popl.	K7	Kwale	4 30 S 39 07 E	Allan
Pangani, popl.	K7	Lamu	2 21.5 S 40 34 E	Hooper & Townsend
Pangani, distr.	T3	Pangani	c.5 35 S 38 45 E	
Pangani, popl., port	T3	Pangani	5 25.5 S 38 59 E	Numerous
Pangani, R.	T2/3	several	3 32 S 37 34 E–5 26 S 38 58 E	Numerous
Pangani Falls	T3	Pangani	5 21 S 38 39 E	Peter, Stuhlmann
Pangani Rapids, falls	T6	Morogoro/Rufiji	7 50 S 37 53 E	Goetze
Pangire, mssn.	T7	Njombe	9 24 S 34 56 E	Wolfe
Pangulidala, Mt.	T7	Njombe	9 15 S 34 22 E	Eusebio
Pangundutani, ln.	T7	Rungwe	c.9 01 S 33 38 E	Brenan & Greenway
[Pangwe] see Pongwe	T3	Tanga	5 07.5 S 38 59 E	Carmichael

[Patta] see Pate	K7	Lamu		c.2 07 S 41 05 E	Greenway & Rawlins
[Patte] see Pate	K7	Lamu		c.2 07 S 41 05 E	
Pawaga, t.a.	T7	Iringa		c.7 08 S 35 30 E	Childs, Ward
Pawe, popl.	U1	Acholi		3 06 N 32 09 E	
Paya, popl.	U3	Mbale		0 50 N 34 00 E	Maitland
[Payida] see Paidha	U1	West Nile		2 25 N 31 00 E	Numerous
[Peet] see Pete	Z	Zanzibar		6 17 S 39 25 E	
Pegi, hill	U3	Busoga		1 23 N 32 52 E	
Pelekech, mt. range	K2	Turkana		c.3 45 N 35 05 E	Dale
Pelekech, peak	K2	Turkana		3 48 N 35 05 E	Buxton
[Peleketch] see Pelekech	K2	Turkana		3 48 N 35 05 E	Kassner
[Pemba] see Cha Shimba	K7	Kwale		c.4 09 S 39 18 E –4 04.5 S 39 32 E	Numerous
Pemba, I.	P	Pemba		c.5 10 S 39 48 E	Busse
Pembamoto, popl.	T5	Mpwapwa		c.6 16 S 36 48 E	Vaughan
Pembe, I.	P	Pemba		5 08 S 39 40.5 E	Tanner
Pembyabwe (near Mwera)	T3	Pangani		near 5 20 S 38 59 E	
[Pendeni] see Mpandeni	T3	Lushotc		5 07.5 S 38 41.5 E	
Pengo, hill, for.	K7	Kwale		4 15 S 39 23 E	Drummond & Hemsley, Magogo & Glover
Peninj, str.	T2	Masai		c.2 20 S 35 58 E	Uhlig
Penny Penn's Farm, area	T7	Iringa		8 14 S 35 07 E	Bidgood, Brummitt & Mwasumbi
Peramiho, mssn.	T8	Songea		10 39 S 35 27 E	Eggeling, Milne-Redhead & Taylor
Perani, t.a., for.	T3	Tanga		c.4 37 S 39 02 E	Drummond & Hemsley
Peremehe, popl.	T5	Dodoma		6 44 S 35 37 E	
Perkerra Irrigation Scheme	K3	Baringo		c.0 28 N 36 02 E	Bogdan
Pero, popl.	T4	Tabora		4 58 S 32 50 E	Hannington
Pete, popl., area	Z	Zanzibar		6 17 S 39 25 E	Vaughan
Peter's Hut = Horombo Hut	T2	Moshi		3 08 S 37 26 E	Numerous
Petete, popl.	U3	Mbale		1 11 N 33 56 E	
P.F.T. Farm	K3	Trans-Nzoia		0 56.5 N 34 47 E	Tweedie
Pian, co.	U1	Karamoja		c.2 00 N 34 40 E	Dyson-Hudson, Kerfoot, J. Wilson

327

Picnic Point	K2	West Suk	1 16 N 35 09 E	Tweedie
Pienaars Heights	T2	Mbulu	4 23 S 35 46 E	Numerous
Piganyonga's, popl.	T8	Songea	9 53 S 35 20 E	Milne-Redhead & Taylor
Pika Pika Rock	K7	Teita	3 50 S 38 53 E	Bally
Piki, popl., area	P	Pemba	5 07 S 39 46 E	Vaughan
Pikurugwe, mt.	T7	Njombe	c.9 05 S 34 02 E	Goetze
Pimbi, hill	T3	Lushoto	5 08 S 38 39 E	Greenway
Pingire, popl., r.c.	U3	Teso	1 23 N 33 22 E	Lind
[Pingiri] see Pingire	U3	Teso	1 23 N 33 22 E	Vaughan
Pink Terraces, ln.	Z	Zanzibar	c.6 06.5 S 39 14.5 E	Tweedie
Pipe Line	K3	Trans-Nzoia	c.1 10 N 34 46 E	A.S. Thomas, Brasnett
Pirre, popl.	U1	Karamoja	3 51 N 34 04 E	Goetze
[Pisaki] ? see Paschiägu	T7	Mbeya	?8 55 S 32 50 E	Richards
Pitiores Gorge	T2	Arusha	prob. c.3 14 S 36 50 E	Bullock
Pito, popl., mssn.	T4	Ufipa	8 07 S 31 37 E	
Plateau, rsta.	K3	Uasin Gishu	0 26 N 35 22 E	Greenway
Poacher's Lookout, ln.	K6	Masai	2 57.5 S 37 58.5 E	
Pokezi = Pokezeni, sw.	Z	Zanzibar	5 56 S 39 19.5 E	
[* Pokot, West] see West Suk	K2	West Suk	c.1 30 N 35 25 E	
(*modern name, West Suk continues to be used for F.T.E.A.)				
Polhill's, Mrs. E.	K3	Naivasha	0 48 S 36 16 E	Hooper & Townsend
Polish Farm	T4	Ufipa	Not traced	Richards
Pommern, mssn.	T7	Iringa	8 06 S 35 46 E	Polhill & Paulo
[Ponda] see Ponde	T3	Lushoto	4 54 S 38 24.5 E	Buchwald
Ponde, popl., hill, str.	T3	Lushoto	4 54 S 38 24.5 E	
Pongolo = Pongola (Uzungwa Mts.)	T7	Iringa	Not traced	Goetze
Pongwe, popl., rsta., mssn.	T3	Tanga	5 07.5 S 38 59 E	Numerous
Pongwe, sisal est.	T3	Tanga	5 06.5 S 38 58 E	
Pongwe, hill, f.r.	T6	Bagamoyo	6 18 S 38 14 E	Harris, Holst
Pongwe, popl., area	Z	Zanzibar	6 02 S 39 24 E	Vaughan

Name	Grid	District/Region	Coordinates	Collector
[Pop Onditi] see Paponditi	K5	Central Kavirondo	0 19 S 34 56 E	Kokwaro
Pori	(Type of vegetation)			Schlieben, Peter
[Poror] see Poro	K1	Northern Frontier	1 15 N 36 37 E	Leakey, Carter & Stannard, J. Bally
Poroto, mts.	K1	Northern Frontier	1 15 N 36 37 E	Numerous
Poroto, popl., sawmill	T7	Rungwe/Mbeya/Njombe	c.9 00 S 33 45 E	Richards
Port Bell, popl.	T7	Mbeya	8 59 S 33 38 E	
[Port Florence] see Kisumu	U4	Mengo	0 17 N 32 39 E	Winkler
Port Tudor, harbour	K5	Kisumu-Londiani	0 06 S 34 45 E	Dale, MacNaughton, Sulemani
Port Victoria, popl., mssn.	K7	Mombasa	4 01 S 39 39 E	Glasgow
Posta, hill	K5	Central Kavirondo	0 06 N 33 58 E	Argyle
Potwe, t.a.	K5	Kericho	0 38 S 35 23 E	Tanner
Potwe, area, sisal est., for.	T3	Pangani	c.5 21 S 38 44 E	Semsei
[Poucha Poucha] see Puchapucha	T3	Tanga	c.5 13 S 38 37 E	J. Bally
[Powys Farm] see Ngare Ndare Ranch	T8	Tunduru	10 58 S 37 42 E	Hooper & Townsend
Pozo Moyo, popl.	K4	North Nyeri	0 07 N 37 25 E	Numerous
[Prison] see Changa	T4	Tabora	5 06 S 31 46 E	
Prison Dam	Z	Zanzibar	6 07 S 39 10 E	Tweedie
Prison Farm	K3	Trans-Nzoia	0 59 N 35 00 E	Tweedie
Puchapucha, popl.	K3	Trans-Nzoia	0 58 N 35 03 E	Milne-Redhead & Taylor
Pugu, f.r.	T8	Tunduru	10 58 S 37 42 E	Numerous
Pugu, hills	T6	Uzaramo	c.6 54 S 39 05 E	Numerous
Pugu, str.	T6	Uzaramo	c.6 55 S 39 02 E	Holtz
Puhi = Pui, for.	T6	Uzaramo	c.6 55 S 39 02 E	Richards & Arasululu
Puma, popl., t.c.	T5	Kondoa	4 26 S 35 46 E	Polhill & Paulo
Punda Milia, est.	T5	Singida	5 00 S 34 44.5 E	
Punda Milia, popl.	K4	Fort Hall	0 52.5 S 37 12 E	Goodhart
Pungatini, popl.	K4	Fort Hall	0 54 S 37 11 E	
Pungu, str./s.	T8	Kilwa	8 26 S 39 02 E	Rammell, Jarrett
Pungulumo, mt.	K4	Machakos	c.2 01 S 37 41 E	
	T7	Mbeya	8 47 S 33 15 E	

Name	Code	Region	Coordinates	Collector
[Pungusi] see Ipunguli	T5	Dodoma	5 51 S 34 23 E	Stuhlmann
[Pungutini] see Pungatini	T8	Kilwa	8 26 S 39 02 E	Braun
Putini, popl.	T3	Tanga	5 13 S 39 03 E	Peter
Pwani, popl.	Z	Zanzibar	6 02 S 39 12.5 E	Faulkner, Greenway
Pwani Mchangani, popl., area	Z	Zanzibar	5 56 S 39 21 E	Greenway
[Pyeda] see Paidha	U1	West Nile	2 25 N 31 00 E	Tothill
Quambo (N. Hanang)	T2	Mbulu	c.4 24 S 35 24 E	Carmichael
[Qwamkembi] see Kwamkembe	T3	Tanga	5 13 S 38 58 E	Faulkner
[Quaraa] see Kwaraha	T2	Mbulu	4 14 S 35 48 E	Carmichael
[Quare] see Kware	T2	Moshi	3 16 S 37 10 E–3 27 S 37 12.5 E	Volkens
[Quarre] see Kware	T2	Moshi	3 16 S 37 10 E–3 27 S 37 12.5 E	Volkens
[Queen Elizabeth Nat. Park] see Rwenzoli	U2	several	c.0 15 S 30 00 E	Lind, Symes
Queen's Cave Waterfall, falls	K4	South Nyeri	0 29 S 36 42 E	Polhill
Quenyasi (Marang Forest)	T2	Mbulu	c.3 42 S 35 40 E	Carmichael
[Quiloa] see Kilwa	T8	Kilwa	c.9 20 S 38 15 E	Kirk
Rabai, hills	K7	Kilifi	c.3 56 S 39 35 E	W.E. Taylor
Rabai, popl.	K7	Kilifi	3 56 S 39 34 E	Numerous
[Rabay] see Rabai	K7 not Z	Kilifi	3 56 S 39 34 E	Sacleux
Rabondo, hills	K5	South Kavirondo	0 33 S 34 35 E	
Rabongo, mt., for.	U2	Bunyoro	c.2 08 N 31 57 E	H.E. Brown, G. Jackson
Ragati, for. stn.	K4	South Nyeri	0 23 S 37 09.5 E	Kerfoot, Kokwaro
Ragem, popl.	U1	Acholi	2 40 N 31 24 E	
Ragwe, popl., area	T3	Tanga	5 00 S 38 58 E	Geilinger
Rainkombe kopje, hill	K4	Meru	0 07 N 38 12.5 E	P. Hamilton
Rakai, popl.	U4	Masaka	0 43 S 31 24 E	
Ramisi, popl.	K7	Kwale	4 32 S 39 24 E	Numerous
Ramisi, R.	K7	Kwale	4 23 S 39 11 E–4 33 S 39 23 E	Drummond & Hemsley
Ramu, p.p., t.c.	K1	Northern Frontier	3 56 N 41 13 E	J. Adamson, Gillett

Name	Code	Region	Coordinates	Collector
[Ramusambi] see Ruamuthambi				
[Rangani] see Rongoni, Ras				
[Rapai] see Arapai				
Rapids	K2	Turkana/West Suk	1 47.5 N 33 37 E	Maitland
Rapogi, mkt., mssn., sch.	K5	South Kavirondo	c.1 29 N 35 02.5 E	Tweedie
Ras Biongwe, pt.	K7	Lamu	0 54 S 34 28 E	Glasgow
Ras Bweni, pt.	T6	Rufiji	2 23 S 40 49 E	Frazier
Ras Chiamboni	Som.		7 41 S 39 52 E	
			1 39 S 41 36 E	J. Adamson
Ras Chokii, pt.	T6	Uzaramo	6 49 S 39 18 E	Frazier
Ras Chukwani, pt.	Z	Zanzibar	6 15 S 39 12.5 E	
Ras Domoni, cape	P	Pemba	5 27 S 39 42.5 E	Greenway
Rasha Rasha, est.	T2	Masai	3 18 S 36 27 E	Ritchie
[Rashitani] see Rishateni				
[Rashitany] see Rishateni	T2	Arusha	3 14 S 36 54.5 E	Richards
Ras Kankadya, pt.	T2	Arusha	3 14 S 36 54.5 E	Procter
[Ras Kasone] see Ras Kazone	T6	Uzaramo	6 43 S 39 16.5 E	Lindeman
Ras Kazone, pt.	T3	Tanga	5 03 S 39 07.5 E	
	T3	Tanga	5 03 S 39 07.5 E	
Ras Kidomoni = English Point	K7	Mombasa	4 03 S 39 41 E	Greenway
Ras Kigomasha, cape	P	Pemba	4 52 S 39 41 E	S.A. Robertson
Ras Kikadini	K7	Kwale	4 10 S 39 38 E	
Ras Kiongwe, pt.	T3	Tanga	4 47 S 39 11 E	Harris
Ras Kiromoni, pt.	T6	Uzaramo	6 37.5 S 39 11.5 E	Greenway, Schlieben
Ras Mbisi, pt.	T6	Rufiji	7 49 S 39 42 E	
Ras Michamvi, pt.	Z	Zanzibar	6 07.5 S 39 30 E	Greenway
Ras Mkumbi, pt.	T6	Rufiji	7 38 S 39 53.5 E	Greenway
Ras Mkumbuu, cape	P	Pemba	c.5 13 S 39 40 E	
Ras Mtangawanda, pt.	K7	Lamu	2 07 S 40 58 E	Polhill & Paulo, Hacker
Ras Ngomeni, pt.	K7	Kilifi	2 59 S 40 14.5 E	Tidbury
Ras Nungwi, pt.	Z	Zanzibar	5 43 S 39 18 E	
Ras Rongoni, pt.	T6	Uzaramo	6 49 S 39 19 E	

331

332

Riwa, mt.	U3	Mbale	1 21.5 N 34 47 E	
Riwa, str.	K2	Turkana	c.1 20 N 34 50 E	
Riwa, area	K2	West Suk	c.1 17 N 34 55 E	Pratt
Roaring Rocks, ln.	K6	Masai	2 56 S 38 07 E	Greenway
[Robaya] see Rubaya				
Robeho, pass	U2	Kigezi	1 25 S 29 57 E	
[Roborogoto] see Ruborogota	T5	Mpwapwa	c.6 34 S 36 17 E	Grant
Roccati Pass	U2	Ankole	0 59 S 30 34 E	Snowden
Rocky Kopjes	U2	Toro	0 26 N 29 54.5 E	Osmaston
[Rogovero] see Rojewero	T2	Masai	prob. c.2 35 S 35 25 E	Greenway
Roi, mkt.	K1/4	N. Frontier/Meru	c.0 16 N 37 56 E – 0 04 S 38 25 E	J. Adamson
Rojewero, R.	K4	Kiambu	0 58 S 36 48 E	
Rojewero Circuit	K1/4	N. Frontier/Meru	c.0 16 N 37 56 E – 0 04 S 38 25 E	J. Adamson, Ament
Roka, area	K4	Meru	0 10 N 37 10 E	Ament
Roka, popl.	K1	Northern Frontier	c.1 33 S 40 27 E	?Dale
Rom, popl., r.h.	K7	Kilifi	3 26 N 39 54 E	Dale
Rom, popl.	U1	Acholi	3 19 N 33 32 E	Eggeling, Purseglove
Rom, mt.	U1	Acholi	3 24 N 33 28 E	
Rom III, mt.	U1	Acholi	3 22 N 33 36 E	Numerous
Rombo, popl.	U1	Acholi	3 24 N 33 37 E	
Rombo, area	K7	Teita	3 03 S 37 42.5 E	Pfennig
Rombo, mssn.	T2	Moshi	c.3 15 S 37 40 E	
[Rombo] see Bombo	T2	Moshi	3 09 S 37 38 E	Haarer
[Romuma] see Lumuma	K6, 7/T2	Masai/Teita/Moshi	c.3 03 S 37 38 E – 3 03 S 37 55 E	D.C. Edwards, Haarer, Volkens
Romwe = Lomwe, popl.	T3	Lushoto/Tanga	c.4 46 S 38 42 E – 4 31 S 38 47 E	Holst
Rondo, R.	T5/6	Mpwapwa/Kilosa	c.7 00 S 36 38 E – 6 39 S 36 42 E	Grant, Peter
Rondo, f.r.	T3	Handeni	5 40 S 37 41.5 E	Semsei & Gane
Rondo Chini, popl.	T7	Iringa	prob. c.8 30 S 35 31 E	Carmichael
Rondo Plateau	T8	Lindi	c.10 09 S 39 15 E	Eggeling, Wigg
	T8	Lindi	10 08 S 39 14 E	
	T8	Lindi	c.10 10 S 39 15 E	Numerous

333

334

Name		Region	Coordinates	Collector/Source
Ruanda, mssn.	T8	Songea	10 33 S 34 57 E	Busse, Milne-Redhead & Taylor
Ruande, f.r.	T1	Mwanza	2 42 S 32 05 E	Gane
Ruaraka, area	K4	Nairobi	c.1 14 S 36 53 E	
Ruasina, f.r.	T1	Bukoba	c.1 20 S 31 42 E	
Rubaare, popl., mssn.	U2	Ankole	1 01 S 30 11 E	Lye
Rubabu, popl., r.c.	U2	Kigezi	0 55 S 29 58 E	Purseglove
Rubafu, popl.	T1	Bukoba	1 02 S 31 50 E	
Rubanda, co.	U2	Kigezi	c.1 15 S 29 50 E	Purseglove
Rubanda, popl.	U2	Kigezi	1 07 S 29 51 E	Purseglove
[Rubare] see Rubaare	U2	Ankole	1 01 S 30 11 E	Katende, Procter
Rubare, f.r.	T1	Bukoba	1 23 S 31 48.5 E	Numerous
Rubaya, popl.	U2	Kigezi	0 47 S 29 46 E	?Purseglove, A.S. Thomas
Rubaya, popl., r.h.	U2	Kigezi	1 25 S 29 57 E	?Purseglove
Rubeho, mts.	T5/6	Mpwapwa/Kilosa	c.6 55 S 36 30 E	B.D. Burtt, Grant
Rubirizi, popl.	U2	Ankole	0 16 S 30 07 E	
Rubogo, sw.	T1	Bukoba	prob. c.1 04 S 31 48 E	Gillman
Rubondo, I.	T1	Mwanza	2 20 S 31 52 E	Gane
Ruboni, str.	U2	Toro	0 20 N 29 58 E–0 21 N 30 02.5 E	Osmaston
Ruborogota, popl.	U2	Ankole	0 59 S 30 34 E	Eggeling
[Ruborogoto] see Ruborogota	U2	Ankole	0 59 S 30 34 E	
Rubuga, popl.	T4	Tabora	5 15 S 33 22 E	Jackson
[Rubugua] see Rubuga	T4	Tabora	5 15 S 33 22 E	Stuhlmann
Rubuguli, mssn.	U2	Kigezi	1 07 S 29 41 E	Purseglove
[Rubugwa] see Rubuga	T4	Tabora	5 15 S 33 22 E	Grant
Rubungo, popl.	T1	Bukoba	1 50 S 31 40 E	
[Rubungu] see Rubungo	T1	Bukoba	1 50 S 31 40 E	Welch
Rubuzigye, popl.	U2	Ankole	0 26 S 30 02 E	Lye
Rubya, f.r.	T1	Mwanza	c.2 03 S 32 50 E	Semkiwa, Carmichael, Procter
Ruchetera, P.W.D.	U2	Kigezi	1 03 S 29 59 E	Norman
[Ruchiga] see Rukiga	U2	Kigezi	c.1 05 S 30 00 E	

Name	Code	Location	Coordinates	Collector
[Ruchigga] see Rukiga				
Rudewa, popl., mssn.	U2	Kigezi	c.1 05 S 30 00 E	Bagshawe
Rudewa, popl.	T6	Kilosa	6 41 S 37 08 E	Swynnerton
[Rudolf] see Turkana, L.	T7	Njombe	10 06.5 S 34 39.5 E	Thulin & Mhoro
Rufiji, distr.	K1/2/Ethiopia	N. Frontier/Turkana	2 22—4 32 N 35 50—36 42 E	Numerous
Rufiji, R.	T6	Rufiji	c.8 00 S 38 45 E	Musk
	T6/8	Ulanga/Morogoro/ Rufiji/Kilwa	c.8 31 S 37 22 E—8 00 S 39 20 E	Numerous
Rufua = Rufuha, R.	U2	Ankole	0 59 S 30 13 E— 1 03 S 30 23 E	Bagshawe
Rugaga, popl.	U2	Ankole	0 50 S 31 01 E	
[Rugalo] see Lugalu	T7	Iringa	7 43 S 35 52 E	Goetze, Stolz
Ruganzo, popl.	T1	Ngara	2 32 S 30 42 E	Tanner
[Rugaro] see Lugalu	T7	Iringa	7 43 S 35 52 E	Goetze
Rugewa = Rujewa	T7	Mbeya	8 42 S 34 23 E	Anderson, Polhill & Paulo, Richards
[Rugeyo] see Rugyeyo	U2	Kigezi	0 54 S 29 49.5 E	Purseglove
Rugezi, str.	U2	Kigezi	1 04 S—1 01 S 29 44 E	
Rugongo, hill	U2	Ankole	0 22 S 30 29 E	Rwaburindore
Rugongo, popl.	T1	Biharamulo	2 55.5 S 31 13 E	Verdcourt
Rugufu, R.	T4	Kigoma	5 44 S 30 40 E—5 21.5 S 29 46 E	
[Rugusu] see Ruguthu	K1/4	several	c.0 03 S 37 22 E—0 17 N 37 33 E	Schelpe
Ruguthu = E. Marania, R.	K1/4	several	c.0 03 S 37 22 E—0 17 N 37 33 E	
[Rugwisi] see Lugungwisi	T4	Mpanda	6 05 S 30 02 E—5 56 S 29 56 E	Richards
Rugyeyo, r.h.	U2	Kigezi	0 54 S 29 49.5 E	
Ruhama, r.h.	U2	Ankole	0 59 S 30 21 E	Eggeling
Ruhamba, peak	T6	Morogoro	c.6 04 S 37 31 E	Drummond & Hemsley, Semsei
Ruhanga, popl.	U2	Ankole	0 52 S 30 11 E	Stuhlmann
[Ruhedje] see Ruhudji	T6/7	Ulanga/Njombe	c.9 26 S 34 34 E—8 52 S 36 01 E	
[Ruhemba] see Buhemba	T1	Musoma		
[Ruhembe] see Luhembe	T6	Kilosa	7 15 S 36 50 E—7 43 S 37 04 E	Goetze
[Ruhengera] see Ruhengere	U2	Ankole	0 22 S 30 45 E	Jarrett
Ruhengere, popl.	U2	Ankole	0 22 S 30 45 E	

Name		Region	Coordinates	Collector(s)
[Ruhidge Falls] see Ruhudji Falls	T7	Njombe	9 21 S 34 45 E	Eggeling
[Ruhidji Falls] see Ruhudji Falls	T7	Njombe	9 21 S 34 45 E	Pollock
[Ruhiji] see Ruhudji	T6/7	Ulanga/Njombe	c.9 26 S 34 34 E—8 52 S 36 01 E	
[Ruhimba] see Luhimba	T8	Songea	c.10 26 S 35 43 E—10 30 S 35 37 E	Purseglove
Ruhinda, popl., r.h.	U2	Kigezi	0 41 S 29 56 E	
Ruhira, popl.	U2	Ankole	0 55 S 30 50 E	
Ruhiza = Luhizha, popl.	U2	Kigezi	1 02 S 29 47 E	Hamilton
[Ruhoho] see Rwoho	U2	Ankole	0 51 S 30 32 E	St. Clair-Thompson
[Ruhudje] see Ruhudji	T6/7	Ulanga/Njombe		
[Ruhudje Falls] see Ruhudji Falls	T7	Njombe	9 21 S 34 45 E	Goetze, Schlieben
Ruhudji, R.	T6/7	Ulanga/Njombe	c.9 26 S 34 34 E—8 52 S 36 01 E	Eggeling
Ruhuhu, R.	T7/8	Njombe/Songea	c.9 30 S 34 48 E—10 31 S 34 34 E	Goetze, Milne-Redhead & Taylor
[Ruhuyango] see Rutenganio	T7	Rungwe	9 21 S 33 38 E	Busse
Ruiga River, f.r.	T1	Biharamulo/Bukoba	c.2 25 S 31 30 E	
Ruimi, str.	U2	Toro	0 26 N 30 00 E—0 20 N 30 17 E	Stolz
[Ruiri] see Ruizi	U2	Ankole	c.0 45 S 30 14 E—0 40 S 30 52 E	Procter, Carmichael
Ruiru, popl., p.p.	K4	Kiambu	1 09 S 36 57.5 E	Eggeling, Osmaston, Scott Elliot
Ruiru, R.	K4	Kiambu	1 02.5 S 36 40 E—1 10 S 37 05 E	Jarrett
[Ruisamba] see George L.	U2	Toro/Ankole	c.0 00 30 12 E	Numerous
Ruisongo, L.	U2	Ankole	near 0 17 S 30 07 E	Bagshawe
Ruiwa, popl., t.c.	T7	Mbeya	8 46 S 33 38 E	Eggeling
Ruizi, R.	U2	Ankole	c.0 45 S 30 14 E—0 40 S 30 52 E	Trapnell, Cribb & Grey-Wilson, Procter
Rujewa, mssn.	T7	Mbeya	8 41 S 34 19 E	Eggeling, Jarrett, G. Taylor
Rujewa, popl., sch.	T7	Mbeya	8 42 S 34 23 E	Anderson
[Rujezi] see Rugezi	U2	Kigezi	1 04 S—1 01 S 29 44 E	St. Clair-Thompson
[Rujwe] see Rujewa	T7	Mbeya		
Rukanga, sch.	K7	Teita	3 49 S 38 38 E	Faden et al.
Rukiga, co.	U2	Kigezi	c.1 05 S 30 00 E	Bagshawe
Rukinga, hill	K7	Teita	3 51.5 S 38 45 E	Vesey-FitzGerald
Rukinga, popl.	T6	Uzaramo	7 23 S 38 45 E	Stuhlmann

338

Name	Code	Region	Coordinates	Collector
Rukiri, popl., r.h., t.c.	U2	Ankole	0 17 S 30 21 E	
[Rukungire] see Rukungiri	U2	Kigezi	0 47 S 29 55 E	
Rukungiri, popl., mssn., t.c.	U2	Kigezi	0 47 S 29 55 E	Eggeling, Purseglove, A.S. Thomas
Rukungwe, R., for. (near Amani)	T3	Lushoto	near 5 06 S 38 38 E	Ali Omare
[Rukungwu] see Rukungwe	T3	Lushoto	near 5 06 S 38 38 E	Ali Omare
Rukwa, L.	T4/7	several	8 00 S 32 25 E	Numerous
Rukwa Rift Valley	T4	Ufipa/Mpanda	c.7 20 S 31 35 E	Numerous
[Rula] ? see Lula	T7	Iringa	c.7 38 S 35 57 E	Stolz
Rulenge, popl., p.p., mkt.	T1	Ngara	2 43 S 30 38 E	Tanner, Eggeling
[Rulongwe] see Bulongwa	T7	Njombe	9 20 S 34 03.5 E	Stolz
Rumakali, R.	T7	Njombe/Rungwe	c.9 06 S 33 52 E –9 20 S 33 55 E	Stolz
[Rumakira] see Rumakali	T7	Njombe/Rungwe	c.9 06 S 33 52 E –9 20 S 33 55 E	Goetze
Rumara	T1	Mwanza	Not traced	Tanner
[Rumbera] see Lumbila	T7	Njombe	9 34.5 S 34 07.5 E	
Rumbhia, popl., mssn.	K4	Embu	0 35 S 37 39 E	
[Rumbia's] see Rumbhia	K4	Embu	0 35 S 37 39 E	
Rumbira, R.	T7	Njombe	c.9 13 S 34 15 E –9 33 S 34 10 E	M.D. Graham
[Rumbira] see Lumbila	T7	Njombe	9 34.5 S 34 07.5 E	Goetze
Rumeno = Lumeno, str.	T6/7	Ulanga/Iringa	c.7 50 S 36 43 E –8 11 S 36 40.5 E	Goetze
[Rumerno] see Rumeno	T6/7	Ulanga/Iringa	c.7 50 S 36 43 E –8 11 S 36 40.5 E	Haerdi
Rumogi, popl.	U1	West Nile	3 33 N 31 22 E	Eggeling
[Rumuma] see Lumuma	T5/6	Mpwapwa/Kilosa	c.7 00 S 36 38 E –6 39 S 36 42 E	
Rumuruti, popl., airstrip	K3	Laikipia	0 15.5 N 36 32 E	Numerous
Runazi, popl., mkt.	T1	Biharamulo	2 47 S 31 28 E	Gane, Procter
Rundi = Lundi, area, popl.	T6	Morogoro	7 04 S 37 50 E	Harris
[Rungaro = Rugaro] see Lugalu	T7	Iringa	7 43 S 35 52 E	
Rungemba, mssn.	T7	Iringa	8 11 S 35 23 E	McLoughlin, Goetze, Eggeling
[Rungulugulu] see Ngulugulu	T7	Rungwe	c.9 25 S 33 31 E	Stolz
Rungwa, popl.	T4	Tabora	6 57 S 33 31.5 E	Boaler
Rungwa, f.r.	T5	Dodoma	c.7 00 S 34 00 E	Carmichael

Name	Code	Region	Coordinates	Collector
[Rungwa] see Rungwa River, g.r.				
Rungwa, popl.	T/4/5/7	Tabora/Dodoma/Chunya	c.7 15 S 34 30 E	Procter
Rungwa, R.	T4	Mpanda	7 20 S 31 40 E	Bullock, Michelmore, Boaler
Rungwa River, f.r.	T4/5/7	several	6 47 S 33 56 E—7 36 S 31 50 E	Richards, Vesey-FitzGerald
Rungwa River, g.r.	T4	Mpanda	c.7 00 S 32 15 E	Carmichael
Rungwa River, g.r.	T4/5/7	Tabora/Dodoma/Chunya	c.7 15 S 34 30 E	Richards, Mdehwa
[Rungwe] see Rungwa	T4/5/7	several	6 47 S 33 56 E—7 36 S 31 50 E	Richards
Rungwe, crater	T7	Rungwe	9 09.5 S 33 40.5 E	
Rungwe, distr.	T7	Rungwe	c.9 20 S 33 30 E	
Rungwe, f.r.	T7	Rungwe	c.9 10 S 33 40 E	Brenan & Greenway, Semsei
Rungwe, mssn.	T7	Rungwe	9 10 S 33 35.5 E	Numerous
Rungwe, mt.	T7	Rungwe	9 08 S 33 40 E	Numerous
Runyenje's, popl.	K4	Embu	0 25 S 37 34 E	W.H. Lewis
Rupa, popl.	U1	Karamoja	2 34 N 34 40 E	J. Wilson, Lye, Harker
Rupa, popl.	T1	Musoma	1 48 S 33 39 E	
Rupa Road	U1	Karamoja	c.2 35 N 34 40 E	Tweedie
Rupia, popl.	T6	Ulanga	8 35 S 37 16 E	Schlieben
Rupia (Ukinga area)	T7	Njombe	Not traced	Braun (label reads Ukinga, Upangu-Ubena)
Rupingazi, R.	K4	S. Nyeri/Embu/Meru	0 10 S 37 19 E—0 43 S 37 28.5 E	F. White, Fries, D. Davis
Ruponda, popl.	T8	Lindi	10 15 S 38 42.5 E	Anderson, Semsei
Rusako, popl.	T6	Bagamoyo	6 26.5 S 38 40 E	Magogo
Rusarus Plateau	K1	Northern Frontier	1 33 N 38 00 E	
[Ruseko] see Rusako	T6	Bagamoyo	6 26.5 S 38 40 E	
Rusende, popl.	T6	Rufiji	7 56 S 38 43 E	Musk, Nicholson
Rusengo (near Muganza)	T1	Ngara	near 2 56 S 30 40 E	Tanner
		(possibly a place in Burundi)		
Rushasha, popl.	U2	Kigezi	1 07.5 S 29 43 E	Lye
[Rushoshi] see Rushozi	U2	Ankole	0 27.5 S 30 43.5 E	Lind
Rushozi, popl.	U2	Ankole	0 28 S 30 43 E	
Rushozi, dam	U2	Ankole	0 27.5 S 30 43.5 E	Tallantire
Rushungi, popl.	T8	Kilwa	9 26 S 39 37 E	Wingfield

Name	Code	Location	Coordinates	Collector
Rusinga, I.	K5	South Kavirondo	0 24 S 34 10 E	Leakey, Opiko
[Rusomo] see Rusumo	T1/Rwanda	Biharamulo	2 23 S 30 47 E	Peter, Busse
Rusotto, str. (near Kwai)	T3	Lushoto	near 4 44 S 38 21 E	Tanner
[Russwisswi] see Luswiswe	T7	Rungwe	c.9 17 S 33 28 E–9 22 S 33 33 E	Procter, Tanner
Rusumo, falls	T1/Rwanda	Biharamulo	2 23 S 30 47 E	Goetze
Rusumo, fy.	T1	Biharamulo/Ngara	2 33 S 30 47 E	
[Ruswiswi] see Luswiswe	T7	Rungwe	c.9 17 S 33 28 E–9 22 S 33 33 E	
Rutamba, popl.	T8	Lindi	10 03 S 39 27 E	Gillman
[Rutambo] ? see Rutamba	T8	Lindi	?10 03 S 39 27 E	
Rutanga, str.	T4	Buha	c.4 37 S 29 37 E	Pirozynski
Rutenga, popl., r.h.	U2	Kigezi	0 59 S 29 51 E	Purseglove, Eggeling
Rutenganio, mssn.	T7	Rungwe	9 21 S 33 38 E	Stolz
Rutengo ? see Rutenganio	T7	Rungwe	?9 21 S 33 38 E	Stolz
Ruti (near Mbarara)	U2	Ankole	near 0 37 S 30 39 E	Eggeling
Rutoma, popl.	U2	Ankole	0 24 S 30 34 E	
[Rutschugi Post] see Uvinza	T4	Kigoma	5 06 S 30 23 E	Procter
[Rutuganjo] see Rutenganio	T7	Rungwe	9 21 S 33 38 E	Stolz
Rutundu, hill	K4	Meru	0 02 S 37 28 E	Schelpe
[Ruvu] see Pangani	T2/3	several	3 32 S 37 34 E–5 26 S 38 58 E	Milne-Redhead & Taylor
Ruvu, f.r.	T6	Morogoro	c.7 01 S 37 51 E	Troll
Ruvu, mssn.	T6	Uzaramo	6 49 S 38 41 E	Grundy, Peter, Vaughan
Ruvu, rsta.	T6	Uzaramo	6 48 S 38 37 E	
Ruvu, R.	T6	Morogoro/Uzaramo/Bagamoyo	6 54 S 37 40 E–6 23 S 38 52 E	Numerous

Name	Code	Location	Coordinates	Collector
[Ruvubu] see Ruvuvu	T1/Burundi	Ngara	3 23 S 30 00 E–2 23 S 30 47 E	Carmichael
Ruvuma, bay	T8/Mozambique	Mikindani	c.10 25 S 40 30 E	Kirk
Ruvuma, R.	T8/Mozambique		c.10 45 S 35 38 E–10 29 S 40 28 E	Numerous
Ruvuvu, R.	T1/Burundi	Ngara	c.3 23 S 30 00 E–2 23 S 30 47 E	Scott Elliot, Eggeling
Ruwalisi = Luwalisi, R.	T7	Rungwe	c.9 16 S 33 38 E–9 30 S 33 50 E	
[Ruwama] see Ruwana	T1	Musoma	c.2 07 S 33 48 E	Gillman

Name	Region	District	Coordinates	Collector
Ruwana, R.	T1	Musoma	c.2 07 S 33 48 E	Greenway
Ruwanda, est.	T7	Rungwe	prob. c.9 20 S 33 45 E	R.M. Davies
[Ruwande] see Ruande	T1	Mwanza	2 42 S 32 05 E	
Ruwarisi = Ruwalisi, R.	T7	Rungwe	c.9 16 S 33 38 E–9 30 S 33 50 E	
Ruwenzori, mts.	U2/Zaire	Toro	c.0 05–0 50 N 29 45–30 25 E	Numerous
Ruwiri, popl.	T5	Dodoma	near 5 42.5 S 34 59 E	B.D. Burtt
[Ruwiri] see Luwila, R.	T5	Dodoma	?5 30 S 34 58 E–5 50 S 35 05 E	B.D. Burtt
[Ruwu] see Ruvu	T6	Morogoro/Uzaramo/Bagamoyo	6 54 S 37 40 E–6 23 S 38 52 E	Brehmer
Ruzhumbura, co.	U2	Kigezi	c.0 35 S 29 50 E	Purseglove
Ruzinga, popl., sw.	T1	Bukoba	1 06 S 31 40 E	Gillman
[Ruzumbura] see Ruzhumbura	U2	Kigezi	c.0 35 S 29 50 E	
Rwabaranda, f.r.	U2	Ankole	Not traced	Greenway & Eggeling
Rwagimba, popl.	U2	Toro	0 29 N 30 06 E	Osmaston, Eggeling
[Rwagimbo] see Rwagimba	U2	Toro	0 29 N 30 06 E	Eggeling
Rwakarindiri	T1	Bukoba	Not traced	Ford
Rwampara, co.	U2	Ankole	c.0 45 S 30 30 E	Snowden
Rwamuchuchu = Rwamucucu, popl.	U2	Kigezi	1 09 S 30 03 E	Purseglove
Rwamunyonyi, mt.	U2	Kigezi	1 05 S 29 50 E	
Rwanyamahembe, gombola, mssn.	U2	Ankole	0 26 S 30 33 E	Lind
Rwashamaire, popl., r.h., t.c.	U2	Ankole	0 50 S 30 08 E	Snowden
[Rwasina] see Ruasina	T1	Bukoba	c.1 20 S 31 42 E	Ford
Rwebisengo, r.h., plains	U2	Toro	1 03 N 30 16 E	Eggeling
Rwensama, popl.	U2	Kigezi	0 24 S 29 47 E	
Rwentobo, popl.	U2	Ankole	1 07 S 30 11 E	
Rwenyaga, hill	U2	Ankole	0 50.5 S 30 39 E	Eggeling
Rwenzoli, Nat. Park	U2	several	0 15 S 30 00 E	
Rwoho, popl.	U2	Ankole	0 51 S 30 32 E	Eggeling
[Saadani] see Sadani	T6	Bagamoyo	6 02.5 S 38 47 E	Peter

Name		Region	Coordinates	Source
Saanane I. (near Mwanza)	T1	Mwanza	c.2 32 S 32 53 E	Carmichael
Saanane, ln.	T3	Lushoto	c.4 48 S 38 30 E	Cribb & Grey-Wilson
[Saanya Juu] see Sanya Juu	T2	Moshi	3 11 S 37 04 E	Cribb & Grey-Wilson
Sabachi, peak	K1	Northern Frontier	0 50 N 37 32 E	Curry
Sabaga, str.	T4	Kigoma	5 55 S 30 47 E–5 50.5 S 30 51.5 E	Kahurananga, Kibuwa & Mungai
[Sabakhi] see Sabaki	K4/7	Kitui/Teita/Kilifi	2 59.5 S 38 31 E–3 10 S 40 08 E	Numerous
Sabaki = Galana, R.	K4/7	Kitui/Teita/Kilifi	2 59.5 S 38 31 E–3 10 S 40 08 E	
[Sabale] see Sabule	K1	Northern Frontier	0 20 N 40 12 E	Gillett & Gachathi
Saba Saba = Thaba Thaba, sw., str.	K4	Fort Hall	0 47 S 36 55 E–0 49 S 37 18 E	Bally
Sabatia, rsta.	K3	Ravine	0 02 S 35 45 E	R.M. Graham
[Sabei] see Kapchorwa	U3	Mbale	1 24 N 34 27 E	Numerous
[Sabei] see Sebei	U3	Mbale	c.1 20 N 34 35 E	Numerous
Sabinio, mt.	U2/Zaire/Rwanda	Kigezi	1 23 S 29 36 E	Purseglove
[Sabinyo] see Sabinio	U2/Zaire/Rwanda	Kigezi	1 23 S 29 36 E	
Saboti, hill	K3	Trans-Nzoia	0 56.5 N 34 51 E	
Saboti, popl.	K3	Trans-Nzoia	0 56.5 N 34 50.5 E	Symes
Sabule, w.h.	K1	Northern Frontier	0 20 N 40 12 E	
Sabwani Turning, ln.	K3	Trans-Nzoia	1 03.5 N 34 55 E	Tweedie
[Sachsenwald] see Mogo f.r.	T6	Uzaramo	c.6 54 S 39 11 E	Engler, Holtz, Busse, Goetze
Sadallah = Zadala's, popl.	T7	Iringa	7 27 S 35 28 E	Zimmermann
Sadani (foot of Pare Mts.)	T3	Pare	near 3 39 S 37 34 E	Engler
Sadani, popl., t.c.	T6	Bagamoyo	6 02.5 S 38 47 E	Peter, Hannington, Procter
Sadani, ln.	T7	Iringa	8 14 S 35 00 E	Childs
[Sadenyi Hill] see Sasenyi	K7	Teita	3 39 S 38 42 E	Bally
Sagala, hills	K7	Teita	c.3 30 S 38 35 E	Bally, Polhill & Paulo, V.G. van Someren
Sagala, peak	K7	Teita	3 27 S 38 34.5 E	
Sagala, popl.	K7	Teita	3 31 S 38 35 E	
Sagala Mbuga = Sagara, L., marsh	T4	Kigoma/Tabora	c.5 13 S 31 06 E	Silungwe

Name		Region	Coordinates	Collector(s)
[Sagalla] see Sagara	T5	Dodoma	6 15 S 36 33 E	Busse
Sagamaganga, ln.	T6	Ulanga	8 03.5 S 36 48 E	B.D. Nicholson
Sagana, popl., rsta.	K4	South Nyeri	0 40 S 37 12.5 E	Bally, C.G. Rogers
Sagana, R.	K4	S. Nyeri/N. Nyeri	0 14.5 S 37 13 E – 0 40 S 37 12 E	Schelpe
Sagana, w.h.	K7	Teita	3 51 S 38 50.5 E	Faulkner
[Sagara] see Segera	T3	Pangani	5 18.5 S 38 33 E	Vesey-FitzGerald
Sagara, L.	T4	Kigoma/Tabora	5 13 S 31 06 E	
Sagara, popl, area	T5	Dodoma	6 15 S 36 33 E	
Sagara, mts.	T5	Mpwapwa	c.6 52 S 36 30 E	
Sagiro, area	T3	Handeni	5 42 S 37 57 E	
Sagitu, I.	U3	Busoga	0 01 S 33 39 E	G.H. Wood
Sai, est., hill	U4	Mengo	c.0 14 N 32 47 E	Dummer
Saint Austins Mission, mssn.	K4	Nairobi	1 16 S 36 47 E	
[Saira] see Sawa	T3	Tanga	5 07.5 S 39 06.5 E	Faulkner
Saisi, R.	T4/7/Zam	Ufipa/Mbeya	8 55 S 31 44 E – 8 53 S 32 16 E	Michelmore
Saiwa, sw., Nat. Park	K3	Trans-Nzoia	1 05 N 35 06 E	Hooper & Townsend, Owen, Gilbert
Saje, ln.	T2	Arusha	3 16 S 36 55.5 E	Greenway
[Saji] see Saje	T2	Arusha	3 16 S 36 55.5 E	Greenway
Saka, popl.	K1	Northern Frontier	0 09 S 39 19.5 E	Sampson, Bally
Sakalilo, popl.	T4	Ufipa	8 12 S 31 58 E	Bullock, Richards
[Sakamalewa] see Sakamaliwa	T4	Nzega	4 11 S 34 05 E	B.D. Burtt
Sakamaliwa, popl.	T4	Nzega	4 11 S 34 05 E	
[Sakanaliwa] see Sakamaliwa	T4	Nzega	4 11 S 34 05 E	
Sakarani, old pl.	T3	Lushoto	c.4 50.5 S 38 24 E	Drummond & Hemsley
Sakare, popl., old pl.	T3	Lushoto	4 59 S 38 26 E	Drummond & Hemsley, Engler, Peter
Sakare Rocks (Monga)	T3	Lushoto	c.5 05 S 38 35 E	Ali Omare
[Sakarre] see Sakare	T3	Lushoto	4 59 S 38 26 E	Busse
Sakila, area, for., sw.	T2	Arusha	c.3 20 S 36 57 E	Richards, Greenway
Sakura, popl.	T3	Pangani	5 37 S 38 52 E	
Sakura, sisal est.	T3	Pangani	5 36 S 38 53 E	Tanner

Name	Grid	District	Coordinates	Collector
Sambret Catchment	K5	Kericho	c.0 22 S 35 23 E	Kerfoot
Samburu, distr.	K1	Northern Frontier	c.1 10 N 37 00 E	Jex-Blake, Newbould
Samburu, Nat. Pk.	K1	Northern Frontier	c.0 35 N 37 30 E	Agric. Dept., Cribb & Grey-Wilson
Samburu, rsta.	K7	Kwale	3 46 S 39 16 E	Numerous
Samburu Lodge	K1	Northern Frontier	0.34.5 N 37 33 E	
Sambwa, pen.	U3	Teso	c.1 24 N 33 20 E	Michelmore
Same, popl., mssn., rsta.	T3	Pare	4 04 S 37 44 E	Numerous
Samia, t.a.	U3	Mbale	c.0 25 N 33 35 E	Snowden
Samia, hills	K5	Central Kavirondo	c.0 18 N 34 09 E	Davidson
Samia Bugwe, co.	U3	Mbale	c.0 30 N 34 00 E	G.H. Wood
Samina (near Geita)	T1	Mwanza	near 2 52 S 32 10 E	Eggeling, Procter
[Samonge] see Ssamunge	T2	Masai	2 09 S 35 42 E	Carmichael
[Samui Hills] see Samuye Hills	T1	Shinyanga	c.3 50 S 33 23 E	B.D. Burtt
Samuye, area	T1	Shinyanga	c.3 50 S 33 25 E	
Samuye, hills	T1	Shinyanga	c.3 50 S 33 23 E	B.D. Burtt, Holtz, Koritschoner
Samuye, popl.	T1	Shinyanga	3 49 S 33 20 E	
[Sanara] see Isunura	T7	Mbeya	8 37 S 34 25 E	Zimmermann
Sand, str./s.	K6/T1	Masai/Musoma	c.1 41 S 35 25 E–1 33 S 35 01 E	Richards
Sandum's Bridge, bdg.	K3	Trans-Nzoia	1 03.5 N 35 03 E	Irwin, Tweedie
Sandy Ridge, rd.	T4	Mpanda	6 38 S 31 08 E–6 49 S 30 56 E	Richards
Sanga, popl., r.h.	U2	Ankole	0 30 S 30 54 E	Eggeling, Harker
Sangaar (near Lolelia)	U1	Karamoja	near 3 19 N 33 56 E	Eggeling
Sangalo, mkt.	K3	Nandi	0 22 N 35 02.5 E	Dale
Sangan, ln.	K3	Ravine	0 02 N 35 32 E	Tweedie
Sangarawe = Sangerawe, old pl.	T3	Lushoto	5 08 S 38 37 E	Numerous
[Sangerawa] see Sangerawe	T3	Lushoto	5 08 S 38 37 E	
Sangerawe, old pl.	T3	Lushoto	5 08 S 38 37 E	Numerous
Sangilwa, popl.	T4	Kahama	3 52 S 32 44 E	Yeoman
Sango, bay	U4	Masaka	0 51 S 31 42 E	Numerous
Sangor, str.	U1	Karamoja	3 30 N 33 57 E–3 28 N 33 48 E	Numerous

[Sangwa] see Songwe	T7	Mbeya		
Sani (or ? Saui)	K4	Machakos	2 03 S 37 48 E	R.M. Davies
Sanituba (near Hosiga)	T3	Lushoto	c.4 50 S 38 14 E	Kassner
Sanje, popl., mssn.	U4	Masaka	0 46 S 31 30 E	Holst
[Sanje] see Saje	T2	Arusha	3 16 S 36 55.5 E	Lye
Sanje, popl.	T6	Ulanga	7 47.5 S 36 54 E	Greenway & Kanuri
Sanje, falls	T7	Iringa	7 46 S 36 54 E	Anderson, Carmichael
Sanje, str.	T6/7	Ulanga/Iringa	7 44 S 36 51.5 E–7 47 S 37 00 E	Polhill, Lovett
Sankuri, popl.	K1	Northern Frontier	0 18 S 39 35 E	Sampson
Sanssu	T1	Bukoba	prob. c.1 04 S 31 48 E	Gillman
Sanswago (Umalila)	T7	Mbeya	Not traced	Leedal
[Santaliya] see Santilya	T7	Mbeya	9 06 S 33 22 E	
Sante, popl.	T7	Mbeya	8 35 S 32 33.5 E	Gillett
Santilya, r.h.	T7	Mbeya	9 06 S 33 22 E	Goetze
Sanya, popl.	T2	Moshi	3 20.5 S 37 07 E	Bally, Endlich, Haarer
Sanya, R.	T2	Moshi	c.3 05 S 37 11 E–3 27 S 37 10 E	Haarer, Peter
Sanya Chini, rsta.	T2	Moshi	3 23 S 37 07 E	Numerous
Sanya Juu, popl., area	T2	Moshi	3 11 S 37 04 E	Numerous
Sao Hill, popl.	T7	Iringa	8 19 S 35 12 E	Firth
Saosa, est.	K5	Kericho	0 25 S 35 18 E	Kerfoot
Saosa = Saoset, str.	K5	Kericho	0 23 S 35 26 E–0 26 S 35 13.5 E	Kelly
Saoset, str.	K5	Kericho	0 23 S 35 26 E–0 26 S 35 13.5 E	
Saponi, R.	K7	Kilifi/Kwale	Not traced	
[Sara] see Sala	U3	Mbale	0 58.5 N 34 22 E–0 58 N 34 18 E	
[Saraka Steppe] see Tharaka	K4	Kitui	c.0 25 S 38 03 E	Balbo
[Sarake Steppe] see Tharaka	K4	Kitui	c.0 25 S 38 03 E	Balbo
Saranda, popl., rsta.	T5	Dodoma	5 42.5 S 34 59 E	B.D. Burtt, Fischer, Peter
[Saranka] see Salanga	T5	Kondoa		
[Sari] see Sare, R.	K5	South Kavirondo	c.0 47 S 34 26 E	Glasgow
Sari Hill (above Lake Kitangiri)	T5	Singida	Not traced	Richards

Saricho, popl.	K1	Northern Frontier	1 08.5 N 39 06 E	
Sasa, str.	U3	Mbale	1 10.5 N 34 28.5 E –1 13 N 34 23.5 E	G.H. Wood
Sasa, hut	U3	Mbale	1 10.5 N 34 28 E	Forbes, G.H. Wood, Lye
Sasamua, dam	K3/4	Naivasha/Kiambu	0 45.5 S 36 40 E	Coe, Lucas, Nattrass
Sasawara, R.	T8	Songea/Tunduru	11 00 S 36 40 E–11 33 S 36 55 E	
Saseni, R.	T3	Pare	4 17.5 S 37 55 E–4 28 S 38 06 E	
Sasenyi, hill	K7	Teita	3 39 S 38 42 E	
[Sasseni] see Saseni	T3	Pare	4 17.5 S 37 55 E –4 28 S 38 06 E	Eggeling
Sasumua = Sasamua, dam	K3/4	Naivasha/Kiambu	0 45.5 S 36 40 E	
Sasumwa = Sasamua, dam	K3/4	Naivasha/Kiambu	0 45.5 S 36 40 E	
Satima = Ol Doinyo Lesatima	K3/4	Laikipia/Naivasha/S. Nyeri	0 19 S 36 37 E	Dale, Fries, Polhill
Sattimma = Ol Doinyo Lesatima, peak	K3/4	Laikipia/Naivasha/S. Nyeri	0 19 S 36 37 E	
[Saube] see Sante	T?	Mbeya	8 35 S 32 33.5 E	Goetze
Saui (or ? Sani)	K4	Machakos	2 03 S 37 48 E	Kassner
Sawa, popl.	T3	Tanga	5 07.5 S 39 06.5 E	Faulkner
Sawago, f.r.	T7	Rungwe	9 01 S 33 38 E	
Sawala, hill, popl.	T7	Iringa	8 30.5 S 35 20.5 E	
[Sayo] see Seya	K1	Northern Frontier	0 50 N 36 44 E–1 30 N 37 06 E	J. Bally
Saza, H.Q. (Buvuma I.)	U4	Mengo	prob. c.0 11.5 N 33 17 E	Dawkins
Saza, mine	T7	Chunya	8 23 S 32 59 E	Richards
S-bends, ln.	K3	Trans-Nzoia	1 10 N 34 57.5 E	Tweedie
Schauri, popl.	T6	Ulanga	c.8 54 S 36 48 E	Schlieben
[Schengena] see Shengena	T3	Pare	4 16 S 37 56 E	Peter
[Schimba] see Shimba	K7	Kwale	c.4 07–4 20 S 39 25 E	Kassner
[Schlesien] see Silesian Mission	T6	Morogoro	near 7 05 S 37 35 E	Schlieben
[Schugulj] see Shuguri	T6/8	Ulanga/Kilwa	8 31 S 37 22 E	Schlieben
[Schume] see Shume	T3	Lushoto	4 40 S 38 13 E	Verdcourt
Seanna, ln.	K7	Kwale	c.4 08 S 39 29 E	Kassner
Sebei, distr.	U3	Mbale	c.1 20 N 34 35 E	Norman

(considered as part of Mbale District for F.T.E.A.)

[Sebei] see Kapchorwa	U3	Mbale	1 24 N 34 27 E	Numerous
[Sebeleni] see Sebleni	Z	Zanzibar	6 00 S 39 13 E	Vaughan
Sebit, sch., t.c.	K2	West Suk	1 23.5 N 35 20 E	Bogdan, Tweedie
Sebleni, popl.	Z	Zanzibar	6 00 S 39 13 E	
[Seboti] see Saboti	K3	Trans-Nzoia	0 56.5 N 34 50.5 E	Tweedie
Sechet (NW. Hanang)	T2	Mbulu	c.4 25 S 35 23 E	Carmichael
Secretariat Hill	K4	Nairobi	1 17 S 36 49 E	Tweedie
Sedia, popl.	T6	Kilosa	c.6 39 S 36 56 E	Busse
[Sedingombe] see Ledingombe	T6	Kilosa	7 04.5 S 36 37.5 E	Meyer
Segera, hill, for.	T3	Pangani	5 18.5 S 38 33 E	Greenway, Faulkner
Segoma, f.r., old pl.	T3	Lushoto	4 58 S 38 45 E	Faulkner, Moreau
Seguru, hill	U3	Busoga	0 28 N 33 38 E	Osmaston
Sekalet = Sikalet, glade, sw.	T2	Masai	c.2 54 S 36 14 E	Carmichael
Sekanyonyi, popl.	U4	Mengo	0 30.5 N 32 08 E	
Seke, area	T1	Shinyanga	c.3 19 S 33 26 E	Lunan, Yeoman
Seke, rsta.	T1	Shinyanga	3 20.5 S 33 31 E	
Sekenke, popl.	T5	Singida	4 16 S 34 10.5 E	Numerous
Sekenwo, for.	K3	Baringo	0 20 N 35 49 E	
Sekerr, mts., for.	K2	West Suk	c.1 40 N 35 23 E	Thorold, Agnew
Sela, hill, escarp.	K2	West Suk	c.1 16 N 35 12 E	Napier
Selambula = Silambula, str.	T4	Mpanda	c.6 20 S 29 54 E–6 25 S 29 52 E	Oxford Univ.
Selander, creek	T6	Uzaramo	6 47 S 39 17 E	Batty
[Selberawawa] see Solberawawa	K1	Northern Frontier	1 12.5 N 38 04 E	
Selebu, rock = Selegu, peak	T7	Iringa	7 29 S 36 10 E	Carmichael
Selelio, ln., w.h.	K1	Northern Frontier	c.2 56 N 39 05 E	
Selem, popl., area	Z	Zanzibar	6 03 S 39 14 E	Vaughan
Selengai, ln., t.c.	K6	Masai	2 11 S 37 10 E	Napper
Selengai, str./s.	K6	Masai	c.1 58 S 37 02 E–2 29 S 37 21.5 E	
Seliman-Mamba (N. Rondo Plat.)	T8	Lindi	Not traced	Busse
Selimweguru (Kungwe Mts.)	T4	Mpanda	c.6 08 S 29 45 E	Newbould & Harley

348

Selous, g.r.	T6/8	Rufiji/Ulanga/Kilwa	c.9 00 S 37 30 E	Nicholson
[Semali] see Semwali	T6	Morogoro	6 08.5 S 37 25.5 E	Schlieben
Sembabule, popl.	U4	Masaka	0 05 S 31 27 E	A.S. Thomas, Lye
Semewani, popl.	P	Pemba	5 11 S 39 47 E	Vaughan
Semini, sw., dam	K3	Naivasha	0 42 S 36 36 E	J.G. Williams
Semliki, R.	U2/Zaire		0 08 S 29 37 E – 1 14 N 30 28 E	Numerous
Sempaya, popl., r.h.	U2	Toro	0 51 N 30 10 E	A.S. Thomas
Sempaya, peak	U2	Toro	0 52 N 30 12 E	Eggeling, St. Clair-Thompson
Sempaya, str.	U2	Toro	c.0 51 N 30 11 E	
[Sempayo] see Sempaya	U2	Toro	0 51 N 30 10 E	Liebenberg
[Sempayu] see Sempaya	U2	Toro	0 51 N 30 10 E	
Semunya, f.r.	U4	Mengo	c.0 09 N 32 24.5 E	Chandler, Dawkins
Semuto, popl.	U4	Mengo	0 37 N 32 19 E	
[Semuye] see Samuye	T1	Shinyanga		Eggeling
Semwali, popl.	T6	Morogoro	6 08.5 S 37 25.5 E	Rounce
Senato (Ngorongoro Crater)	T2	Masai	c.3 10 S 35 35 E	Greenway
Senato, L.	T2	Arusha	3 16 S 36 54 E	Richards
[Seneto] see Senato	T2	Arusha	3 16 S 36 54 E	Richards
Senga, popl.	T4	Tabora	5 45 S 32 02.5 E	Procter
Sengale (near Longuza)	T3	Lushoto	near 5 03 S 38 42 E	Greenway
Sengani (near Simba) ? see Kwa Sengiwa	T3	Pare	?3 44 S 37 45 E	Engler
[Senganya Dya] see Mzingani–Idya	P	Pemba		R.O. Williams
Sengerenu ?see Sangerawe	T3	Lushoto	?5 08 S 38 37 E	Engler
[Sengina] see Kwa Sengiwa	T3	Pare	3 44 S 37 45 E	Uhlig
Sentema, popl.	U4	Mengo	0 22 N 32 25 E	
Sera, hill, str.	K1	Northern Frontier	1 05 N 37 52 E	
[Seranera] see Seronera	T1	Musoma	2 26 S 34 49 E	Eggeling
Serawani (near Kipumbwi)	T3	Pangani	near 5 38 S 38 53.5 E	Tanner
[Serenera] see Seronera	T1	Musoma	2 26 S 34 49 E	Eggeling
Serengeti National Park, g.r.	T1/2	Musoma/Maswa/Masai	c.2 20 S 34 50 E	Numerous

349

Sezibwa, R.	U4	Mengo	0 16 N 33 02 E–1 22 N 32 45 E	
Shaba, hill	K4	Meru	0 29 N 37 50 E	Bally
Shaba Dogo = Shaptiga, hill	K4	Meru	0 28.5 N 37 50 E	Bally & Smith
Shabal Taragwa, for.	K6	Masai	c.0 30 S 35 49 E	Glover et al.
Shabal Taragwa, str.	K6	Masai	0 30 S 35 49 E–0 41 S 35 40 E	
Shabal Taragwa, t.c.	K6	Masai	0 33 S 35 47.5 E	
Shabal Tarakwa = Shabal Taragwa, t.c.	K6	Masai	0 33 S 35 47.5 E	
Shabele, popl. (near Shaba Hill)	K4	Meru	near 0 29 N 37 50 E	Bally
Shafa Dika, w.h.	K1	Northern Frontier	0 40 N 37 55 E	T. Adamson
[Shagai] see Shagayu	T3	Lushoto	4 31 S 38 17.5 E	Drummond & Hemsley, Semsei, Procter, Eggeling, Willan
Shagayu, f.r.	T3	Lushoto	c.4 30 S 38 18 E	Procter, Mgaza, Carmichael
Shagayu, Mt., area	T3	Lushoto	c.4 31 S 38 17.5 E	Drummond & Hemsley
Shagein, peak	T3	Lushoto	4 31 S 38 17.5 E	
Shaitani, for.	K7	Kwale	4 20 S 39 34 E	Brenan
Shakababo, L.	K7	Tana River	2 25 S 40 11 E	Hooper & Townsend
Shamata, hill, for. stn.	K3	Laikipia	0 07 S 36 33 E	Mabberley
[Shambali] see Shombole	K6	Masai	2 08 S 36 05.5 E	Greenway
Shambangeda (near Amani)	T3	Lushoto	c.5 06 S 38 38 E	Faden
Shambarai, area	T2	Masai	3 39 S 36 54 E	Welch
Shanga, popl.	T1	Ngara	2 32.5 S 30 34 E	Tanner
Shangani, popl., reef	T8	Mikindani	10 15 S 40 12 E	Richards
Shankala, str.	T4	Mpanda	c.6 18 S 31 08 E	Groome
Shanwa, popl.	T1	Maswa	3 10 S 33 46 E	B.D. Burtt
Shanzu, beach	K7	Mombasa	3 58 S 39 45 E	Irwin, Greenway
Shaptiga, hill	K4	Meru	0 28.5 N 37 50 E	Bally & Carter
Shara, R.	T2	Arusha	prob. c.3 14 S 36 50 E	Greenway
[Shaurimayo] see Shauri Moyo	T2	Mbulu	3 55 S 35 39.5 E–3 55.5 S 35 42 E	Carmichael
Shauri Moyo, str.	T2	Mbulu	3 55 S 35 39.5 E–3 55.5 S 35 42 E	Carmichael
[Shegejuu] see Shengejuu	P	Pemba	5 04.5 S 39 48 E	

351

352

Name	Region	District	Coordinates	Collector
[Shegejuu] see Shengejuu				
Shekalage, ln.	T3	Lushoto	c.4 48 S 38 30 E	Cribb & Grey-Wilson
Shela, popl., area, sand-dunes	K7	Lamu	2 17.5 S 40 54.5 E	Magogo & Glover
Sheldrick's Falls	K7	Kwale	4 17 S 39 25.5 E	Greenway & Rawlins
[Shella] see Shela	K7	Lamu	2 17.5 S 40 54.5 E	Saville
Shelui, popl.	T5	Singida	4 20 S 34 17 E	Eggeling
Shema, co.	U2	Ankole	c.0 40 S 30 20 E	Greenway
Shengejuu, popl.	P	Pemba	5 04.5 S 39 48 E	Leechman, Eggeling
Shengena, peak	T3	Pare	4 16 S 37 56 E	Greenway
Shesheda (N.E. Mt. Hanang)	T2	Mbulu	4 26 S 35 24 E	Leedal
Sheyo, popl.	T7	Rungwe	9 20.5 S 33 22 E	Tanner
Shigara, popl.	T1	Mwanza	2 24 S 33 38 E	Peter
Shigatini, popl., mssn.	T3	Pare	3 40 S 37 38.5 E	Davies
Shikurufumi, f.r.	T6	Morogoro	7 10 S 37 31 E	Drummond & Hemsley, Lucas et al.
Shimba, for.	K7	Kwale	c.4 07–4 20 S 39 17–39 30 E	Numerous
Shimba, hills	K7	Kwale	c.4 07–4 20 S 39 25 E	Bally & Smith
Shimo la Tewa, fy.	K7	Mombasa	3 57.5 S 39 44.5 E	Birch, Drummond & Hemsley, Whyte
Shimoni, popl., for.	K7	Kwale	4 39 S 39 23 E	Geilinger
Shinyanga, distr.	T1	Shinyanga	c.3 40 S 33 30 E	Numerous
Shinyanga, popl.	T1	Shinyanga	3 40 S 33 26 E	
Shinyanga, Old., popl.	T1	Shinyanga	3 33.5 S 33 24.5 E	
Shira Plateau, area	T2	Moshi	c.3 02 S 37 14 E	Haarer, Salt, Volkens
Shiraha, sch.	K5	North Kavirondo	0 12 N 34 35 E	Birch
Shirati, popl., r.h., t.c., pier	T1	North Mara	1 08 S 33 59 E	Gane, Winkler
Shisaki, area	T1	Musoma	c.2 00 S 33 50 E	Gaertner
Shishiye, area	T2	Mbulu	c.4 26 S 35 24.5 E	Carmichael
Shombole, hill	K6	Masai	2 08 S 36 05.5 E	Bally, Milne-Redhead & Taylor
[Shomboli] see Shombole	K6	Masai	2 08 S 36 05.5 E	Greenway
Shuguri, falls	T6/8	Ulanga/Kilwa	8 31 S 37 22 E	Schlieben
[Shuguru] see Shuguri	T6/8	Ulanga/Kilwa	8 31 S 37 22 E	Schlieben

Name	Code	Region	Coordinates	Collector(s)
Shumba, hill (N. Rukiga)	U2	Kigezi	c.1 00 S 30 00 E	Purseglove
Shume, area, for.	T3	Lushoto	c.4 40 S 38 13 E	Numerous
Shume, popl.	T3	Lushoto	4 42 S 38 13 E	Numerous
Shume-Lands, f.r.	T3	Lushoto	c.4 39 S 38 10 E	
Shume-Magamba, f.r.	T3	Lushoto	c.4 40 S 38 15 E	Mgaza, Eggeling
Shungi, popl., area	P	Pemba	5 16 S 39 44.5 E	Greenway
Shungubweni, popl.	T6	Uzaramo	7 11 S 39 23 E	Procter
Shungu-Mbili, I.	T6	Rufiji	7 42 S 39 41 E	Frazier
Siabei = Seyabei, gorge	K6	Masai	c.0 37 S 35 59 E –1 20 S 35 59.5 E	Bally
Siakago, popl.	K4	Embu	0 35 S 37 38 E	Braun
[Siangira] see Liangira	T7	Njombe	c.9 15 S 34 42 E	
Siapei = Seyabei, R.	K6	Masai	c.0 37 S 35 59 E –1 20 S 35 59.5 E	
Siavona, hill	U3	Busoga	0 15 N 33 51 E	
Siaya, distr.	K5	Central Kavirondo	c.0 00 34 15 E	
Siba, f.r.	U2	Bunyoro	c.1 40 N 31 25 E	G.H. Wood, Dale
Siba, str.	U2	Bunyoro	1 37 N 31 39 E –1 42 N 31 23.5 E	Eggeling, Greenway, Sangster
Sibaga, hill	T4	Mpanda	6 01 S 29 59 E	C M. Harris
Sibu ("Nandi Country")	K3/5		Not traced	
[Sidilike] see Sitalike				
Siedentopf's Farm	T4	Mpanda	6 38 S 31 08 E	Evan James
Siende (near Mwatate)	T2	Masai	3 09 S 35 36.5 E	Richards
Sifala, popl.	K7	Teita	near 3 31 S 38 24 E	Bally
Siga Caves (near Tanga)	P	Pemba	5 10 S 39 47 E	Verdcourt & Polhill
Sigi, for.	T3	Tanga	near 5 04 S 39 06 E	Vaughan
Sigi, popl.	T3	Lushoto	c.5 06 S 38 39 E	Bally, Geilinger, Peter
Sigi, hill	T3	Lushoto	5 07 S 38 40 E	Pegler, Renvoize
Sigi, popl.	T3	Lushoto	5 00 S 38 45.5 E	Numerous
Sigi, R.	T3	Lushoto/Tanga	5 00.5 S 38 48 E	Faulkner, Greenway
Sigi Chini, pl.	T3	Lushoto/Tanga	4 55 S 38 42 E –5 02.5 S 39 04 E	Numerous
Sigi Chini, pl.	T3	Lushoto	5 06 S 38 39 E	Greenway
Sigirari, old area	K6/T2	several	2 45 –3 30 S 37 00 E	Fischer

Name	Region	District	Coordinates	Collector
Sigi Segoma, est.	T3	Lushoto/Tanga	4 58 S 38 46 E	
Sigi Singali, area	T3	Lushoto	5 06 S 38 39 E	Verdcourt
Signal Hill	U1	Karamoja	2 28 S 34 40 E	
Signal Hill = Mwangea, hill	K7	Teita	3 23 S 38 33 E	Napier
Signal Hill, Iringa	T7	Iringa	c.7 46 S 35 42 E	St. Clair-Thompson
Sigor = Sigorr, sch., p.p.	K2	West Suk	1 29.5 N 35 28.5 E	Bogdan, Carter & Stannard, Tweedie
Sigulu, I.	U3	Busoga	c.0 06 N 33 48 E	
Sigutioi (near Sitoton)	K3	Nakuru	near 0 31 S 35 38 E	Bally
Sikalet = Sekalet, sw., glade	T2	Masai	c.2 54 S 36 14 E	Carmichael
Sikanki, I.	U2	Toro	0 04 S 30 13 E	Lock
Sikh Sawmills	T3	Tanga	?5 02 S 38 53 E	Faulkner
[Sikilike] see Sitalike				
Sikitiko, ln.	T4	Mpanda	6 38 S 31 08 E	Richards
Sikoke, popl.	T4	Mpanda	6 40 S 31 16 E	Procter
Sikonge, f.r.	T4	Mpanda	6 12 S 29 48 E	Newbould & Jefford
Sikonge, popl.	T4	Tabora	c.5 20 S 32 57 E	Shabani
Silai, popl.	T4	Tabora	5 38 S 32 46 E	Lindeman
Silambula, popl.	T3	Lushoto	4 57 S 38 23.5 E	Holst
Silambula = Selambula, str.	T4	Mpanda	6 25 S 29 52 E	Oxford Univ. Exped.
Silesian Mission (near Lugongo)	T4	Mpanda	c.6 20 S 29 54 E–6 25 S 29 52 E	Oxford Univ. Exped.
Silkcub Highlands	T6	Morogoro	near 7 05 S 37 35 E	Peter
Silule = Siluti camp (L. Kitangiri)	T4	Mpanda	c.6 13 S 31 18 E	Richards
Silversands, bay	T5	Singida	c.4 04 S 34 19 E	Richards
Silver Sea, L.	K7	Kilifi	3 16 S 40 07.5 E	Rawlins
Sima, area	T2	Arusha	c.3 14 S 36 52 E	Richards
Sima (near Mkwaja)	T1	Mwanza/Kwimba	c.2 35 S 33 15 E	Procter
Simambaya, I.	T3	Pangani	near 5 47 S 38 51 E	Tanner
Simambaya, popl.	K7	Lamu	c.1 51 S 41 25 E	Greenway
Simambwe, area, sw.	K7	Lamu	1 53 S 41 22 E	Greenway & Rawlins
	T7	Mbeya	8 58.5 S 33 36.5 E	
Simanjiro, plain, mbuga	T2	Masai	c.4 00 S 36 30 E	Procter

Name		District	Coordinates	Collector
Simba, hill	K4	Machakos	1 49 S 37 35.5 E	Edwards
Simba, rsta.	K4	Machakos	2 09 S 37 36 E	Bogdan, Edwards, Kassner
Simba, peak	K7	Kwale	4 09.5 S 39 26 E	Kassner, Bally, W.E. Taylor
Simba Kopjes, hills	T1	Maswa	2 42 S 34 55 E	Greenway
Simba, R. = Mto wa Simba	T2	Masai/Mbulu	3 18 S 35 48.5 E –3 26 S 35 49.5 E	Greenway
[Simba] see Campi ya Simba	T3	Pare	3 36 S 37 42 E	Engler
[Simba] see Zimba	T4	Ufipa	7 51.5 S 31 49.5 E	
Simbah = Zimbo, popl.	T4	Tabora	5 36 S 33 50 E	Grant
Simbawanga	T8	Kilwa	Not traced	Busse
Simbili, popl.	T3	Lushoto	5 07 S 38 22 E	Holst
[Simbili] see Zimbili	T3	Lushoto	4 49 S 38 45.5 E	Pócs
[Simbini] see Sumbini	T6	Morogoro	6 55.5 S 37 44 E	Semsei
Simbo, popl.	T4	Mpanda	6 21 S 30 42 E	Richards
Simbo, popl.	T4	Nzega	4 40 S 33 27 E	Carmichael
Simbo, f.r.	T4	Tabora	4 57 S 33 02 E	Numerous
Simbo, popl.	T4	Tabora	4 57 S 33 02 E	B.D. Burtt
Simbo, area	T5	Kondoa	c.5 00 S 35 40 E	B.D. Burtt
[Simbo hill] see Simbu	T5	Kondoa	5 06 S 35 32 E	B.D. Burtt
Simbu, hill	T5	Kondoa	5 06 S 35 32 E	Williams
Simini's Farm, S. Kinangop	K3	Naivasha	c.0 44 S 36 39 E	Geilinger
[Simipala] see Sinipale	T7	Chunya	8 18 S 33 10 E	Fischer
[Simii] see Simiyu	T1	Maswa/Kwimba/Mwanza	c.2 55 S 34 48 E –2 33 S 33 25 E	Tanner, Fischer
Simiyu, R.	T1	Maswa/Kwimba/Mwanza	c.2 55 S 34 48 E –2 33 S 33 25 E	Tweedie
Simpson's Rocks	K3	Trans-Nzoia	c.1 10.5 N 34 49.5 E	Greenway
[Simuyu] see Simiyu	T1	Maswa/Kwimba/Mwanza	c.2 55 S 34 48 E –2 33 S 33 25 E	Greenway
[Simyuyu] see Simiyu	T1	Maswa/Kwimba/Mwanza	c.2 55 S 34 48 E –2 33 S 33 25 E	Volkens
Sinas Boma	T2	Moshi	3 14 S 37 20 E	Harris
Sinda, Is.	T6	Uzaramo	6 49 S 39 24 E	Numerous
Sindeni, popl., t.c., hill	T3	Handeni	5 21 S 38 14 E	Buechner
Sindikwa, triangulation point	U2	Toro	0 55.5 N 30 21 E	

355

Sinero (Selous Game Reserve)	T6	Ulanga	Not traced	Rees
[Singale] see Sengale	T3	Lushoto	near 5 03 S 38 42 E	Greenway
[Singhiso] see Thingithu	K4	Meru	c.0 04 S 37 29 E–0 11 S 37 59 E	Balbo
Singida, distr.	T5	Singida	c.4 35 S 34 40 E	Numerous
Singida, L.	T5	Singida	4 47 S 34 45 E	Numerous
Singida, popl., r.h., airstrip	T5	Singida	4 49 S 34 45 E	Geilinger
[Singide] see Singida	T5	Singida		
Singino, hill	T8	Kilwa	8 47 S 39 24 E	Busse, Braun
Singo, co.	U4	Mengo	c.0 45 N 31 50 E	Eggeling, Langdale-Brown, Snowden
Singoni ? see Singino	T8	Kilwa	?8 47 S 39 24 E	Braun
Sinipale, mt.	T7	Chunya	8 18 S 33 10 E	Geilinger
Sino, hill	T2	Mbulu	3 56.5 S 35 44 E	Haarer
Sinoia, popl.	K7	Kilifi	3 10 S 40 06 E	
Sinyori, area	K7	Lamu	c.2 16.5 S 40 53.5 E	Sangai
Sipa, sawmill	T4	Mpanda	c.7 12 S 32 20 E	Carmichael
Sipi, falls	U3	Mbale	1 20 N 34 22.5 E	Lewys Lloyd
Sipi, popl., hill, mssn.	U3	Mbale	1 20 N 34 22 E	Numerous
Sira, R.	T7	Chunya	8 40 S 33 10 E–8 28 S 33 00 E	
Siria, escarp.	K6	Masai	1 08 S 35 05 E	Taiti
Siriba, popl., Govt. Training Centre	K5	Central Kavirondo	0 0.5 S 34 36 E	Abraham
Sirimon Track	K4	North Nyeri	0 03.5 N 37 12 E–0 04.5 S 37 18.5 E	Bernardi, Verdcourt, Hepper & Field
Siroko, R.	U3	Mbale	1 07 N 34 31 E–1 28.5 N 34 14 E	Snowden, Tothill, Eggeling
Sirowa = Serewa, popl.	K2	West Suk	1 20 N 35 00 E	Tweedie
[Sirra] see Sera	K1	Northern Frontier	1 05 N 37 52 E	Ritchie
Sirwan = Siruan, peak	K1	Northern Frontier	1 47.5 N 37 06 E	Newbould
Sisa, popl., r.h.	U4	Mengo	0 11 N 32 30 E	Chandler
Sisaga, peak	T4	Mpanda	6 09 S 29 49 E	Oxford Univ. Exped.
Sisal Research Station, Thika	K4	Fort Hall	1 00 S 37 05 E	Bogdan
[Sissa] see Sisa	U4	Mengo	0 11 N 32 30 E	Chandler
[Sit] see Siti	U3	Mbale	1 11 N 34 33 E–1 26.5 N 34 43.5 E	Eggeling, Dale

357

Name		Region	Coordinates	Collector
Sogo, area	K6	Masai	0 50 S 35 36 E	
[Sogolemon] see Sogolimen	U1	Karamoja	2 58 N 34 32 E	
[Sogolime] see Sogolimen	U1	Karamoja	2 24 N 34 43 E	
Sogolimen, mt.	U1	Karamoja	2 58 N 34 32 E	J. Wilson
Sogolimen, mt.	U1	Karamoja	2 24 N 34 43 E	
[Sogoo] see Sogo	K6	Masai	0 50 S 35 36 E	
Soitayai, ln.	T2	Masai	c.2 28 S 35 20 E	Kratz
[Soitick] see Sotik	K5	Kericho	0 41.5 S 35 06.5 E	Greenway, Paulo
Soit Ololol = Isuria, escarp.	K6	Masai	c.1 00 S 35 10 E –1 25 S 34 47 E	
[Sokdek] see Moroto mt.	U1	Karamoja	2 32 N 34 46 E	J. Wilson
[Sokeni] see Nzukini	K4	Fort Hall	1 04.5 S 37 28 E	Bally
Soko, hill	U2	Kigezi	1 12.5 S 30 02 E	Purseglove
Soko, hill	T2	Moshi	3 27 S 37 32 E	Bally
Sokoke, for. stn.	K7	Kilifi	3 30 S 39 50 E	Numerous
Sokon, popl.	T2	Arusha	3 22 S 36 42 E	Milne-Redhead & Taylor
Sokorte Dika, crater	K1	Northern Frontier	2 18.5 N 37 58.5 E	Bally & Smith
Sokorte Guda = L. Paradise	K1	Northern Frontier	2 16 N 37 56 E	Bally & Smith
[Sokota] see Esokota	K6	Masai	c.2 19 S 36 46 E –2 32 S 36 56 E	Bally
[Sokota Dika] see Sokorte Dika	K1	Northern Frontier	2 18.5 N 37 58.5 E	Bally & Smith
Sokoto Drift	K6	Masai	2 26 S 36 50.5 E	Bally
Solai, escarp.	K3	Nakuru	c.0 02 N 36 10 E	Gardner
Solai, hill	K3	Nakuru	0 06 N 36 12 E	Gardner
Solai, popl., rsta.	K3	Nakuru	0 01 N 36 09 E	
Solai, p.p., stores	K3	Nakuru	0 03 S 36 07.5 E	
Solai Lake, sw., airstrip	K3	Nakuru	0 03 N 36 09 E	Baillie, Bally, MacDonald
Solberawawa, w.h.	K1	Northern Frontier	1 12.5 N 38 04 E	Richardson
Solo, str.	U3	Mbale	c.0 28 N 34 06 E –0 35 N 33 59 E	
Sololo, w.h.	K1	Northern Frontier	1 41 S 40 25 E	
Sololo, w.h., p.p.	K1	Northern Frontier	3 33 N 38 39.5 E	Bally, Gillett
Solulu, popl.	T6	Ulanga	8 00 S 36 51 E	Haerdi

Somerville's Road	K3	Trans-Nzoia	1 09 N 34 50 E	Tweedie
[Sonau] see Sondu				
Sondang, mt.	K5	Kavirondo/Kericho	0 23 S 35 00.5 E	Mabberley & McCall, Thorold
Sondu, popl., t.c., p.p.	K2	West Suk	1 24 N 35 24 E	
	K5	Kavirondo/Kericho	0 23 S 35 00.5 E	
Songa, popl.	Zaire *not* U1		1 55 N 30 16 E	Stuhlmann
Songa, popl., est.	T3	Tanga/Pangani	5 16 S 38 39 E	Paulo
Songambele, beach	T4	Ufipa	c.8 27.5 S 31 08 E	Wingfield
Songe, mt.	T2	Arusha	3 08.5 S 36 47.5 E	Richards
Songea, distr.	T8	Songea	c.10 35 S 35 50 E	
Songea, popl., airstrip	T8	Songea	10 41 S 35 39 E	Numerous
Songhor, popl	K5	Kisumu-Londiani	0 03.5 S 35 13 E	Bally, Green, Harvey
Songoro, hill	T2	Arusha	3 19 S 36 49 E	Willan
Songot, hills	K2	Turkana	c.4 02 N 34 30 E	Champion
Songot, mt.	K2	Turkana	4 03 N 34 27 E	Champion
[Songwa] see Songwe				
Songwe, popl.	T7	Mbeya	8 57 S 33 32 E	R.M. Davies, Feb.–Mar. 1933
Songwe, R.	T7	Mbeya/Chunya	c.9 10 S 33 23 E–8 26 S 32 50 E	Procter, Cribb & Grey-Wilson
Songwe, R.	Zambia/T7	Rungwe/Mbeya	c.9 08 S 33 12 E–9 43 S 33 56 E	B.D. Burtt, Goetze, Boaler
Soni, popl., mssn., r.h., falls	T3	Lushoto	4 50.5 S 38 22 E	Richards, Stolz, Leedal
Sonjo, popl.	T2	Masai	2 17 S 35 42 E	Numerous
[Sonjo] see Sanje				
Sonso, R.	T6	Ulanga	7 47 S 36 54 E	Verdcourt, Bally, Carmichael, Merker
Sonso, popl.	U2	Bunyoro	1 40 N 31 35 E–1 53 N 31 24 E	Harris
Sonta, plain	U2	Bunyoro	1 39.5 N 31 35 E	Eggeling, Harris
Sopa, popl.	T4	Mpanda	c.7 40 S 31 46 E	Eggeling
Soroti, popl., mssn.	T4	Ufipa	7 30 S 31 21 E	Richards
Soroto Forest, Milo	U3	Teso	c.1 45 N 33 30 E	Richards, Wingfield
[Sosawara] see Sasawara	U3	Teso	1 43 N 33 37 E	Numerous
Sosian = Sosio, R.	T7	Njombe	c.9 53 S 34 39 E	Richards
	T8	Songea/Tunduru	11 00 S 36 40 E–11 33 S 36 55 E	Busse
	K5	North Kavirondo	1 06 N 34 34.5 E–0 45 N 34 46 E	Tweedie

Name	Code	Region	Coordinates	Collector
Soysambu, hill	K6	Masai	2 08 S 37 23 E	Bally & Carter
Sozi, pt.	U4	Masaka	0 20 S 32 19.5 E	Numerous
Speke, Mt.	U2/Zaire	Toro	0 24 N 29 53 E	
Speke Gulf, bay	T1	Mwanza/Kwimba/Musoma	c.2 20 S 33 20 E	B.D. Burtt
Springfield, rsta.	K3	Uasin Gishu	0 48 N 35 08 E	Busse
Ssagassa, popl.	T3	Handeni	5 43 S 37 23 E	
[Ssalanga] see Salanga	T5	Kondoa		
[Ssamuje] see Samuye	T1	Shinyanga	c.3 50 S 33 25 E	Bally
Ssamunge, popl.	T2	Masai	2 09 S 35 42 E	Bally
[Sseke] see Seke	T1	Shinyanga	c.3 19 S 33 26 E	Peter
[Sseria] see Serya	T5	Kondoa	4 56 S 35 41 E	Stuhlmann
[Ssonga] see Songa	Zaire *not* U1		1 55 N 30 16 E	Carmichael
[Stanbrush] see Steinbruch	T3	Tanga	5 06 S 39 00.5 E	
Stanley, Mt.	U2	Toro	0 23 N 29 52.5 E	Ludanga
[Starike] see Sitalike	T4	Mpanda	6 38 S 31 08 E	Hooper & Townsend
[Stavilike] see Sitalike	T4	Mpanda	6 38 S 31 08 E	Cribb & Grey-Wilson
[Steinbrach] see Steinbruch	T3	Tanga	5 06 S 39 00.5 E	Eggeling, Faulkner
Steinbruch, f.r., gorge	T3	Tanga	5 06 S 39 00.5 E	Vollesen
Stieglers Gorge	T6	Rufiji	7 48 S 37 53 E	Polhill & Verdcourt
Stone Bridge	K4	Meru	0 10 N 37 52.5 E	Trapnell
[Stoni Athi] see Stony Athi	K4	Machakos	1 36 S 37 00 E	
Stony Athi, rsta.	K4	Machakos	1 36 S 37 00 E	
Stony Athi = Engare Olduroto, R.	K4/6	Machakos/Masai	1 46 S 37 02 E – 1 25.5 S 37 00 E	Bogdan
Stream Head, ln.	K3	Trans-Nzoia	1 01 N 35 00 E	Tweedie
Stuhlmann Pass	U2	Toro	0 23.5 N 29 53 E	Hedberg
Stutchbury, fm.	K3	Trans-Nzoia	1 10 N 34 47.5 E	Tweedie
Suam, R.	U3/K2, 3	several	1 09 N 34 32 E – 1 54 N 35 19.5 E	Numerous
Suam, saw mills	K3	Trans-Nzoia	1 10 N 34 44 E	Gillett, Hedberg, Tweedie
Suam Bridge, bdg.	U3/K3	Mbale/Trans-Nzoia	1 13.5 N 34 44 E	
Suam Estates	K3	Trans-Nzoia	c.1 11 N 34 50 E	Tweedie

Name	Code	Region	Coordinates	Collector
Suma, popl.	T7	Rungwe	9 12 S 33 43 E	Cribb & Grey-Wilson
[Sumara-Mpika] see Kimarampaka	T8	Songea	c.10 41 S 35 33 E	Richards
Sumbagulo, R.	T6	Ulanga	c.7 42 S 36 54 E	Carmichael
Sumbawanga, popl., airstrip	T4	Ufipa	7 57.5 S 31 37 E	Numerous
Sumbini, popl.	T6	Morogoro	6 55.5 S 37 44 E	Pócs
Summit	T2	Masai	Not traced	Child
Suna, popl.	K5	South Kavirondo	1 05 S 34 26.5 E	Napier
[Suna] see Issuna	T5	Singida	5 23 S 34 46 E	
[Sunda] see Sundu	T4	Ufipa	8 31.5 S 31 38.5 E	Richards
Sundu, L.	T4	Ufipa	8 31.5 S 31 38.5 E	Richards, Vesey-FitzGerald
Sunga, popl.	U4	Mubende	0 54 N 30 53 E	
Sunga, popl.	T3	Lushoto	4 31.5 S 38 14.5 E	Drummond & Hemsley, Eichinger, Procter
Sungira, hill	U4	Mengo	1 19 N 32 27 E	Langdale-Brown
[Sungwe] see Sungwi	T3	Lushoto	4 43 S 38 14 E	Drummond & Hemsley
Sungwi, mt.	T3	Lushoto	4 43 S 38 14 E	Eggeling, Semsei, Procter
[Sunta's] ? see Nsundas	T7	Mbeya	?9 12 S 33 09 E	Goetze
[Sunura] see Isunura	T7	Mbeya	8 37 S 34 25 E	Zimmerman
Supuko Looltian, area	K3	Naivasha/Nakuru	0 02–0 18 S 36 15–36 21 E	Piers
Sura, for.	T2	Arusha	c.3 15 S 36 52 E	Richards
[Surgilo] see Sagiro	T3	Handeni	5 42 S 37 57 E	Semsei
Surima, Mt. Kulal	K1	Northern Frontier	c.2 43 N 36 55 E	J. Adamson
Surungai, area	T5	Dodoma	6 08 S 35 14 E	Peter
Suse, Mt.	K1	Northern Frontier	1 49 N 37 16 E	Carter & Stannard
Susu, hill	K7	Teita	3 27 S 38 20 E	Gardner
Suswa, mt., crater	K3/6	Naivasha/Masai	c.1 09 S 36 21 E	Bally, Glover et al.
Suthie, I.	K7	Lamu	c.1 45 S 41 30 E	Gillespie
Swagaswaga, mt.	T5	Kondoa	4 52 S 35 32 E	B.D. Burtt
[Swagilo] see Sagiro	T3	Handeni	5 42 S 37 57 E	Semsei
[Swala] see Sawala	T7	Iringa	8 30.5 S 35 20.5 E	Carmichael

Name	Code	Region	Coordinates	Collector
Tambani, popl.	T6	Uzaramo	6 59 S 39 12.5 E	Stuhlmann
Tamburu, str.	T6	Rufiji	8 23 S 39 04 E –8 12 S 39 13 E	Busse
Tame, R.	T6	Kilosa/Morogoro	6 18.5 S 37 00.5 E –6 32 S 37 27 E	
Tamkal Valley	K2	West Suk	1 39 N 35 28 E	Tweedie
Tamota, popl.	T3	Lushoto	5 04 S 38 23 E	
Tamota, popl.	T3	Handeni	5 35.5 S 37 33 E	Semsei
[Tamta] see Taveta	K7	Teita	3 24 S 37 40.5 E	Sampson
Tana, R.	K1/4/7	several	0 25 S 37 56 E –2 30 S 40 30 E	Numerous
[Tana] see Tawa	T6	Morogoro	7 01 S 37 44 E	Stuhlmann
Tana Falls = Kora Falls	K1/7	N. Frontier/Tana River	0 04 S 38 43 E	Sampson
Tanana, area	T6	Morogoro	7 01 S 37 39.5 E	E.M. Bruce, Schlieben
Tanangozi, popl.	T7	Iringa	7 55 S 35 35.5 E	Numerous
Tana River, distr.	K7	Tana River	c.1 30 S 39 30 E	
Tana River Reservoir, dam	K4	Embu/Kitui	c.0 49 S 37 47 E	Robertson
Tandahimba, L., popl.	T8	Newala	10 45 S 39 38 E	Gillman
Tandala, popl.	T5	Kondoa	4 58 S 36 13 E	B.D. Burtt, C. Jackson
Tandala, popl.	T7	Mbeya	9 10 S 33 12.5 E	Leedal
Tandala, mssn.	T7	Njombe	9 23 S 34 14 E	McLoughlin, Stolz, Leedal
Tandangongoro, L.	T8	Lindi	10 02.5 S 39 30.5 E	Busse
[Tandaua] see Tundaua	T3	Tanga	5 14 S 39 03.5 E	
Tandega, ln.	T2	Arusha	c.3 16 S 36 53.5 E	
[Taneka] see Taveta	K7	Teita	3 24 S 37 40.5 E	Sampson
Tanga, bay	T3	Tanga	c.5 02 S 39 06 E	Greenway
Tanga, distr.	T3	Tanga	c.5 00 S 38 50 E	
Tanga, popl., airport	T3	Tanga	5 04 S 39 06 E	Numerous
[Tangalbei] see Tangulbei	K3	Baringo	0 48 N 36 17 E	Powys
Tanganyika, L.	T4/Zaire/Bur./Zam.	several	3 20 –8 45 S 29 00 –31 15 E	
Tangata, hill	T3	Tanga	5 10.5 S 39 04 E	
Tangata, popl.	T3	Tanga	5 13.5 S 39 04 E	Peter, Zimmermann

Name	Region	District	Coordinates	Authority
[Tangawanda] see Ras Mtangawanda				
Tangul, popl.	K7	Lamu	2 07 S 40 58 E	Rawlins
Tangulbei, popl.	K3	Elgeyo	1 09.5 N 35 30 E	Townsend
[Tannana] see Tanana				
Tapu (Image Mt., W.)	K3	Baringo	0 48 N 36 17 E	Meyerhoff
Tarakia, str.	T6	Morogoro	7 01 S 37 39.5 E	E.M. Bruce
Tarambas Hill, for.	T7	Iringa	c.7 28 S 36 08 E	Carmichael
[Tarana] see Tanana				
Tarangire, g.r.	T2	Moshi	3 04 S 37 28 E–3 00 S 37 34 E	Dale
Tarangire, popl.	K3	Baringo	c.0 29 N 35 48 E	E.M. Bruce
Tarangire, str.	T6	Morogoro	7 01 S 37 39.5 E	Richards, Vesey-FitzGerald
Tarangire mbuga, sw.	T2	Mbulu	c.3 40–4 18 S 36 07 E	Numerous
Tarangire Ridge, hills	T2	Mbulu	3 41.5 S 35 57 E	Vesey-FitzGerald
Tarasa, str.	T2/5	Masai/Mbulu/Kondoa	4 25 S 36 05 E–3 43 S 35 51 E	Vesey-FitzGerald
Tarbaj, t.c.	T2	Mbulu	c.3 40 S 35 58 E	Hooper & Townsend
Tare = Tale, hills	T2	Masai	c.4 15 S 36 00 E	Bally & Radcliffe-Smith
[Tareme] see Tarime	K7	Tana River	c.2 26 S 40 10 E	Peter
Tarik, mt.	K1	Northern Frontier	2 13 N 40 07 E	Lynes
Tarime, f.r.	T4	Buha	4 45 S 29 51 E	Bancroft, Watkins
Tarime, hill	T1	North Mara	1 21 S 34 23 E	Gillman, Carmichael
Tarime, popl., r.h., t.c.	T7	Iringa	Not traced	Richards
[Taringire] see Tarangire	T1	North Mara	c.1 21 S 34 21 E	Kassner
[Taro] see Taru	T1	North Mara	1 20 S 34 24 E	Kassner
Tarohora	T1	North Mara	1 21 S 34 23 E	Elliot
Tarosero, mt.	T2/5	Masai/Mbulu/Kondoa	3 46 S 39 07 E	H. Elliot
Tarosero-Burko ridge	K7	Kwale	4 45 S 39 08 E	Numerous
Taru, hill	T3	Tanga	3 12 S 36 22 E	Greenway
Taru, rsta.	T2	Masai	c.3 15 S 36 20 E	Tweedie
Taru Desert, area	T2	Masai	3 46 S 39 07 E	Kassner
Taru Plains	K7	Kwale	3 44 S 39 09 E	
	K7	Kwale	3 44 S 39 09 E	
	K7	Kwale	c.3 45 S 39 10 E	
	K7	Kwale	c.3 45 S 39 10 E	

Name	Code	Region	Coordinates	Authority
Tasini, popl.	P	Pemba	5 25 S 39 41 E	Greenway
Tatanda, mssn.	T4	Ufipa	8 32 S 31 29 E	Sturtz, Hooper & Townsend
[Tatohora] see Mtotohovu				
Tanga	T3	Tanga	4 42 S 39 10 E	Kassner
[Taua] see Tawa				
Taveta, popl.	T6	Morogoro	7 01 S 37 44 E	Stuhlmann
Taveta, mssn.	K7	Teita	3 24 S 37 40.5 E	Numerous
Tavu, I.	T6	Ulanga	9 01 S 35 37 E	Haerdi
Tawa, area	U4	Mengo	0 02 S 32 42 E	Dawkins
Tawatawa, ln.	T6	Morogoro	7 01 S 37 44 E	
[Tchamtei] see Maji ya Chumvi	T8	Kilwa	8 38 S 38 35 E	
Tegetero, mssn.	K7	Kwale/Kilifi	3 48 S 39 23 E	Hildebrandt
Teita, distr.	T6	Morogoro	6 56 S 37 43 E	Drummond & Hemsley, Stuhlmann
Teita, hills	K7	Teita	c.3 35 S 38 30 E	Numerous
[Telek] see Talek	K7	Teita	3 25 S 38 20 E	Bally, Kirrika
Teleki Valley	K6	Masai	1 24 S 35 18 E–1 26 S 35 04 E	Hedberg, Schelpe
[Telel] see Tulel	K4	North Nyeri	c.0 10 S 37 14–37 18 E	Dale
Tembo, hill	U3	Mbale	1 21.5 N 34 42 E	Haarer
Temi = Themi, R.	T6	Kilosa	6 07 S 37 07 E	Carmichael
Tendaguru, area	T2	Arusha	3 29 S 36 44 E–3 31 S 36 47 E	Hennig, Schlieben, Migeod
[Tendhimba] see Tandahimba	T8	Lindi	9 45 S 39 20 E	
Tenende, fy.	T8	Newala	10 45 S 39 38 E	Rees
Tengeni, popl., rsta.	T6/8	Ulanga/Kilwa	8 28 S 37 28 E	Numerous
Tengeru, rsta.	T3	Tanga	5 11 S 38 45 E	Numerous
Tengeru, str.	T2	Arusha	3 23 S 36 50 E	
Tenges, popl.	T2	Arusha	c.3 21.5 S 36 48 E–3 26 S 36 51 E	Tweedie
Tengulingi = Tengulinye	K3	Baringo	0 19 N 35 48 E	Goetze
Tenwek, popl., mssn.	T7	Iringa	near 7 55 S 35 15 E	Ivens
Te'okoto, popl.	K5	Kericho	0 45 S 35 21 E	
Terego, co.	U1	Acholi	2 21 N 31 36 E	Purseglove
Terego, popl.	U1	West Nile	c.3 10 N 31 10 E	Eggeling, Hazel
	U1	West Nile	3 11 N 31 02 E	

367

Name	Code	Region	Coordinates	Authority
[Thererekue] see Thiririka	K4	Kiambu	c.0 52 S 36 35.5 E–1 12 S 37 07.5 E	
Thessalia, popl.	K5	Kisumu-Londiani	0 10.5 S 35 10 E	Johansen
Theta, str., sw.	K4	Kiambu	0 55 S 36 42 E–1 07.5 S 37 00 E	
Thiba, R.	K4	S. Nyeri/Embu	0 18.5 S 37 15.5 E–0 49.5 S 37 41 E	Battiscombe, Kirrika
Thika, distr.	K4	Fort Hall/Kiambu/Machakos	c.1 00 S 37 20 E	Numerous
Thika, Hort. Res. Stn.	K4	Fort Hall	1 00 S 37 04 E	Njogu
Thika, falls	K4	Fort Hall	1 01.5 S 37 04 E	Pegler, Napier
Thika, popl., rsta.	K4	Fort Hall/Kiambu	1 02 S 37 05 E	Numerous
Thika, str.	K4	Fort Hall/Kiambu/Machakos	0 40 S 36 43 E–0 53.5 S 37 29 E	Faden, Battiscombe
Thika Road House, inn	K4	Nairobi	1 15.5 S 36 51 E	Verdcourt
Thikuni, hill	K4	Machakos	1 49.5 S 37 31 E	Trapnell & Birch
Thingithu, R.	K4	Meru	c.0 04 S 37 29 E–0 11 S 37 59 E	
Thiririka, str., falls	K4	Kiambu	c.0 52 S 36 35.5 E–1 12 S 37 07.5 E	Kirrika, C.G. van Someren
[Thitabi] see Thitani	K4	Kitui	1 03 S 37 55 E	
Thitani, sch., dam	K4	Kitui	1 03 S 37 55 E	Bally & Smith
[Thlawa = Thlawi] see Thawi	T2/5	Kondoa	4 43 S 35 44 E	B.D. Burtt
Thompson, est.	K4	Nairobi	c.1 17.5 S 36 46 E	Napier
[Thompson's Falls] see Nyahururu	K3	Nakuru	0 02.5 N 36 22 E	Hyde
[Thomsons Falls] see Nyahururu	K3	Nakuru	0 02.5 N 36 22 E	Numerous
Thowa, R.	K4/7	Kitui/Tana River	1 21 S 38 08 E–1 30 S 40 03 E	Lindblom
Thua = Thowa, R.	K4/7	Kitui/Tana River	1 21 S 38 08 E–1 30 S 40 03 E	Edwards
Thua, R. (near Mombasa)	K7	Mombasa	c.4 01 S 39 44 E	Gillett, D. Davis
Thuchi, R.	K4	Embu	0 11 S 37 21 E–0 21 S 37 51.5 E	Gardner
Thura, popl., for.	K4	Meru	0 03 N 37 43 E	
Thuuri, for.	K4	Meru	0 05 N 37 53 E	
Tiati, hill	K3	Baringo	1 19 N 35 56.5 E	Beentje
[Tibirizi] see Tibirinzi	P	Pemba	5 14.5 S 39 44.5 E	Vaughan
Tibirinzi, popl.	P	Pemba	5 14.5 S 39 44.5 E	
Tigeri = Bargera, R.	K3	Baringo	c.0 11 N 35 39 E–0 32 N 36 05 E	Bally
[Tiggeri] see Tigeri	K3	Baringo	c.0 11 N 35 39 E–0 32 N 36 05 E	Bally

Name		District	Coordinates	Collector
Tiwi, popl.	K7	Kwale	4 14 S 39 34.5 E	Battiscombe, Jex-Blake
[Tjuni] see Tyuni				
Tlahara (Nou Forest)	T7	Njombe	8 57 S 34 04 E	
Tlambi (Marang Forest)	T2	Mbulu	c.4 05 S 35 30 E	Carmichael
Tlawi, L.	T2	Mbulu	c.3 42 S 35 40 E	Carmichael
Tlawi, popl., mssn.	T2	Mbulu	3 54 S 35 28 E	B.D. Burtt, Moreau
Tobolwa Rock	T2	Mbulu	3 55 S 35 29 E	B.D. Burtt
Todenyang, p.p.	K3	Nandi	0 23.5 N 34 58.5 E	Wye
[Togaba] see Takabba	K2	Turkana	4 31 N 35 56 E	Martin
Togoro Plains	K1	Northern Frontier	3 23 N 40 13.5 E	Dale
Tokwe, sw.	T1	Musoma	c.2 10 S 34 55 E	Greenway
[Tola] see Tula	U2	Toro	0 50 N 30 02 E	Mukasa
[Tomati] see Tumati	K7	Tana River	c.0 42 S 38 45 E–0 50 S 39 51 E	
[Tome] see Temi	T2	Mbulu	4 02 S 35 31 E	Carmichael
Tomoko, hills	T2	Arusha	3 29 S 36 44 E–3 31 S 36 47 E	Lindeman
Tomondo, popl.	T5	Kondoa	4 47 S 36 03 E	
Tona	Z	Zanzibar	6 12 S 39 14 E	
[Tondahua] see Tundaua	T3	Pare	near 4 06 S 37 47 E	Peter
Tondola, popl., r.h.	T3	Tanga	5 14 S 39 03.5 E	Peter
Tondwa, popl.	U4	Mengo	0 10 N 31 54 E	Maitland
Tongaren, str.	T6	Rufiji	7 51.5 S 39 46 E	Greenway
Tongo, w.h.	K3	Trans-Nzoia	0 56 N 34 51 E–0 51 N 34 55 E	
[Tongolo] see Pongolo	T2	Masai	prob. c.2 45 S 36 15 E	Richards
Tongoni, popl., r.h.	T7	Iringa	Not traced	
[Tongoren] see Tongaren	T3	Tanga	5 13 S 39 03.5 E	Lindeman
Tongwe, popl., mssn.	K3	Trans-Nzoia	0 56 N 34 51 E–0 51 N 34 55 E	Tweedie
Tongwe, hill, f.r.	T3	Tanga	5 06.5 S 38 42.5 E	Greenway, Faden
Tongwe, area	T3	Pangani	5 18 S 38 44 E	Numerous
Tonto, r.c.	T4	Kigoma	c.5 25–6 00 S 30 00 E	Grant, Toyoshima
Tonya, popl.	T3	Pare	4 18 S 37 54 E	Eggeling
	U2	Bunyoro	1 35 N 31 05 E	Eggeling

371

Name		Location	Coordinates	Collector
Toola, popl.	T7	Mbeya	?c.8 40 S 32 35 E	Goetze
[Top Camp, Elgon] see Mudangi	U3	Mbale	1 10 N 34 26 E	Tothill
Top of Road, ln.	K3	Trans-Nzoia	1 07 N 34 40.5 E	Tweedie
Toro, distr.	U2	Toro	c.0 30 N 30 30 E	
Toro, g.r.	U2	Toro	c.0 55 N 30 25 E	Buechner
Toroma, popl., t.c.	U3	Teso	1 45 N 33 57 E	Fiennes
Toroma, mssn.	U3	Teso	1 47 N 33 58 E	
Toror, mt.	U1	Karamoja	2 50 N 34 12 E	Numerous
Tororo, popl., mssn., rsta.	U3	Mbale	0 42 N 34 11 E	Numerous
Tororo, hill	U3	Mbale	0 41 N 34 11 E	A.S. Thomas
Torosei, str.	K6	Masai	1 53.5 S 36 43.5 E–1 53.5 S 36 38 E	Glover & Cooper
Tosamaganga, mssn.	T7	Iringa	7 50 S 35 36 E	Mathias & Taylor, Greenway
[Tosamanga] see Tosamaganga	T7	Iringa	7 50 S 35 36 E	Mathias & Taylor
Tot, popl.	K3	Elgeyo	1 12 N 35 39 E	Bally, Tweedie
[Totohovo] see Mtotohovu	T3	Tanga	4 42 S 39 10 E	
[Totohovu] see Mtotohovu	T3	Tanga	4 42 S 39 10 E	
Towa, f.r.	U4	Masaka	c.0 21 S 32 14 E	A.S. Thomas, Eggeling, Purseglove
[Trakimboga] see Trekimboga	T7	Mbeya	7 35 S 34 47 E	Greenway
Trans Mara Masai = area W. of Mara R.	K6	Masai		
Trans-Nzoia, distr.	K3	Trans-Nzoia	c.1 00 N 35 00 E	
Trappe Farm = Momela Farm	T2	Arusha	3 15 S 36 51 E	Richards, Greenway
[Trapper Farm] see Trappe Farm	T2	Arusha	3 15 S 36 51 E	Richards
Trekimboga, ln.	T7	Mbeya	7 53 S 34 47 E	Greenway & Kanuri
[Trikanboga] see Trekimboga	T7	Mbeya	7 53 S 34 47 E	Greenway & Kanuri
[Trikimboga] see Trekimboga	T7	Mbeya	7 53 S 34 47 E	Greenway
Tsagwa, ln.	K7	Kilifi	3 50 S 39 40 E	Gilbert & May
Tsansingewe = Tsausingwe, area	T7	Njombe	c.9 34 S 34 41 E	Goetze
Tsausingwe, area	T7	Njombe	c.9 34 S 34 41 E	Kassner
[Tsavi] see Tsavo	K4/7	Machakos/Teita	2 59 S 38 28 E	
Tsavo, rsta.	K4/7	Machakos/Teita	2 59 S 38 28 E	Numerous

Tsavo Park (East), Nat. Pk., g.r.	K4/7	Machakos/Kitui/Teita	c.2 45 S 39 00 E	Numerous
Tsavo Park (West), Nat. Pk., g.r.	K4/6/7	Machakos/Masai/Teita	c.3 30 S 38 15 E	Numerous
Tsavo Royal National Park, g.r.	K4/6/7	several	c.3 00 S 38 45 E	Numerous
[Tschaia] see Chaya	T5	Dodoma	5 37 S 34 03.5 E	Stuhlmann
[Tschaja] see Chaya	T5	Dodoma	5 37 S 34 03.5 E	Peter
[Tschali] see Chali	T5	Dodoma	6 15 S 35 15 E	Peter
[Tschamtei] see Maji ya Chumvi	K7	Kwale/Kilifi	3 48 S 39 23 E	Hildebrandt
Tschamtuara, ? hill	T1	Bukoba	prob. c.2 04 S 31 30 E	Stuhlmann
Tschaworo, distr.	U2	Kigezi	c.1 00 S 29 50 E	Stuhlmann
[Tschaya] see Chaya	T5	Dodoma	5 37 S 34 03.5 E	Peter
[Tschenzema] see Chenzema	T6	Morogoro	7 07 S 37 35 E	Schlieben
[Tschueni] see Chuini	Z	Zanzibar	6 03 S 39 13.5 E	Stuhlmann
[Tschugulu Fälle] see Shuguri Falls	T6/8	Ulanga/Kilwa	8 31 S 37 22 E	Schlieben
[Tschukuani] see Chukwani	Z	Zanzibar	6 15 S 39 12.5 E	Stuhlmann
Tshak'henge, popl.	T6	Uzaramo	6 58 S 38 53 E	Busse
[Tshumo] see Chumo	T8	Kilwa	8 33 S 39 02 E	Busse
[Tshunyo] see Chunyu	T5	Mpwapwa	6 18 S 36 20 E	Busse
[Tsimba] see Shimba	K7	Kwale	c.4 07–4 20 S 39 25 E	Kassner
Tubila, rsta.	T4	Kigoma	c.5 02 S 30 10 E	Procter
Tubugwe, popl.	T5	Mpwapwa	6 20 S 36 37 E	B.D. Burtt, Hornby
Tubugwe, str.	T5	Mpwapwa	c.6 20 S 36 36 E	B.D. Burtt
Tugen = Kamasia, hills	K3	Baringo	c.0 23 N 35 48 E	Gillett
[Tugwell] see Turkwel	K1/2	N. Frontier/Turkana/ W. Suk	c.1 54 N 35 20 E–3 05 N 36 09 E	Wellby
[Tukin] see Tugen				
Tukuyu, popl.	K3	Baringo	c.0 23 N 35 48 E	Numerous
Tula = Lac Tula, R.	T7	Rungwe	9 15 S 33 38.5 E	L.C. Edwards, Bally
Tula Drift, ln.	K7	Tana River	c.0 42 S 38 45 E–0 50 S 39 51 E	Faden
Tulel Valley	K7	Tana River	0 39 S 39 17 E	Tweedie
Tulia, popl.	U3	Mbale	1 21.5 N 34 42 E	Polhill & Paulo
	T5	Singida	4 07 S 34 20 E	

Name	Region	District	Coordinates	Collector
Tulia, str.	T5	Singida	c.4 05 S 34 19 E	
[Tuliani] see Turiani	T6	Morogoro	6 09 S 37 35 E	Schlieben
Tulimani, for.	K4	Machakos	1 32 S 37 24 E	
[Tulisia] see Tulusia	T2	Arusha		Greenway
Tulu Dimtu, crater	K1	Northern Frontier	3 10 N 37 35 E	Bally
[Tululusie] see Tulusia	T2	Arusha	3 12.5 S 36 54.5 E	Greenway & Kanuri
Tulusia, hill, gorges, str.	T2	Arusha	c.3 15 S 36 50 E	Richards
Tulusia, L.	T2	Arusha	3 12.5 S 36 54.5 E	Richards
Tumati, area	T2	Mbulu	c.4 02 S 35 31 E	
Tumati, sawmill	T2	Mbulu	4 01 S 35 28 E	Eggeling, Carmichael
Tumba, camp	T4	Mpanda	7 31 S 31 39.5 E	Numerous
Tumbako, popl., area	T6	Morogoro	6 52 S 37 47 E	Pócs & Lungwecha
Tumbatu, I.	Z	Zanzibar	5 49 S 39 13 E	Vaughan
Tumbe, popl., area	P	Pemba	4 57 S 39 47 E	Vaughan
Tumbi, popl.	T4	Tabora	5 04 S 32 44 E	Lindemann, Lunan, Joseph
Tumbiri, str.	T7	Mbeya	c.9 03 S 33 18 E	Goetze
Tumbisi = Tumbiri, str.	T7	Mbeya	c.9 03 S 33 18 E	
[Tumbwa] see Tumba	T4	Mpanda	7 31 S 31 39.5 E	Richards
Tumu Tumu, area	K4	South Nyeri	0 30 S 37 05 E	
Tundaua, popl.	T3	Tanga	5 14 S 39 03.5 E	Vaughan, Lyde
Tundaua, popl.	P	Pemba	5 15.5 S 39 42 E	Rodgers, Ludanga
Tundu, hills	T8	Kilwa	c.8 50 S 38 25 E	Numerous
Tunduma, popl.	T7	Mbeya	9 18 S 32 46 E	
Tunduru, distr.	T8	Tunduru	c.11 00 S 37 25 E	Numerous
Tunduru, popl.	T8	Tunduru	11 07 S 37 21 E	Carmichael
Tungamalenga = Tungamarenga, hill	T7	Iringa	7 46 S 35 00 E	Vaughan
[Tungau] see Tunguu	Z	Zanzibar	6 12 S 39 19 E	
Tunguli, hill	T3	Handeni	5 55 S 37 19 E	Busse
Tunguu, popl., area	Z	Zanzibar	6 12 S 39 19 E	Oxtoby, Vaughan
Tungwe	Z	Zanzibar	Not traced	Greenway

Tununguo, mssn.	T6	Morogoro	7 03 S 37 55.5 E	Stuhlmann, Boaler
Tunya, hill, popl.	T3	Lushoto	4 48 S 38 38.5 E	Peter
Tura, rsta., popl.	T4	Tabora	5 31 S 33 50.5 E	Peter, Stuhlmann, Grant
Turbi, hills	K1	Northern Frontier	c.3 20 N 38 22 E	Gillett, Bally
Turbo, popl., p.p.	K3	Uasin Gishu	0 38 N 35 03 E	Brodhurst Hill, Newton, Gosnell
Turi, popl., p.p.	K3	Nakuru	0 16.5 S 35 45 E	Bogdan
Turiani, popl.	T3	Handeni	5 35.5 S 37 59 E	
Turiani, popl., mssn., falls	T6	Morogoro	6 09 S 37 35 E	Numerous
[Turkan] see Kamasia	K3	Baringo	c.0 23 N 35 48 E	
Turkana, L.	K1/2/Ethiopia	N. Frontier/Turkana	2 22−4 32 N 35 50−36 42 E	
Turkana, distr. (E. of R. Turkwell)	K1	Northern Frontier	c.2 30 N 36 00 E	
Turkana, distr. (W. of R. Turkwell)	K2	Turkana	c.3 00 N 35 00 E	
Turkana Escarpment	U1/K2	Karamoja/Turkana	4 10 N 34 00 E−2 50 N 34 40 E	Pole Evans, J. Wilson
[Turkin] see Kamasia	K3	Baringo	c.0 23 N 35 48 E	
Turkwel, R.	K1/2	N. Frontier/Turkana/ West Suk	c.1 54 N 35 20 E−3 05 N 36 09 E	Delamere, Wellby
Turkwell = Turkwel, R.	K1/2	N. Frontier/Turkana/ West Suk	c.1 54 N 35 20 E−3 05 N 36 09 E	Bally
Turkwell Gorge	K2	Turkana	1 55 N 35 20 E	J. Wilson
Turoka, rsta.	K6	Masai	1 55 S 36 39 E	
Turtle Bay, hotel	K7	Kilifi	3 21.5 S 40 00 E	van Someren
Turu, t.a.	T5	Singida/Dodoma	c.5 10 S 34 45 E	Stuhlmann, Peter, von Trotha, Grant
Tuso, popl., mssn.	K4	Fort Hall	0 40 S 36 50 E	Balbo, Napier
Tuso, str.	K4	Fort Hall	c.0 39 S 36 47 E−0 41 S 36 51 E	
[Tusu] see Tuso	K4	Fort Hall	c.0 39 S 36 47 E−0 41 S 36 51 E	Balbo
[Tusu] see Tuso	K4	Fort Hall	0 40 S 36 50 E	Gwynne
Tutho = Tuso, popl., mssn.	K4	Fort Hall	0 40 S 36 50 E	Greenway
[Tutishi] see Titushi	T1	Musoma	c.2 37 S 34 46 E	Tweedie
Tuyabei Farm	K3	Trans-Nzoia	1 07 N 34 50 E	Langdale-Brown
Tweyanze, popl.	U4	Mengo	1 18 N 32 04 E	

Name		Code	Region	Coordinates	Authority
Twiga, beach, hotel		K7	Kwale	c.4 13 S 39 37 E	Verdcourt
Two Tarn Col		K4	North Nyeri	0 09 S 37 18 E	Hedberg
Tyack's Bridge		K3	Trans-Nzoia	0 58 N 35 05 E	Tweedie
Tyuni, mt.		T7	Njombe	8 57 S 34 04 E	Goetze
Tyunia, popl.		K2	West Suk	1 24 N 35 16 E	Tweedie
Uafiwa = Uhafiwa, area		T7	Iringa	c.8 25 S 35 45 E	Carmichael
Uaraguess = Warges, mt.		K1	Northern Frontier	0 57 N 37 23.5 E	Newbould
Uasin Gishu, distr.		K3	Uasin Gishu	c.0 30 N 35 15 E	
[Uaso Narok] see Ewaso Narok		K3	Laikipia	0 00 36 22 E–0 32 N 36 52 E	Carter & Stannard
Uaso Nyiro (N) = Ewaso Ngiro, R.		K1/3/4	several	0 19 36 39 E–0 27 N 39 55 E	Adamson, Dalton, Edwards, Gardner, Pratt, Tweedie, Aosta
Uaso Nyiro (S.), R.		K6	Masai	c.0 39 S 35 44 E–2 04 S 36 07 E	M.S. Evans, Glover, Bally, Mearns, Milne-Redhead & Taylor
Uassi, area		T5	Kondoa	4 32 S 35 45 E	Peter
[Ubana] see Ubena		T6/7	Ulanga/Njombe	c.9 15 S 35 20 E	Schlieben
[Ubangala] see Mbangala		T6	Ulanga	8 36 S 36 44 E	Numerous
Ubena, t.a.		T6/7	Ulanga/Njombe	c.9 15 S 35 20 E	Procter
Ubenazomozi, popl.		T6	Bagamoyo	6 38 S 38 10 E	Hooper & Townsend
Ubende, t.a., plateau		T4	Mpanda	c.6 20 S 30 50 E	Greenway
Uberi = Ubili, area, grasslands		T3	Lushoto/Tanga	c.5 07 S 38 33 E	
Ubili, area		T3	Lushoto/Tanga	c.5 07 S 38 33 E	Greenway
Ubili, peak		T3	Lushoto/Tanga	5 07 S 38 35 E	
Ubiri = Ubili		T3	Lushoto/Tanga		
Ubungo, popl.		T6	Uzaramo	6 47.5 S 39 12 E	Mwasumbi
[Ubungo] see Uwungu		T7	Chunya	c.8 20 S 33 00 E	Goetze
Uchindire, ln.		T7	Iringa	c.8 38 S 35 18 E	Procter
Uchunga, area		T1	Shinyanga	3 35 S 33 45 E	B.D. Burtt
Udagaje		T6	Ulanga	prob. c.8 31 S 35 57 E	Carmichael
Udehei		T2	Mbulu	Not traced	Peter

Name	Region	District	Coordinates	Collector
Udekwa, popl.	T7	Iringa	7 38 S 36 25 E	Rodgers & Hall
Udigo, t.a.	T3	Tanga	c.4 50 S 39 00 E	Peter
Udiko, hill	U3	Busoga	c.0 30 N 33 55 E	G.H. Wood
Udinde, popl.	T7	Chunya	8 06 S 32 37 E	Sanane
Udoe, t.a.	T6	Bagamoyo	c.6 15 S 38 35 E	Stuhlmann
Uduhe, t.a.	T1	Shinyanga	c.3 35 S 33 50 E	B.D. Burtt
[Uduhi] see Uduhe	T1	Shinyanga	c.3 35 S 33 50 E	B.D. Burtt
Udupi, popl.	U1	West Nile	3 18 N 31 10 E	
[Udzungwa] see Uzungwa	T6/7	Ulanga/Iringa		
[Uera] see Nera	T1	Kwimba	c.2 57 S 33 05 E	Mabberley, Mwasumbi
Ufana, popl., area, str.	T2	Mbulu	4 15 S 35 21 E	Carmichael
[Ufiguru] see Ifuguru	T7	Iringa	7 36 S 35 03 E	B.D. Burtt
Ufiome, f.r.	T2	Mbulu	4 14 S 35 48 E	Greenway
Ufiome, t.a.	T2	Mbulu	c.4 15 S 35 50 E	Carmichael
Ufiome = Kwaraha, mt.	T2	Mbulu	4 14 S 35 48 E	B.D. Burtt, F.G. Smith, Peter
Ufipa, distr.	T4	Ufipa	c.8 00 S 31 30 E	
Ufipa, plateau	T4	Ufipa	c.8 00 S 31 30 E	Webb
[Ufiume] see Ufiome	T2	Mbulu	c.4 15 S 35 50 E	B.D. Burtt
Uflaga, plain	T4	Kigoma	4 42 S 31 35 E	
Ufuagi, area	T7	Iringa	c.8 32 S 35 00 E	Goetze
Ufufuma, area	Z	Zanzibar	c.6 10 S 39 20 E	Vaughan, Greenway
Ufukumi (Nou Forest)	T2	Mbulu	c.4 05 S 35 30 E	Carmichael
[Ufwangano] see Mfangano	K5	South Kavirondo	0 28 S 34 00 E	Napier
Ugaga, area	T4	Buha	c.4 35 S 30 05 E	
[Ugala Mbuga] see Ugalla Mbuga	T4	Tabora/Mpanda	c.5 46 S 32 03 E–5 08 S 30 42 E	Shabani
Ugalla, area	T4	Tabora/Mpanda	c.5 00–6 15 S 31 30 E	Boehm
Ugalla, g.r.	T4	Tabora/Mpanda	5 50 S 31 50 E	
Ugalla, popl., rsta.	T4	Tabora/Mpanda	5 47 S 31 09 E	
Ugalla, R., mbuga (marsh)	T4	Tabora/Mpanda/Kigoma	5 46 S 32 03 E–5 08 S 30 42 E	Numerous
Ugalla River, f.r.	T4	Mpanda	6 10 S 32 00 E	

Uganda Pass	U3	Mbale	?c.1 09 N 34 34 E	Hedberg
[Ugano] see Umgano	T8	Songea	10 58 S 34 55 E	Zerny
Ugenya, area	K5	Central Kavirondo	c.0 15 N 34 15 E	Davidson
Ugogo, t.a.	T5	Dodoma/Mpwapwa	6 15 S 35 45 E	Braun, Stuhlmann
[Ugueno] see Ugweno	T3	Pare	c.3 36 S 37 38 E	H. Meyer, Volkens
[Uguja Ukoo] see Unguja Ukuu	Z	Zanzibar	6 18.5 S 39 22 E	Faulkner
Ugunda, f.r.	T4	Tabora	5 50 S 32 30 E	
[Ugungu] ? see Ugunja	K5	Central Kavirondo	0 11 N 34 18 E	Davidson
[Ugunja] see Ogunja, popl.	K5	North Kavirondo	0 11.5 N 34 17.5 E	Davidson
Ugunja = Ogunja, mkt.	K5	Central Kavirondo	0 11 N 34 18 E	
Uguwe ? = Ugwe, popl.	T4	?Mpanda	?6 30 S 30 55 E	
Ugweno, area	T3	Pare	c.3 36 S 37 38 E	Swynnerton
[Ugweno Gebirge] see North Pare, mts. (part)	T3	Pare (does not include the part sometimes known as Middle Pare)	c.3 30–3 50 S 37 33—37 44 E	Davies, Meyer, Volkens
[Uha] see Buha (part of)	T4	Buha	c.3 30 S 31 10 E	Peter
Uhafiwa, area	T7	Iringa	c.8 25 S 35 45 E	
Uhanyana, mssr., area	T7	Njombe	8 56 S 34 52 E	
Uhehe, t.a.	T7	Iringa/Mbeya	c.8 00 S 35 30 E	Goetze, Prince, von Prittwitz
Uhimbe = Uhimba, t.a.	T1	Bukoba	2–2 30 S 31 00 E	
Uhimbila (Image Mt.)	T7	Iringa	c.7 31 S 36 10 E	von Trotha
Uhinga	T7	Njombe	c.9 12 S 34 05 E	Carmichael
Uhoka, mt.	T1	Mwanza	2 56 S 32 47 E	Goetze
Uhuru Point = Kaiser Wilhelm Spitze, peak	T2	Moshi	3 04.5 S 37 21 E	Stuhlmann
[Uippi, escarp.] see Uleppi	U1	West Nile	2 45 N 31 03 E	Greenway & Eggeling
Ujamba, popl.	T4	Mpanda	6 13.5 S 29 51 E	Newbould & Jefford, Mahinda
Ujashi, t.a.	T1	Mwanza	c.2 30 S 33 05 E	Tanner
Ujiji, popl.	T4	Kigoma	4 55 S 29 41 E	Numerous
Ujiji, popl.	T6	Ulanga	8 43 S 36 45 E	Haerdi
Uka	T7	Rungwe	prob. c.9 25 S 33 40 E	Stolz
Ukaguru, mts.	T5/6	Mpwapwa/Kilosa	c.6 25 S 36 50 E	Mabberley, Cribb & Grey-Wilson

Name	Prov.	District	Coordinates	Recorded by
[Ukama] ? = Bukama	T4	Nzega	?4 09.5 S 33 49 E	B.D. Burtt
Ukamba = Ukambani, area	K4	Machakos/Kitui	c.0 40—3 00 S 35 00—38 30 E	Hildebrandt, Scott Elliot, Fischer
Ukambani, area	K4	Machakos/Kitui	c.0 40—3 00 S 35 00—38 30 E	Scheffler, Scott Elliot, Bally

(old area bounded by Tana R., Athi R., corresponding roughly to present-day Machakos and Kitui Districts)

Name	Prov.	District	Coordinates	Recorded by
Ukami, area	T6	Morogoro	c.7 00 S 37 30—38 30 E	Stuhlmann, Peter
[Ukanga] see Ukangu		Njombe	9 32 S 34 08 E	
Ukangu, mt.	T7	Njombe	9 32 S 34 08 E	Goetze
Ukano, area, mt. (?)	T7	Iringa	c.7 59 S 35 18 E	Goetze
Ukara, I.	T1	Mwanza	1 50 S 33 03 E	McLoughlin
[Ukasi] see Ukazzi	K4	Kitui	0 49 S 38 32.5 E	
Ukazzi, area, hill	K4	Kitui	0 49 S 38 32.5 E	Gillett & Gachathi
[Ukera] see Ukira	T1	North Mara	c.1 25 S 34 00 E	
Ukereni, popl., area, hill	T3	Tanga	5 00 S 39 03 E	Fischer
Ukerewe, f.r.	T1	Mwanza	c.2 03 S 32 52 E	Peter
Ukerewe, I.	T1	Mwanza	c.2 03 S 33 00 E	Conrads, Rounce, Uhlig
Ukidi, t.a.	U1	Acholi/Lango	c.2 15 N 32 00—33 00 E	Grant
Ukimbu, t.a.	T5/7	Dodoma/Chunya	c.7 10 S 33 25 E	Schlieben
Ukindu ? = Kindu	T6	Ulanga	?7 54 S 37 41 E	Numerous
Ukinga, t.a.	T7	Njombe	c.9 10 S 34 15 E	Fischer, Feb. 1886
Ukira, old area	T1	North Mara	c.1 25 S 34 00 E	Numerous
[Ukirigulu] see Ukiriguru		Mwanza	2 43 S 33 01 E	
Ukiriguru, popl.	T1	Mwanza	2 43 S 33 01 E	Stuhlman
Ukome, popl.	T1	Biharamulo	2 50 S 31 48 E	Vaughan, Harris
Ukonga, popl.	T6	Uzaramo	6 52.5 S 39 11 E	Numerous
Ukonongo, t.a.	T4	Mpanda/Tabora	c.6 30 S 32 10 E	
Ukunda, area	K7	Kwale	c.4 19 S 39 32 E	
Ukunda, popl.	K7	Kwale	4 17 S 39 33.5 E	
Ukuti, popl.	U1	Acholi	3 39 N 33 32 E	Eggeling
Ukutu, area	T6	Morogoro/Rufiji	c.7 30 S 37 45 E	Goetze
Ukwala, popl.	K5	Central Kavirondo	0 12 N 34 11 E	Davidson

379

Name	Code	District	Coordinates	Reference
Ulumwa, f.r.	T4	Tabora	c.5 20 S 32 57 E	Shabani
[Ulunji] see Ulenje	T7	Mbeya	8 54 S 33 42 E	Leedal
[Uluwira] see Uruwira	T4	Mpanda	6 27 S 31 21.5 E	
Uluzi	T7	Iringa	near 7 55 S 36 05 E	Carmichael
Ulyankulu, popl., f.r.	T4	Tabora	4 45 S 32 19 E	Manolo, Procter
Umalila, t.a.	T7	Mbeya/Rungwe	c.9 05 S 33 15 E	Goetze, Procter, Stolz
Umalila, f.r.	T7	Mbeya	9 12 S 33 15 E	Procter
[Umanda] see Manda	T7	Chunya	7 58 S 32 26 E	Goetze
Umani Pools	K4	Machakos	2 28 S 37 55 E	Timberlake
Umba, R.	K7/T3	Kwale/Lushoto/Tanga	4 31 S 38 16 E–4 39 S 39 13 E	Numerous
Umba Steppe, area	T3	Lushoto/Tanga	c.4 30 S 38 45 E	Geilinger, Greenway, Kassner
Umba Steppe, g.r.	T3	Lushoto/Tanga	c.4 32 S 38 45 E	
[Umbugwe] see Mbugwe	T2	Mbulu	c.3 55 S 35 50 E	Merker
[Umbulu] see Mbulu	T2	Mbulu	3 51 S 35 32 E	Geilinger
[Umbure] ? see Umbwe	T2	Moshi	(this could quite easily be a mis-reading of Umbwe)	
Umbwe, popl.	T2	Moshi	3 13.5 S 37 17.5 E	
Umbwe, R.	T2	Moshi	3 05 S 37 20 E–3 20 S 37 19 E	Greenway, Bie
[Umeroke] see Bubu	T5	Dodoma	6 18 S 35 54 E–7 03 S 35 49 E	Lynes
[Umfwe] see Umbwe	T2	Moshi	3 05 S 37 20 E–3 20 S 37 19 E	Greenway
Umgano, mssn.	T8	Songea	10 58 S 34 55 E	Zimmer, Zerny
[Umia-Aneyma] see Omiya Anyma	U1	Acholi	3 16 N 33 16 E	
[Umia-Anyema] see Omiya Anyma	U1	Acholi	3 16 N 33 16 E	
[Umiya Anyema] see Omiya Anyma	U1	Acholi	3 16 N 33 16 E	Eggeling
[Umpeke] see Bumpeke	T1	Mwanza	3 05 S 32 28 E	Stuhlmann
Umuamba, old distr.	T7	Rungwe	c.9 20 S 33 45 E	Goetze
Umuhenya (Gombe Stream Reserve)	T4	Buha	Not traced	Pirozynski
Una, str.	T2	Moshi	3 14 S 37 31 E–3 19 S 37 31 E	Volkens
Unangwa, hill	T8	Songea	c.10 42 S 35 42 E	Milne-Redhead & Taylor
Undali = Bundali, t.a.	T7	Mbeya/Rungwe	c.9 10–30 S 33 15–30 E	R.M. Davies, Cribb & Grey-Wilson

Name	Region	District	Coordinates	Authority
Undussuma, area, r.c.	Zaire		c.1 25 N 30 18 E	Stuhlmann
Ungoni, t.a.	T8	Songea	c.10 45 S 36 00 E	Busse
[Ungu] see Nguru	T6	Morogoro	6 00 S 37 30 E	Fischer
[Unguja Kuu] see Unguja Ukuu	Z	Zanzibar	6 18.5 S 39 22 E	
Unguja Ukuu, popl., area	Z	Zanzibar	6 18.5 S 39 22 E	Greenway, Faulkner
Unguru, t.a.	T6	Morogoro	c.6 20 S 37 45 E	
Ungwana Bay	K7	several	2 45 S 40 20 E	Frazier
Uni, f.r.	U4	Mengo	0 23 N 32 19 E	Dawkins
[Unianembe] see Unyanyembe	T4	Tabora	c.5 15 S 32 45 E	Stuhlmann
[Uniyiha] see Unyiha	T7	Mbeya/Chunya	c.8 50 S 33 00 E	
[Unkana] see Nkanka	T7	Mbeya	c.9 03 S 32 58 E –9 04 S 32 41 E	Goetze
[Unnabach] see Una (stream)	T2	Moshi	3 14 S 37 31 E –3 19 S 37 31 E	Volkens
[Untali] see Undali	T7	Rungwe	c.9 30 S 33 30 E	Goetze
Unyakyusa, area	T7	Rungwe	c.9 15 S 33 40 E	Leedal
Unyamwanga, area	T7	Mbeya	c.8 55 S 32 30 E	R.M. Davies
Unyamwezi, t.a.	*T4	Tabora/Nzega/Kahama	c.5 00 S 33 00 E	Peter, Stuhlmann, W.E. Taylor, Grant

(*mainly, with small parts of Mwanza, Biharamulo and Dodoma Districts)

Name	Region	District	Coordinates	Authority
Unyangwira, popl., t.a.	T5	Dodoma	5 48 S 35 04 E	Stuhlmann
Unyanyembe, distr.	T4	Tabora	c.5 15 S 32 45 E	Stuhlmann, Peter
Unyiha, t.a., plateau	T7	Mbeya/Chunya	c.8 50 S 33 00 E	R.M. Davies, Goetze
[Unyika] see Unyiha	T7	Mbeya/Chunya	c.8 50 S 33 00 E	Goetze
[Unyoro] see Bunyoro	U2	Bunyoro	c.1 40 N 31 30 E	Dawe, Grant
[Upanga] see Upangwa	T7	Njombe	c.10 00 S 34 40 E	
[Upangu] see Upangwa	T7	Njombe	c.10 00 S 34 40 E	Braun
Upangwa, t.a.	T7	Njombe	c.10 00 S 34 40 E	Eggeling, Hill
[Upare] see Pare	T3	Pare	c.3 30 –4 35 S 37 33 –38 03 E	Haarer
Upe, co.	U1/K2	Karamoja/Turkana	c.1 55 N 35 00 E	J. Wilson, Philip
Upenja = Upenya, popl.	T6	Rufiji	7 51 S 39 47.5 E	Greenway
Upenja, popl., area	Z	Zanzibar	5 59 S 39 20 E	Vaughan, Mosha
Upenya, popl.	T6	Rufiji	7 51 S 39 47.5 E	

Name	Region	District	Coordinates	Collector
Upinda Hill, f.r.	T7	Iringa	c.8 20 S 35 55 E	Gane
Uplands, rsta.	K4	Kiambu	1 02.5 S 36 38 E	Numerous
Upogoro, t.a.	T6	Ulanga	c.8 50 S 36 45 E	
Uponela = Uponera, popl.	T6	Kilosa	6 23 S 37 02 E	Mabberley
Uponera, popl.	T6	Kilosa	6 23 S 37 02 E	Busse
Upper Kiwira Fishing Camp	T7	Rungwe	9 02 S 33 38 E	Greenway, Napper
Ura, str.	K4	Meru	c.0 03 S 38 18 E	J. Adamson
[Uragess] see Uaraguess	K1	Northern Frontier	0 57 N 37 23.5 E	
[Uraguess] see Uaraguess	K1	Northern Frontier	0 57 N 37 23.5 E	
Urambo, rsta.	T4	Tabora	5 04 S 32 03 E	Numerous
Uramboni = Ulamboni, val.	T8	Songea	c.10 41 S 35 33 E	
Uranzi, area	T6	Rufiji	c.7 47 S 39 52 E	Greenway
Urema = Karema. t.a.	T4	Mpanda	c.6 45 S 30 35 E	C.H.B. Grant
[Urigi See] see Burigi Lake	T1	Bukoba	2 07 S 31 16 E	Braun, Scott Elliot
Urima, t.a.	T1	Mwanza	c.2 52 S 33 05 E	Rounce, Tanner
Uroa, popl., area	Z	Zanzibar	6 05.5 S 39 25.5 E	Werth
Uru, mssn., popl.	T2	Moshi	3 15 S 37 21 E	Carmichael
Uruma, f.r.	T4	Tabora	5 10 S 32 48 E	Greenway & Hoyle
Uruma, popl.	T4	Tabora	5 10 S 32 45 E	
Urumandi Hut	K4	Meru	0 08.5 S 37 25 E	Le Pelley
[Urumwa] see Uruma	T4	Tabora	5 10 S 32 48 E	Carmichael
Urungu, t.a.	T4/Zambia	Ufipa	c.8 10–8 55 S 31 00–31 25 E	Muenzner
Uruwira, popl., mssn.	T4	Mpanda	6 27 S 31 21.5 E	Numerous
Usa = Usa River, popl.	T2	Arusha	3 22 S 36 51.5 E	Numerous
Usa, rsta.	T2	Arusha	3 23.5 S 36 52 E	
[Usafua] see Usafwa	T7	Mbeya/Chunya	c.8 50 S 33 40 E	
Usafwa, t.a.	T7	Mbeya/Chunya	c.8 50 S 33 40 E	R.M. Davies, Goetze, Stolz
Usafwa Forest Reserve (N.)	T7	Mbeya	c.8 50 S 33 32 E	Procter, Eggeling
Usagara, popl.	T1	Mwanza	2 41 S 33 00 E	Tanner
Usagara = Sagara, mts.	T5	Mpwapwa	c.6 52 S 36 30 E	Kirk

384

Usagara, t.a.	T5/6	Mpwapwa/Kilosa	c.6 30 S 36 30 E	Numerous
Usagari, popl.	T4	Tabora	4 53 S 32 29 E	Jones
Usake (Kaabong—Kamion)	U1	Karamoja	3 31 N 34 08 E—3 43 N 34 12 E	Eggeling
Usambara Mts.	T3	Lushoto/Tanga	c.4 25—5 15 S 38 10—38 50 E	Numerous
Usambiro, t.a., mssn.	T1	Mwanza	c.3 05 S 32 40 E	Stuhlmann
Usanda, peak	T1	Shinyanga	3 49 S 33 14 E	B.D. Burtt, Koritschoner
Usanga, mt.	T7	Mbeya	c.9 09 S 33 09 E	Goetze
Usangi, popl., mssn.	T3	Pare	3 42 S 37 39.5 E	Haarer
Usangu Flats, plains	T7	Mbeya	c.8 30 S 34 15 E	Richards
Usangu Plain = Usangu Flats	T7	Mbeya	c.8 30 S 34 15 E	B.D. Burtt, Richards
Usankara, popl.	T4	Tabora	5 04 S 31 52 E	
[Usaramo] see Uzaramo	T6	Uzaramo	c.7 10 S 38 50 E	Stuhlmann
Usa River = Usa, popl.	T2	Arusha	3 22 S 36 51.5 E	Greenway, Willan
Usasi (Mkalama area)	T5	Singida	near 4 06 S 34 38 E	B.D. Burtt
Uschiri, area	T2	Moshi	c.3 11 S 37 36 E	Volkens
[Usega] ? see Usige	?Burundi		?c.3 30 S 29 25 E	Scott Elliot
Useguha, old area	T3/6	several	c.6 00 S 38 15 E	Peter
Usekhe, area	T5	Dodoma	c.6 25 S 35 00 E	Grant
[Usenga] see Usanga	T7	Mbeya	c.9 09 S 33 09 E	
Usengi = Wusengi, pt.	K5	Central Kavirondo	0 05 S 34 02 E	
Usense, area	T4	Mpanda	c.6 25 S 31 10 E	Richards
Useri = Usseri, area	T2	Moshi	c.3 06 S 37 36 E	Haarer, Volkens
Useri, old camp	K6*	Masai	c.3 03.5 S 37 41 E	
		(*on T2 Moshi border)		
Ushetu, popl.	T4	Tabora	4 10 S 32 16 E	Akiley
Ushetu, t.a.	T4	Tabora	c.4 10 S 32 06 E	
Ushirombo, popl., mssn.	T4	Kahama	3 29 S 31 58 E	Boaler, B.D. Burtt
[Ushola] see Ushora	T5	Singida	4 41 S 34 15 E	B.D. Burtt
Ushongo = Ushongo Mtoni	T3	Pangani	5 31 S 38 58 E	Tanner
Ushongo Mabaoni, popl.	T3	Pangani	5 33 S 38 58 E	

Name	Region	District	Coordinates	Collector
Ushongo Mtoni, popl.	T3	Pangani	5 31 S 38 58 E	
Ushora, popl.	T5	Singida	4 41 S 34 15 E	B.D. Burtt
Usige, area	Burundi		c.3 30 S 29 25 E	
Usiha, area	T1	Shinyanga	c.3 45 S 33 40 S	Fischer
[Usinda] see Uzinza	T1	Biharamulo/Mwanza	c.3 00 S 32 00 E	
[Usindja] see Uzinza	T1	Biharamulo/Mwanza	c.3 00 S 32 00 E	Stuhlmann
[Usindscha] see Uzinza	T1	Biharamulo/Mwanza	c.3 00 S 32 00 E	Stuhlmann
Usinge, rsta.	T4	Tabora	5 05 S 31 18 E	Bullock, Peter
Usinge, sw.	T4	Kigoma	5 25 S 31 10 E	Shabani, Michelmore
[Usinja] see Uzinza	T1	Biharamulo/Mwanza	c.3 00 S 32 00 E	Stuhlmann
[Usoga] see Busoga	U3	Busoga	c.0 45 N 33 30 E	Scott Elliot
[Usongora] see Busongora	U2	Toro	c.0 05 N 30 00 E	
[Ussagara] see Usagara	T5/6	Mpwapwa/Kilosa	c.6 30 S 36 30 E	
[Ussambiro] see Usambiro	T1	Mwanza	c.3 05 S 32 40 E	Stuhlmann
Ussangu, area	T7	Njombe	c.9 00 S 34 10 E	Goetze, von Prittwitz & Gaffron
[Ussangu Plain] see Usangu	T7	Mbeya	c.8 30 S 34 15 E	Richards
Ussare, str. (near Ngolai str.)	T2	Masai	c.3 21 S 35 20 E	Peter
Usseke = Usekhe, area	T5	Dodoma	c.6 25 S 35 00 E	
[Ussenge] see Usinge	T4	Tabora	5 05 S 31 18 E	
[Ussenge Swamp] see Usinge	T4	Kigoma	5 25 S 31 10 E	
Usseri, area, R.	T2	Moshi	c.3 06 S 37 36 E	Volkens, Schlieben
[Ussinge] see Usinge	T4	Tabora	5 05 S 31 18 E	
[Ussinge Swamp] see Usinge	T4	Kigoma	5 25 S 31 10 E	
Ussui = Usui, old area	T1	Biharamulo	c.2 25 S 31 15 E	Braun
Ussure, popl.	T5	Singida	4 39 S 34 23 E	Fischer
Usui, old area	T1	Biharamulo	c.2 25 S 31 15 E	Grant
Usuku, co.	U3	Teso	c.2 00 N 34 00 E	Fiennes, Eggeling
Usukuma, area	T1	Mwanza	c.2 27 S 33 08 E	Geilinger, Fischer
[Usula] see Usule	T1	Shinyanga	c.3 45 S 33 13 E	Fischer
Usule, area	T1	Shinyanga	c.3 45 S 33 13 E	Lindeman

386

Entry	Region	District	Coordinates	Collector
[Utwani Ndogo] see Utwani	K7	Lamu	c.2 22 S 40 30 E	
[Uukana] see Nkanka	T7	Mbeya	c.9 03 S 32 58 E–9 04 S 32 41 E	
Uvidunda, area	T6	Kilosa	c.7 30 S 37 00 E	
[Uvindje] see Uvinje	T6	Bagamoyo	5 58 S 38 47.5 E	Fischer
[Uvingi] see Uvinje	T6	Bagamoyo	5 58 S 38 47.5 E	
Uvinje, popl.	T6	Bagamoyo	5 58 S 38 47.5 E	
[Uvinsa] see Uvinza	T4	Kigoma	5 06 S 30 23 E	Bullock, Peter
Uvinza, f.r.	T4	Kigoma	5 10 S 30 20 E	Procter
Uvinza, popl., rsta.	T4	Kigoma	5 06 S 30 23 E	Numerous
Uvinza, t.a.	T4	Kigoma	c.5 10 S 30 45 E	
Uwarungu, mt.	T7	Rungwe	9 18 S 33 19 E	Goetze
Uwemba, mssn.	T7	Njombe	9 28 S 34 46 E	Childs, Eggeling, Gilli
Uwemba, hill	Z	Zanzibar	5 51 S 39 17 E	Vaughan
Uwungu, t.a.	T7	Chunya	c.8 20 S 33 00 E	Goetze
Uyansi, old area	T5	Dodoma	c.5 45 S 34 30 E	Peter
Uyogo, t.a.	T4	Tabora	c.4 10 S 32 22 E	?Braun
Uyogo, t.a.	T4	Tabora	c.4 50 S 32 05 E	?Braun
Uyombo, sw.	K7	Kilifi	3 23 S 39 57 E	MacNaughton
Uyui, popl., hill	T4	Tabora	4 55 S 32 52 E	W.E. Taylor, Hannington
Uyui, t.a.	T4	Tabora	4 55 S 33 05 E	
Uzaramo, distr.	T6	Uzaramo	c.7 10 S 38 50 E	
Uzia, popl.	T4	Ufipa	7 42.5 S 31 36.5 E	Vesey-FitzGerald
Uzigua, area	T6	Bagamoyo	c.6 08 S 38 22 E	Sacleux
Uzigua, f.r.	T6	Bagamoyo	c.6 00 S 38 10 E	
Uzini, popl., area, plain	Z	Zanzibar	6 05 S 39 20 E	Vaughan
Uzinza, t.a.	T1	Biharamulo/Mwanza	c.3 00 S 32 00 E	B.D. Burtt, Tanner, Wigg
Uzinza, popl., area	T1	Mwanza	2 32 S 32 44 E	Tanner
Uzungwa, mts.	T6/7	Ulanga/Iringa	c.8 15 S 35 50 E	
Uzungwa Scarp, f.r.	T6/7	Ulanga/Iringa	c.8 20 S 35 58 E	
Uzuruga camp (Uzinza area)	T1	Mwanza	prob. c.2 30 S 32 49 E	B.D. Burtt

387

Name		Region	Coordinates	Collector
Vanga, admin. post	K7	Kwale	4 39.5 S 39 13 E	Fischer, R.M. Graham, Smith
[Varaguess] see Uaraguess	K1	Northern Frontier	0 57 N 37 23.5 E	B.D. Burtt
[Vawa] see Vwawa	T7	Mbeya	9 07 S 32 54.5 E	Kassner
[Vena] see Mwena	K7	Kwale	c.4 24 S 39 00 E–4 35 S 39 15 E	Greenway
Verani, popl.	P	Pemba	4 55 S 39 41 E	
[Veruni] see Vuruni	T3	Lushoto	c.4 58 S 38 23 E	Koritschoner
[Viansi] see Vianzi	T6	Bagamoyo	6 27 S 38 41 E–6 30 S 38 49 E	
Vianzi, str.	T6	Bagamoyo	6 27 S 38 41 E–6 30 S 38 49 E	Stuhlmann
[Vibongoje] see Vitongoge	P	Pemba	5 14 S 39 49.5 E	
Vibura, popl.	T6	Uzaramo	7 02 S 39 17 E	
Victoria, L. = Nyanza	U3/4, K5, T1	several	c.1 00 S 33 00 E	Procter
Victoria Falls, Foweira	U2	Bunyoro	near 2 10 N 32 19 E	Bagshawe
Victoria Gardens, park	Z	Zanzibar	6 10 S 39 11 E	Vaughan
Victoria Nile, R.	U1/2/3/4	several	0 25 N 33 12 E–2 14 N 31 26 E	
[Victoria Nyanza] see Victoria L.	U3/4, K5, T1	several	c.1 00 S 33 00 E	
Vidani, mt. (S. Pare Mts.)	T3	Pare	Not traced	Greenway
Vidunda, mssn., mts.	T6	Kilosa	7 35 S 37 02 E	Goetze
[Viehboma la Ngombe] see Boma la Ngombe	T2	Moshi	3 20 S 37 10 E	Uhlig
Vigai (on Magunga est.)	T3	Lushoto/Tanga	c.5 08 S 38 33 E	Faulkner
Vigoka	T7	Iringa	near 8 00 S 36 00 E	Carmichael
Vigola	T7	Iringa	prob. 7 48 S 36 21 E	Ede
Vigude	T6	Kilosa	prob. c.7 30 S 37 00 E	Semsei
Vihiga, sch.	K5	North Kavirondo	0 18 N 34 56.5 E	Hooper & Townsend
[Vijansi] see Vilanzi	T6	Morogoro	6 46 S 37 34 E–6 42 S 37 28 E	Busse
Vikana, ln.	T7	Iringa	c.8 03 S 36 01 E	Carmichael
[Vikindo] see Vikindu	T6	Uzaramo	6 59 S 39 17 E	Stuhlmann
Vikindu, popl., f.r.	T6	Uzaramo	6 59 S 39 17 E	Numerous
[Vilansi] see Vilanzi	T6	Morogoro	6 46 S 37 34 E–6 42 S 37 28 E	
Vilanzi, R.	T6	Morogoro	6 46 S 37 34 E–6 42 S 37 28 E	Stuhlmann
Villa Maria, popl.	U4	Masaka	0 14 S 31 44 E	Stuhlmann

Name	Code	Area	Coordinates	Collector
Vindili, popl.	T6	Morogoro	6 58.5 S 35 52 E	Stuhlmann
Vingo, popl., pl.	T3	Lushoto	5 02 S 38 30 E	
Vipingo, popl.	K7	Kilifi	3 49 S 39 47 E	Bally, Verdcourt
Vipingoni, popl.	K7	Kilifi	3 49 S 39 48 E	
[Viransi] see Vilanzi				
Viratsi, R.	T6	Morogoro	6 46 S 37 34 E–6 42 S 37 28 E	Stuhlmann
Virunga, mts.	K5	N./C. Kavirondo	0 12 N 34 38 E–0 09 N 34 18 E	Davidson
	U2/Zaire/ Rwanda	Kigezi	c.1 20 S 29 30 E	Numerous
Visoi, rsta.	K3	Ravine	0 06.5 S 35 47.5 E	Graham, Tweedie
[Vite] see Wite	T3	Lushoto	4 39.5 S 38 13.5 E–4 40 S 38 17.5 E	Eggeling
Vitengeni, popl.	K7	Kilifi	3 22 S 39 43 E	Faden
[Viti] see Wite	T3	Lushoto	4 39.5 S 38 13.5 E–4 40 S 38 17.5 E	Carmichael, Parry
Vitonga, popl.	T6	Morogoro	6 55 S 37 35 E	Schlieben
Vitongoge = Vitongoji, popl.	P	Pemba	5 14 S 39 49.5 E	Greenway, Vaughan, Taylor
[Vitongoje] see Vitongoge	P	Pemba	5 14 S 39 49.5 E	Vaughan
Vitongoji, popl.	P	Pemba	5 14 S 39 49.5 E	
[Vlugani] see Ndugani	T1	Maswa	3 15 S 34 50 E	Greenway
V.N.R. = Victoria Nile River	U1–4	several	0 25 N 33 12 E–2 14 N 31 26 E	Snowden
Voi, Nat. Pk. gate	K7	Teita	3 23 S 38 35 E	Greenway
Voi, popl., rsta.	K7	Teita	3 23 S 38 35 E	Numerous
Voi = Goshi, R.	K7	Teita/Kilifi	3 22 S 38 23 E–3 21 S 39 40 E	Bally
Voo, ln.	K4	Kitui	1 40 S 38 19.5 E	Kimani
[Vudea] see Vudee				
Vudee, area	T3	Pare	c.4 15 S 37 50 E	Greenway
Vudee, str.	T3	Pare	4 14 S 37 53 E–4 22 S 37 50 E	
Vuga, popl.	K7	Kwale	4 11 S 39 30 E	
Vuga, popl., mssn., r.h.	T3	Lushoto	4 53 S 38 20 E	Drummond & Hemsley
Vugiri, popl.	T3	Lushoto	5 04 S 38 27.5 E	Archbold
Vukula, popl.	U3	Busoga	0 57 N 33 36 E	
Vuma, hill, area	T6	Kilosa	7 26 S 37 07 E	Greenway & Kanuri

389

Name		Region	Coordinates	Authority
Wajir Bor, w.h.	K1	Northern Frontier	1 45 N 40 32 E	Gillett
[Wakara] see Ukara	T1	Mwanza	1 50 S 33 03 E	Staples
Wakepele Spring	T4	Ufipa	prob. c.8 10 S 31 25 E	Richards
Waki, popl.	U2	Bunyoro	1 48 N 31 19 E	Purseglove, Eggeling
Waki, R.	U2	Bunyoro	1 29 N 31 23.5 E–1 47.5 N 31 19 E	Bagshawe, Purseglove, Eggeling
Wakyato, popl., r.h.	U4	Mengo	0 52.5 N 32 13 E	Langdale-Brown
Wala, R.	T4	Tabora	c.5 00 S 33 10 E–4 46 S 32 03 E	Boehm, Lindeman, Procter
Walasi, hill	U3	Mbale	1 11 N 34 13 E	Snowden, A.S. Thomas, Tothill
[Walasi] see Buwalasi	U3	Mbale	1 11 N 34 14 E	
Walegga	Zaire			Stuhlmann
Walezo, popl., area	Z	Zanzibar	6 09.5 S 39 14 E	Vaughan
Walker's Point (Ngorongoro Crater)	T2	Masai	?c.3 08 S 35 40 E	Greenway
[Walla] see Wala	T4	Tabora	c.5 00 S 33 10 E–4 46 S 32 03 E	Carmichael
[Wallasi] see Buwalasi	U3	Mbale	1 11 N 34 14 E	A.S. Thomas
Wallers Camp (near Voi)	K7	Teita	near 3 23 S 38 34 E	Trapnell
[Walleso] see Walezo	Z	Zanzibar	6 09.5 S 39 14 E	Stuhlmann
Walufumbo = Wabufumbo, f.r.	U4	Mengo	0 11 N 32 25 E	Dawkins
Walugogo, f.r.	U3	Busoga	0 37 N 33 28 E	G.H. Wood
Walulumba, f.r.	U3	Busoga	c.0 29 N 33 20 E	G.H. Wood
Wamba, popl., p.p.	K1	Northern Frontier	0 59 N 37 20 E	Gardner, Jex-Blake, Newbould
Wamba, str.	T5	Singida	c.5 20 S 34 37 E–5 04 S 34 01 E	B.D. Burtt
Wambabya, f.r.	U2	Bunyoro	c.1 26 N 31 08 E	
[Wambogo's] see Wambugu's				
Wambugu's, popl.	K4	South Nyeri	0 35 S 37 02 E	Battiscombe, Mearns
Wambugu, area	K4	South Nyeri	0 35 S 37 02 E	Engler
Wami, hill	T3	Lushoto	c.4 40 S 38 15 E	
Wami, R.	K4	Machakos	1 39 S 37 08 E	Numerous
Wamiland, area	T6	Kilosa/Morogoro/Bagamoyo	6 30 S 36 57 E–6 08 S 38 49 E	Kirk
[Wamsuku] see Wamusuku	T6	Bagamoyo	c.6 12 S 38 42 E	Oxford Univ.
Wamuhu, str.	T4	Mpanda	6 12 S 29 52 E	
	K4	Fort Hall	c.0 41 S 36 40.5 E	Kerfoot

Name	Code	Region	Coordinates	Collector
Waterloo Farm	K3	Trans-Nzoia	1 10 N 34 49 E	Tweedie
Wati, mt.	U1	West Nile	3 13 N 31 02 E	Chancellor
Waturuma	K7	Kwale	c.4 20 S 39 13 E	Kassner
Wawi, popl., area	P	Pemba	5 14 S 39 47 E	Vaughan
Wazo, hill	T6	Uzaramo	near 6 48 S 39 15 E	Batty, Harris
Weani, popl.	P	Pemba	5 04 S 39 47.5 E	Greenway
Webuye, falls	K5	North Kavirondo	0 36 N 34 48 E	
Webuye, popl., rsta.	K5	North Kavirondo	0 36 N 34 46 E	
[Weisoke] see Waisoke				
Weiwei, R.	U2	Bunyoro	1 47.5 N 31 42 E –1 56 N 31 24 E	Eggeling
	K2	West Suk	1 21 N 35 28.5 E –1 37.5 N 35 32 E	Bogdan
Welene, popl.	P	Pemba	5 16 S 39 48 E	Greenway
[Welere] see Welene	P	Pemba	5 16 S 39 48 E	Greenway
[Wema] see Bwema	U4	Mengo	0 06 N 33 06 E	Bagshawe
Wema, popl., area	K7	Tana River	2 13 S 40 10.5 E	Greenway & Kibuwa
Wembere, R., plains	T4/5	Nzega/Tabora/Singida	4 58 S 33 56 E –4 10 S 34 11 E	Numerous
[Weni] see Weru	T7	Iringa	7 51 S 35 30 E	Goetze
[Wentzel-Heckmann] see Ngozi, L.	T7	Rungwe	9 00.5 S 33 33 E	B.D. Burtt, Menzies, St. Clair-Thompson
Wera, popl., t.c.	U3	Teso	1 52 N 33 45 E	W.H. Lewis, Harker
Wera, swamp	U3	Teso	prob. c.1 52 N 33 43 E	Lind
[Wera] ? see Nera (Uera)	?T1	?Kwimba	?2 57 N 33 05 E	Braun
Wergidhudhi, area	K1	Northern Frontier	2 15 N 40 20 E	Bally & Smith
Werther's Peak	T2	Mbulu	4 27 S 35 23 E	Greenway
Weru, popl.	T7	Iringa	7 51 S 35 30 E	Goetze, Zimmermann
Weru Weru, popl.	T2	Moshi	3 19 S 37 15.5 E	Greenway
Weru Weru, R., gorge	T2	Moshi	3 05 S 37 17 E –3 24.5 S 37 19 E	Numerous
Wesha, popl.	P	Pemba	5 14 S 39 44 E	Vaughan
West Budama, co.	U3	Mbale	c.0 45 N 34 05 E	
West Bugwe, f.r.	U3	Mbale	c.0 32 N 34 00 E	Dawkins, Drummond & Hemsley, G.H. Wood
West Chyulu, game conserv. area	K6	Masai	c.2 40 S 37 50 E	

Wite, str.	T3	Lushoto	4 39.5 S 38 13.5 E–4 40 S 38 17.5 E	Eggeling
Witu, for.	K7	Lamu	c.2 22 S 40 30 E	Mohamed Abdullah, Rawlins
Witu, popl., p.p.	K7	Lamu	2 23 S 40 26 E	Numerous
[Wituland] see Lamu distr.	K7	Lamu	c.2 05 S 41 00 E	
Wituruma = Waturuma	K7	Kwale	c.4 20 S 39 13 E	Kassner
Wobulenzi, popl., r.h.	U4	Mengo	0 43.5 N 32 32 E	
[Wodakurea] see Wogakuria	T1	Musoma	1 39 S 35 00 E	Greenway
Wogakuria, hill, g.p.	T1	Musoma	1 39 S 35 00 E	Greenway & Turner
[Worgess] see Uaraguess	K1	Northern Frontier	0 57 N 37 23.5 E	J. Adamson
World's End, ln.	T7	Mbeya	8 49 S 33 32 E	Cribb & Grey-Wilson, Wingfield
World's View, ln.	T3	Lushoto	4 42 S 38 13 E	Drummond & Hemsley, Richards, Eggeling
Worssera Lookout	K7	Teita	3 20 S 38 33 E	Greenway
[Wuawesi] see Wuwawesi	T8	Songea	c.10 41 S 35 35 E	
Wudee = Wudei, mssn.	T3	Pare	4 06 S 37 47 E	Peter
[Wugu] see Iwugu	T7	Rungwe	9 28 S 33 37 E	
Wundanyi, popl., p.p.	K7	Teita	3 24 S 38 22 E	Bally, Drummond & Hemsley
Wusengi = Usengi, pt.	K5	Central Kavirondo	0 05 S 34 02 E	Norman
Wusi, mssn.	K7	Teita	3 27 S 38 21 E	Drummond & Hemsley, Napier, Bally
[Wusso Nyiro] see Uaso Nyiro	K1/3/4	several	0 19 S 36 39 E–0 27 N 39 55 E	J. Adamson
Wuthei (Mt. Debasien)	U1	Karamoja	c.1 45 N 34 43 E	Eggeling
Wuwawesi, str.	T8	Songea	c.10 41 S 35 35 E	Milne-Redhead & Taylor
[Wuwawezi] see Wuwawesi	T8	Songea	c.10 41 S 35 35 E	Milne-Redhead & Taylor
[Wuzi] see Wusi	K7	Teita	3 27 S 38 21 E	Drummond & Hemsley
Yabicho = Yabichu, ln.	K1	Northern Frontier	c.3 56 N 41 11 E	Gillett
Yaida, popl.	T2	Mbulu	3 51 S 35 09 E	
Yaida, str.	T2	Mbulu	c.3 57 S 35 07 E	Richards
Yaida, sw.	T2/5	Mbulu/Singida	3 55 S 35 05 E	Richards
Yaida, val., escarp.	T2/5	Mbulu/Singida	c.4 00 S 35 00 E	Richards
Yaka (Uaso Nyiro), ln.	K1	Northern Frontier	prob. c.0 45 N 38 18 E	J. Adamson

Yala, R.	K3/5	Nandi/C. & N. Kavirondo	0 03.5 N 34 10 E–0 08.5 N 35 00 E	Numerous
Yala, sw.	K5	Central Kavirondo	c.0 03 N 34 05 E	Tweedie, Gillett
Yala River Bridge	K5	North Kavirondo	0 10 N 34 44.5 E	Drummond & Hemsley
Yale, peak	K7	Teita	c.3 26 S 38 18 E	Rauh
Yama y Mota	K1	Northern Frontier	Not traced	
Yamba, popl.	T3	Lushoto	4 47 S 38 21 E	Magogo
[Yamba] see Niamba	T7	Mbeya	c.8 52 S 32 40 E–8 25 S 32 37 E	Goetze
Yambe, I.	T3	Tanga	5 06 S 39 09.5 E	Faulkner
[Yambe] see Niamba	T7	Mbeya	c.8 52 S 32 40 E–8 25 S 32 37 E	Goetze
Yambe, Little, I.	T3	Tanga	5 05 S 39 10 E	
Yamimbi = Djamimbi	T7	Njombe	9 40 S 34 18 E	Goetze
Yange Yange, str.	T6	Morogoro	c.6 48 S 38 01 E–6 46 S 38 03 E	Busse
Yangwani, popl.	T8	Lindi	9 53 S 39 37 E	Busse
[Yariri] see Yawiri	T7	Njombe	c.9 26 S 34 04 E	Goetze
Yasini = Jasini, popl.	T3	Tanga	4 40 S 39 11 E	Kassner
Yatta Furrow, str.	K4	Fort Hall/Machakos	c.1 11 S 37 34 E–1 06 S 37 22 E	Bally
Yatta Gap, pass	K4	Kitui	2 11 S 38 05 E	Bally & Smith
Yatta Plateau	K4	Machakos/Kitui	1 00–3 00 S 37 25–38 35 E	Numerous
Yawiri, mts.	T7	Njombe	c.9 26 S 34 04 E	Goetze
Yawnae, mts.	T7	Njombe	c.9 04 S 34 02 E	Goetze
Yawuanda, mt.	T7	Njombe	c.9 35 S 34 10 E	Goetze
Yawuanga, mt.	T7	Njombe	c.9 09 S 34 02 E	Goetze
Yegea, popl.	T6	Uzaramo	7 21 S 38 45 E	Stuhlmann
Yemit, popl.	K3	Elgeyo	0 59 N 35 27.5 E	Townsend
Yihirini = Yirihini, popl., hill	T3	Tanga	4 56.5 S 38 55 E	Peter
[Yila] see Njila	T7	Chunya	8 17 S 32 46 E	Goetze
[Yilihini] see Yihirini	T3	Tanga	4 56.5 S 38 55 E	
[Yilihori] ? see Yirihini	T3	Tanga	?4 56.5 S 38 55 E	Peter
Yirihini = Yihirini, popl., hill	T3	Tanga	4 56.5 S 38 55 E	
[Yjunga] see Ijunga	T7	Mbeya	9 04 S 33 11 E	Goetze 25 10 1899

Name	Code	Place	Coordinates	Collector
Yovi, R.	T6	Kilosa	7 06 S 36 39 E–7 35 S 36 48 E	Thulin & Mhoro
[Yozani] see Jozani	Z	Zanzibar	6 16 S 39 25.5 E	Greenway
Yumba ya Nguaro, cave	T2	Moshi	2 59 S 37 25 E	Volkens
Yumbe, popl. t.c., r.h.	U1	West Nile	3 28 N 31 15 E	Eggeling, Langdale-Brown
[Yungururu] see Chungururu	T7	Rungwe	9 18 S 33 52 E	Goetze
Zadala's = Sadallah, popl.	T7	Iringa	7 27 S 35 28 E	Bally
Zagana = Sagana, w.h.	K7	Teita	3 51 S 38 50.5 E	Swynnerton
Zagayu, popl.	T1	Maswa	2 57 S 33 45 E	Eggeling
Zaipi, popl.	U1	West Nile	3 23.5 N 31 59 E	Tanner
Zanaki, area	T1	Musoma	c.1 42 S 33 53 E	Tanner
Zanaki, popl., r.h.	T1	Musoma	1 43 S 33 59 E	
Zanzibar, I.	Z	Zanzibar	c.6 10 S 39 20 E	Numerous
Zanzibar, popl., harbour	Z	Zanzibar	6 10 S 39 11 E	Procter
Zaraninge Plateau	T6	Bagamoyo	prob. 6 20 S 38 30 E	Faden
Zawadi, est.	K4	North Nyeri	0 24 S 36 59 E	
Zebiel, Mt. Debasien	U1	Karamoja	c.1 45 N 34 43 E	A.S. Thomas
[Zeio] see Zeu	U1	West Nile	2 32 N 30 48 E	Eggeling
[Zeo] see Zeu	U1	West Nile	2 32 N 30 48 E	Eggeling
Zeu, hill	U1	West Nile	2 32 N 30 46 E	
Zeu, popl., r.h.	U1	West Nile	2 32 N 30 48 E	Chancellor, Eggeling
Zevigambo, peak	T3	Lushoto	4 33.5 S 38 20 E	Drummond & Hemsley
[Zibwesa] see Kibwesa	T4	Mpanda	6 30 S 29 57 E	Oxford Univ.
[Zigi] see Sigi	T3	Lushoto/Tanga	4 55 S 38 42 E–5 02.5 S 39 04 E	
Zika = Ziku, for.	U4	Mengo	0 08 N 32 31.5 E	Pegler, Dawkins
Ziku, for.	U4	Mengo	0 08 N 32 31.5 E	Pegler, Dawkins
Zimba, popl.	T4	Ufipa	7 51.5 S 31 49.5 E	Bullock, Michelmore, Vesey-FitzGerald
Zimbili, popl., area	T3	Lushoto	4 49 S 38 45.5 E	
Zimbili, popl.	T3	Lushoto	4 55 S 38 58 E	
Zimbili, popl.	T3	Lushoto	5 07 S 38 42.5 E	?Holst

Zimbili, popl.	T3	Lushoto	5 13 S 38 43 E	?Holst
Zimbo, popl.	T4	Tabora	5 36 S 33 50 E	Grant
Zimro Rocks (Kisiwani)	T3	Lushoto	c.5 06 S 38 40 E	Ali Omare
[Zindeni] see Sindeni	T3	Handeni	5 21 S 38 14 E	B.D. Burtt
[Zingout Hills] see Songot Hills	K2	Turkana	c.4 02 N 34 30 E	Champion
[Zingout Mt.] see Songot Mt.	K2	Turkana	4 03 N 34 27 E	
Zingwe, Zingwe, popl.	Z	Zanzibar	6 00 S 39 13.5 E	Greenway
Zintengese = Zintengeze, popl.	U4	Mengo	0 30 N 32 59.5 E	Dummer
Ziwa Chokwe, sw.	Z	Zanzibar	c.5 58 S 39 21 E	Vaughan
Ziwa Kavu, sw.	T2	Arusha	3 15 S 36 52.5 E	Greenway
Ziwa Kibokwa, sw.	Z	Zanzibar	5 58 S 39 20 E	Vaughan
[Ziwa Kibokwe] see Ziwa Kibokwa	Z	Zanzibar	5 58 S 39 20 E	
Ziwa la Mbogo	T2	Arusha	c.3 13.5 S 36 54 E	Richards
Ziwani, popl.	Z	Zanzibar	6 19 S 39 17 E	Vaughan
Ziwani, popl.	P	Pemba	5 10.5 S 39 46 E	Vaughan
Ziwa Pokezeni = Pokezi, sw.	Z	Zanzibar	5 56 S 39 19.5 E	Vaughan
Ziwa Ziwa, sawmill	T4	Tabora	c.5 20 S 32 42 E	Carmichael
Zoka, (= Oju) for.	U1	West Nile	c.3 05 N 31 39 E	Eggeling, Leggat, A.S. Thomas
Zombe, popl.	T6	Rufiji	7 48 S 38 19 E	Anderson, Vesey-FitzGerald
Zulia, mt.	U1	Karamoja	4 07 N 33 59 E	Dawkins
Zulia, f.r.	U1	Karamoja	c.4 00 N 34 00 E	
[Zungomera] see Ngomero	T6	Morogoro	c.7 27 S 37 35 E	
[Zungomero] see Ngomero	T6	Morogoro	c.7 27 S 37 35 E	